Perspectives in Flow Control and Optimization

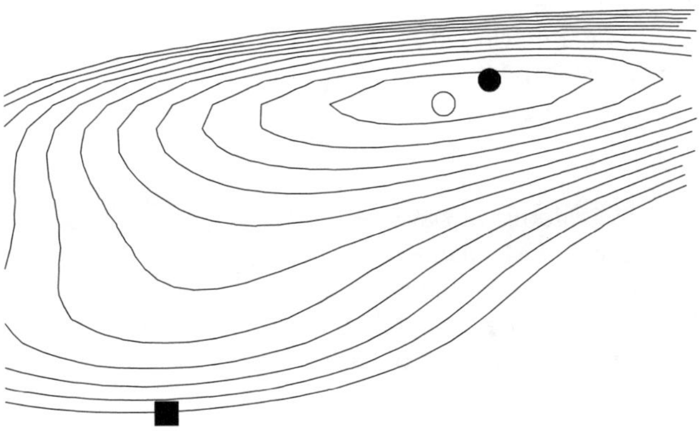

Advances in Design and Control

SIAM's Advances in Design and Control series consists of texts and monographs dealing with all areas of design and control and their applications. Topics of interest include shape optimization, multidisciplinary design, trajectory optimization, feedback, and optimal control. The series focuses on the mathematical and computational aspects of engineering design and control that are usable in a wide variety of scientific and engineering disciplines.

Series Volumes

Gunzburger, Max D., *Perspectives in Flow Control and Optimization*
Delfour, M. C. and Zolésio, J.-P., *Shapes and Geometries: Analysis, Differential Calculus, and Optimization*
Betts, John T., *Practical Methods for Optimal Control Using Nonlinear Programming*
El Ghaoui, Laurent and Niculescu, Silviu-Iulian, eds., *Advances in Linear Matrix Inequality Methods in Control*
Helton, J. William and James, Matthew R., *Extending H^∞ Control to Nonlinear Systems: Control of Nonlinear Systems to Achieve Performance Objectives*

Perspectives in Flow Control and Optimization

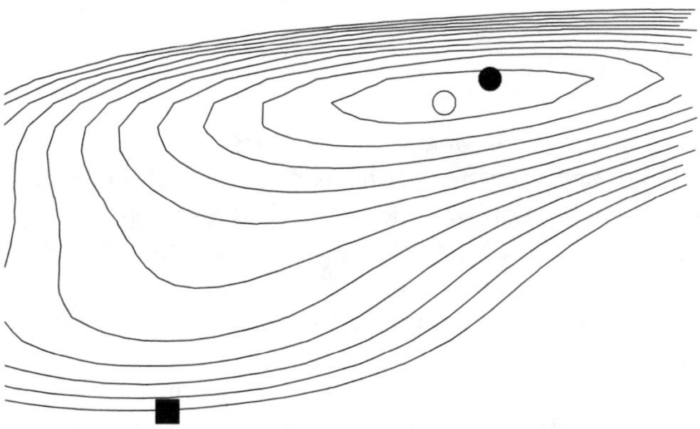

Max D. Gunzburger
Iowa State University
Ames, Iowa
and
Florida State University
Tallahassee, Florida

siam
Society for Industrial and Applied Mathematics
Philadelphia

10 9 8 7 6 5 4 3 2 1

Library of Congress Cataloging-in-Publication Data
Gunzburger, Max D.
Perspectives in flow control and optimization / Max Gunzburger.
p. cm. — (Advances in design and control)
Includes bibliographical references and index.
ISBN 0-89871-527-X
1. Fluid dynamics—Mathematics. 2. Numerical analysis. 3. Mathematical optimization.
I. Title. II. Series.

TA357 .G88 2002
620.1'064'0151—dc21 2002029213

To Lu Ting and George Fix

Contents

Preface

Flow control and optimization is at once an old and a new subject of mathematical and engineering interest. It has fascinated many great minds, among them those of Newton and Prandtl. It has been an important part of experimental flow science throughout the last century. As computational fluid dynamics (CFD) became a mature subject of research and as CFD codes came into routine use for the simulation of fluid flows, it was natural for mathematicians and engineers to examine the use of CFD algorithms and codes for optimization and control problems for fluid flows. Thus, for the last 20 years or so, there has been much activity in the use of sophisticated CFD codes for the solution of flow control and optimization problems. Early on, this activity did not always involve the use of sophisticated optimization algorithms. However, in the last 10 years or so, the marriage of mature CFD methodologies with state-of-the-art optimization methods has become the center of activity in computational flow control and optimization. Today, it is a rather large-scale endeavor in terms of participants, funding, publications, etc.

This book deals with flow control and optimization as a subdiscipline of computational mathematics and computational engineering. My first goal for the book is to present an introduction to the development and analysis of several approaches for solving flow control and optimization problems through the use of modern CFD and optimization methods. A second goal is to discuss some of the many issues that arise, e.g., choices that can be made and difficulties that can be encountered, in the practical implementation of algorithms for flow control and optimization. My hope is that the book will provide the reader with a good idea of what types of flow control and optimization problems can be solved, how to develop effective algorithms for solving such problems, and what to watch for when implementing the algorithms.

The intended audience for the book includes engineers, applied mathematicians, and other scientists interested in computational methods for flow control and optimization. Both those interested in developing new algorithms and those interested in the application of existing algorithms to engineering problems should find useful information in the book. Anyone contemplating offering a course in flow control and optimization should be able to use the book as a resource for preparing lectures and for student readings. The book should be even more useful to those that are entering the field of flow control and optimization either as developers or practitioners. Except for Chapter 6, the great majority of the book should be comprehensible to anyone with a solid background in calculus and a little familiarity with partial differential equations; knowledge about fluid mechanics, computational fluid dynamics, calculus of variations, control theory, or optimization would certainly be beneficial but is not essential. Chapter 6, on the other hand, requires a substantially higher level of mathematical knowledge, most notably in the areas of functional analysis, numerical analysis, and partial differential equations. Fortunately, Chapter 6 is completely independent of the other chapters so that the latter may be well understood even if Chapter 6 is not.

The title of the book includes the word "perspectives." A title that is more descriptive of the contents would have included the modified wording "personal perspectives." As I already mentioned, flow control and optimization is by now a big-time occupation. A comprehensive book covering all aspects and all important contributions would easily be 1000 pages long and would probably require three or four authors. Thus, I make no pretense to be exhaustive or to have written the definitive treatise on the subject of flow control and optimization. Instead, I write mostly from personal experience. As a result, many important topics and results are not found in this book and others are given only superficial treatment. I sincerely hope that other members of the flow control and optimization community will also write books giving their perspectives on the subject.

Plan of book

The first chapter is an attempt to describe the environment in which the huge explosion in interest in systematic and complex approaches to flow control and optimization developed. It also provides, in very broad terms, a discussion of what constitutes a flow control and optimization problem. In Chapter 2, several strategies for solving optimization and control problems are introduced, developed, and discussed in the context of simple[1] example problems. In Chapter 3, the approaches considered in Chapter 2 are illustrated through the consideration of concrete flow control and optimization problems. Chapters 4 and 5 provide a discussion of a number of issues that arise in the application of the algorithms discussed in Chapters 2 and 3 to practical flow control and optimization problems. Chapter 6, although the longest in the book, only provides a rather short and superficial examination of the analysis and numerical analysis of flow control and optimization problems. We include this chapter to illustrate the large body of theoretical work that has been developed and to illustrate its influence on algorithmic developments. Chapter 7 provides a brief examination of some computational techniques for feedback-control problems for fluid flows; the techniques are presented in the context of concrete feedback-control problems.

Although the bibliography seems very comprehensive, it is by no means a complete listing of all available literature on mathematical and computational aspects of flow control and optimization. Although omissions from the bibliography are completely unintentional, I apologize to anyone whose work is not properly represented. The bibliography is imprecisely divided into several groups: groups [A] to [K] contain references to the flow control and optimization literature, grouped according to broad categories such as optimal control, shape optimization, etc. Not all references listed are cited in the text of the book. Indeed, one motivation for separating the bibliography into groups is to provide the reader with the opportunity to locate material in particular categories not considered in the book. Certainly many, if not most, papers in any particular group could justifiably have been included in one or more other groups. Thus, there is a certain amount of arbitrariness and imprecision with regard to which group a paper is assigned to. As a result, a reader who is looking for materials about a particular subject is urged to look at other related groups in the bibliography.

[1] "Simple" is in comparison to flow control and optimization problems.

Group [L] lists references to the literature in optimization, control, automatic differentiation, and other subjects; the citations included in this group are not directly concerned with flow control and optimization problems. Nevertheless, they are closely related and important to the understanding and development of theories and algorithms for flow control and optimization. Group [M], the final group, contains the references cited in the text that are outside the scope of the previous groups.

Acknowledgments

This book grew out of notes for a series of short courses on flow control and optimization that I either presented or was a part of. First, I was one of several presenters at a short course held in 1997 at the von Karman Institute for Fluid Dynamics in Brussels and a workshop held in 1998 at the Centre de recherche en calcul appliqué in Montreal. I thank R. A. Van den Braembussche and M. Manna, the lecture series directors at the VKI, and R. Camarero, A. Garon, D. Pelletier, M. Reggio, and J.-Y. Trepanier, the organizers of the CERCA workshop, for inviting me to participate. However, the main precursor to this book was the notes I prepared for a short course I delivered in 2000 at the Canadian CFD Conference in Montreal. I thank Dominique Pelletier for providing me the opportunity to deliver the short course; without the notes I developed for that course, this book would never have come about. I also prepared a shorter and slightly updated version of the Montreal notes for a short course I delivered in 2001 at the Fifth SIAM Conference on Control and Its Applications in San Diego. I want to thank the SIAM staff and the Organizing Committee of that conference, chaired by Anthony Block and William Hager, for inviting me to that event.

My own entry into research in flow control and optimization was through the Ph.D. studies of L. Steven Hou in the late 1980s. When Steve started his dissertation, I knew very little about flow control and optimization so that he, Thomas Svobodny (who was an equal partner in those early efforts), and I learned a lot together. I, in fact, learned more from them than they did from me; for this, I am truly grateful. Steve's remarkable dissertation (Carnegie Mellon University, 1989) still remains the underpinning for all my work in flow control and optimization. Then, a very fortuitous occurrence took place—I moved to Virginia Tech just at the time that other faculty in the Interdisciplinary Center for Applied Mathematics were becoming interested in flow control and optimization. The intellectual and social atmosphere provided there by Terry Herdman, the director of ICAM, John Burns and Eugene Cliff, the leaders of the flow control and optimization efforts at ICAM, and Jeffrey Borggaard, a crucial participant in our efforts, made my research not only possible, but also very enjoyable. I sincerely thank them for their collaboration and, even more so, for their friendship. I was very lucky to have Janet Peterson and several postdocs (John Burkardt, Ersin Ozugurlu, Ajit Shenoy, Lazarus Tenek, and Xiaonan Wu) work with me on flow control problems at Virginia Tech and/or Iowa State University. I have also been extremely fortunate, in my efforts in flow control and optimization, to work with and be profoundly influenced by many other outstanding mathematicians and engineers. These are too numerous to mention and if I try to do so, I am sure to leave out someone who should definitely be included. In any case, they all know who they are and I thank them all!

For many years my research in flow control and optimization was funded by the Office of Naval Research and, to a much greater extent, by the Air Force Office of Scientific Research. The funds my colleagues, my students, and I received from these agencies were crucial to the success of our research efforts.

I thank Marianne Will for being an ideal editor; she has provided encouragement and advice and, at just the right times, the gentle prodding necessary to get this book completed. I also thank the reviewers for their very helpful comments, especially the reviewer who thoroughly read the manuscript and found numerous small errors. The SIAM book staff, especially Sara Murphy, were also very helpful.

The final two sets of acknowledgments are the most important. First, I have been truly privileged and blessed with the opportunity to work with outstanding students; in many cases, I received at least as much advice from them as they received from me. Without their effort, enthusiasm, and ability, my own work on flow control and optimization would be miniscule. In addition to Steve Hou at Carnegie Mellon University, my students whose dissertations involved control and optimization problems include (in chronological order) Hong-Chul Kim, Hyung-Chun Lee, John Burkardt, Yanzhao Cao, Justin Appel, Sandro Manservisi, and Hyesuk Kwon Lee at Virginia Tech, and Jeehyun Lee, Konstantinos Chrysafinos, and Wenxiang Zhu at Iowa State University. One merely has to look through this book to see the paramount role these students played in my work on flow control and optimization.

Finally, this book is dedicated to my own advisor, Lu Ting, and to George Fix. To this day, I follow the approaches, the discipline, and, most of all, the principles Lu Ting taught me about doing good research. He taught me how to have good taste, not only in Chinese food and Danish butter cookies, but also in all professional activities. George Fix greatly facilitated my entry into numerical analysis and finite element methods. But, beyond that, his enthusiasm and drive taught me about dedication and perseverance; he also helped refine my tastes, not only in my professional activities, but also with respect to great beer and great music. What I am today in my profession I owe in large part to Lu and George, and for that I am profoundly grateful.

Max Gunzburger

Chapter 1

Introductory Comments

1.1 A brief history of flow control and optimization

Flow control and optimization is an ancient *practice* of man. For example, any dam, sluice, canal, levee, irrigation ditch, valve, duct, pipe, pump, hose, vane, etc., is an exercise in flow control or optimization, i.e., an attempt to

> *control the mechanical state, e.g., the rate and direction of motion, and/or the thermodynamic state, e.g., the temperature, of a fluid in order to achieve a desired purpose.*

Even the animal kingdom has examples, such as beaver dams, of attempts at flow control.

Flow control and optimization has also been an *experimental science* for at least 100 years. For example, consider flow past a circular cylinder for which fluid is sucked through a slit on the top side of the cylinder; see Figure 1.1. The objective of the suction is to reduce the drag by delaying separation on the suction side of the cylinder. This was the subject of Prandtl's first paper in 1904 (see [443]) and marked the start of the experimental science of boundary layer control. Here, we are interested in flow control and optimization as a *computational science.* Despite their importance in numerous settings and despite the many successes on the experimental side of the science, it is only recently that effective computational means for the design and control of flows have emerged.

In practice, flow control and optimization has been, for the most part, effected without the use of sophisticated fluid models and/or without the use of sophisticated optimization techniques. In spite of this, substantial successes have been achieved. On the other hand, sophisticated current and future uses of flow controls require a more systematic approach to these problems, and, in particular, will require the use of sophisticated optimization techniques in conjunction with sophisticated flow models. This need has been recognized for quite some time even in the popular organs; see, for example, the January 1993 issue of *Popular Mechanics*, in which the injection of fluid near the nose of an aircraft in order to steer the aircraft in stall environments is discussed. Another example is the March 1, 1993 issue of *Aviation Week & Space Technology*, in which the need for flow control theories involving thousands of degrees of freedom to replace the then current ones involving 10

Figure 1.1. *A depiction of Prandtl's 1904 experiment. Reprinted by permission from* Boundary Layer Theory, *8th revised and enlarged English edition, H. Schlichting and K. Gersten, Springer-Verlag, Berlin, 2000, Figure 11.3, page 293.*

degrees of freedom is discussed.

The main goal of this chapter is to briefly review some of the past attempts at flow control and optimization. We also discuss why the time is now right for the incorporation of sophisticated fluid models and sophisticated optimization techniques into practical flow control and optimization methodologies. We also make some remarks about the structure of flow control and optimization problems and list some examples of interesting objective functionals and control mechanisms. We close the chapter with some remarks concerning some philosophical differences between simulation and control problems.

Lest one think that flow control and optimization is a recent quest among mathematicians, engineers, and scientists, consider the following drag minimization problem:

> *What is the shape that a surface of revolution (moving at constant velocity in the direction of its axis) must have if it is to offer the least resistance to the motion?*

An example body is sketched in Figure 1.2. After making certain assumptions about the flow, one can show that this problem is equivalent to finding a $y(x)$, $x_1 \leq x \leq x_2$, that minimizes

$$\mathcal{J}(y) = \int_{x_1}^{x_2} \frac{y(x)[y'(x)]^3}{1+[y'(x)]^2}\, dx\,.$$

A "solution" for the optimal shape in terms of parametric equations is given by

$$x = \frac{c}{p}(1+p^2)^2 \qquad \text{and} \qquad y = a + c\left(-\ln p + p^2 + \frac{3}{4}p^4\right).$$

This is the *first significant problem in the calculus of variations* [435] and was posed (1687) and solved[2] (1694) by Newton!

1.1.1 Flow control without fluids

By *flow control without fluids* we mean attempts to control a fluid motion and thermodynamic state without the utilization of sophisticated fluid models involving partial differential

[2]Newton's solution is incorrect—his mathematics was impeccable, but his fluid model was incorrect.

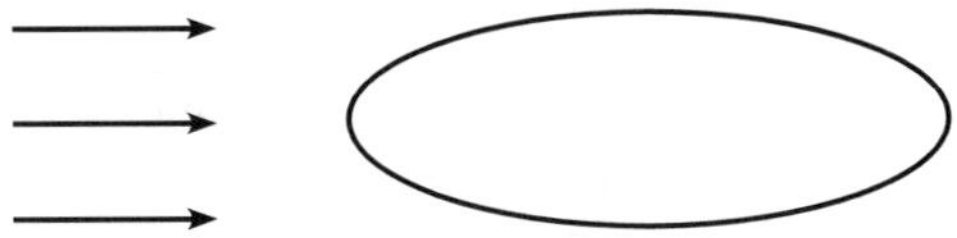

Figure 1.2. *Flow about a body of revolution.*

equations such as the Navier–Stokes, Euler, or potential flow equations. For the examples of ducts, pumps, etc., mentioned above, flow control is effected without any attempt to solve such fluid equations. It is certainly unlikely that beavers solve the Navier–Stokes equations as they build a dam!

An example of a very successful application of flow control without an accurate modeling of the fluid flow is the *design of the heating and cooling system in a building.* Here, one designs a system of ducts, fans, registers, vanes, sensors, actuators, heat pumps, furnaces, air conditioners, etc., so that the temperature throughout a building is close to a uniform, comfortable value and so that the heating/cooling costs are as low as possible. In the design process, the air flow is not computed using sophisticated models involving partial differential equations. Rather, one simply uses empirical rules for determining the flow rates necessary for carrying out the design. One also assumes that pumps, fans, furnaces, etc., move the air at constant flow rates through the ducts, registers, etc. Heat and temperature losses are also determined in an empirical manner.

Perhaps the most spectacular example of successful flow control without fluids is that of *aerodynamic controls.* Here, one determines a position of the rudder, wing flaps, elevators, ailerons, throttle, etc., so that an aircraft executes a desired maneuver. To some extent, all modern aircraft employ automatic controls, i.e., controls that are not determined by the pilot, but perhaps by a computer. The extreme example in this regard is the Grumman X-29 airplane which uses such automatic control to keep the plane from becoming "unstable." Typically, aerodynamic controls are set by solving a small system of ordinary differential equations. The influence of the fluid flow on the controls appears as functions or constants in the differential equations. These functions and constants are determined a priori, very often using an empirical process. When the control settings are being determined, no attempt is made to solve partial differential equations for the fluid flow.

In these and numerous other examples, no attempts are made to employ sophisticated fluid models such as those involving partial differential equations. The flow of the fluid is modeled by a few constants, or at best functions of time, appearing in systems of ordinary differential equations that determine the optimal control settings, or by using a Bernoulli equation to relate mass flow and pressure, or, most often, by assuming constant mass flow rates. In this sense, one may view these efforts as constituting *flow control without fluids.*

1.1.2 Flow optimization without optimization

By *flow optimization without optimization* we mean attempts to control a fluid motion and thermodynamic state in order to meet a desired objective without the utilization of sophisticated optimization techniques such as Lagrange multiplier methods, quasi-Newton methods, etc. In many cases, including the ones described below, although sophisticated optimization algorithms are not involved, a high-fidelity description of the fluid motion and thermodynamic state is employed. Such a description can be determined by experimental measurements, or analytical solutions, or computational simulations.

For the first example of flow optimization without optimization, we mention again the large body of experimental work and somewhat smaller body of analytical work on *boundary layer control* [443]. Here, the size, shape, formation, etc., of a boundary layer is to be affected, e.g., controlled, in order to meet a desired objective. Control mechanisms that have been considered are the movement of solid walls such as for a rotating cylinder, the injection and suction of fluid through orifices or slits, and shape variations such as camber, thickness, and flap adjustments, etc. Objectives that have been considered are maximizing lift, minimizing drag, preventing separation, preventing or facilitating transition to turbulence, etc. For example, consider the question Prandtl addressed in his first paper: Can the drag on a body be lowered by the suction of fluid through a narrow slit? Here, we have a cylinder in a uniform stream and we have fluid sucked through a slit on the back side of the cylinder; see Figure 1.3. Prandtl found (see [443]), through experimentation, that his question had an affirmative answer.

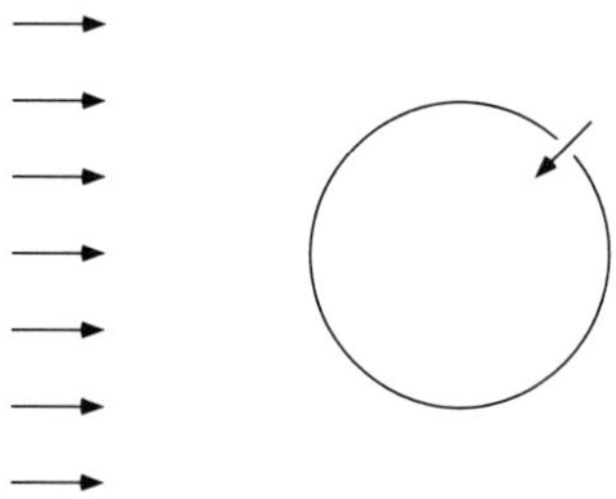

Figure 1.3. *A cylinder in uniform flow with suction through a slit.*

Another example is provided by attempts toward the *cancellation of wave drag.* The Busemann biplane (1930) was an attempt to design a wing shape in order to reduce wave or shock drag; see Figure 1.4 and [436]. The left-hand figure shows the shock waves under design conditions; the wedge angles are exactly those needed to cancel out the outgoing waves. The right-hand figure shows off-design conditions for which the outgoing waves are not completely cancelled.

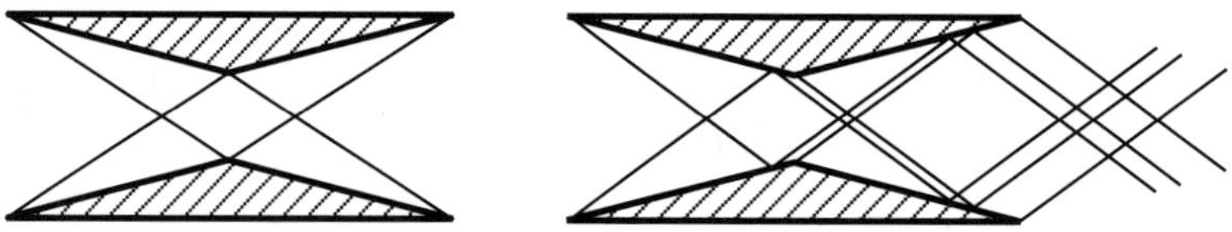

Figure 1.4. *The Busemann biplane at design conditions (left) and at off-design conditions (right).*

More recently, Bauer, Garabedian, and Korn [416], and others, have designed transonic airfoil shapes that generate shock-free flows. Again, under design conditions, there is no shock present at the rear of the supersonic bubble on the upper side of the airfoil; at off-design conditions, a weak shock is present there.

In these examples, and many others as well, sophisticated flow models were used in experiments, analyses, or computations of optimal designs. However, no attempt was made to employ sophisticated optimization algorithms. Solutions were obtained by doing experiments or solving equations for a (small) set of configurations and then comparing results. In essence, optimization, e.g., minimization, is effected by variants of the following algorithm which, for simplicity, we describe in the case of only a single design parameter: Given a cost (or objective or performance) functional $\mathcal{J}(\alpha)$ depending on a design parameter α,

1. choose n distinct values $\{\alpha_1, \alpha_2, \ldots, \alpha_n\}$ of the parameter;
2. evaluate $\mathcal{J}(\alpha_i)$ for $i = 1, \ldots, n$;
3. examine the set $\{\mathcal{J}(\alpha_1), \ldots, \mathcal{J}(\alpha_n)\}$ and choose a value α_j such that $\mathcal{J}(\alpha_j) \leq \mathcal{J}(\alpha_i)$ for $i = 1, \ldots, n$.

Graphically, this is equivalent to plotting the values of $\mathcal{J}(\alpha_i)$, $i = 1, \ldots, n$, as in Figure 1.5, and then choosing the parameter value that yields the minimal value of $\mathcal{J}$ among the plotted values. Each time one evaluates $\mathcal{J}(\alpha_i)$, one has to solve the flow equations or conduct another flow experiment. In essence, optimization, e.g., minimization of the functional, is effected by a "brute force" search through parameter space. If there are multiple parameters, say K of them, one has to sample values in K-dimensional parameter space. One may view such efforts as *flow optimization without optimization.*

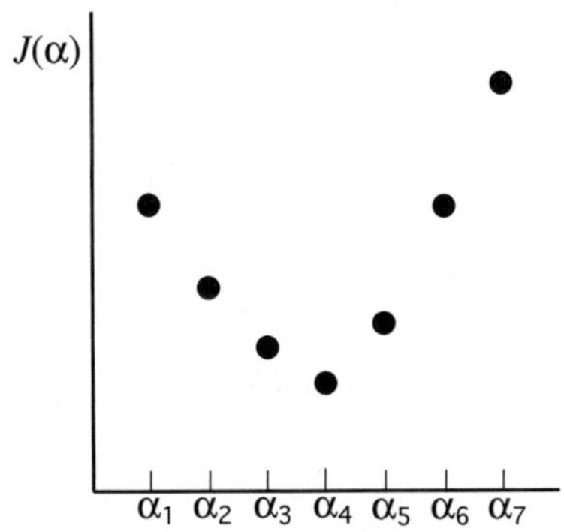

Figure 1.5. *Graphically, the "best value" for the parameter is* α_4.

1.1.3 Flow control without objectives

By *flow control without objectives* we mean attempts to use control and optimization ideas in a fluids setting, not to have the fluid flow meet some desired optimization objective, but for some other independent reason.

As an example, we consider the work reported in [420][3] and related papers on the use of optimization ideas to generate *incompressible computational fluid dynamics* (*CFD*)

[3]Other examples are provided by [426, 431, 437].

algorithms. The connection between a CFD algorithm and flow optimization is made as follows. If $(\mathbf{u}, p)$ is a solution of the Navier–Stokes equations

$$-\nu\Delta\mathbf{u} + \mathbf{u}\cdot\nabla\mathbf{u} + \nabla p = \mathbf{f} \quad \text{in } \Omega\,,$$

$$\nabla\cdot\mathbf{u} = 0 \quad \text{in } \Omega\,,$$

and

$$\mathbf{u} = \mathbf{0} \quad \text{on } \Gamma$$

in some region Ω with boundary Γ, then $\mathbf{u}$ minimizes the functional

$$\mathcal{J}(\mathbf{v}) = \frac{\nu}{2}\int_\Omega |\nabla\Phi(\mathbf{v})|^2\, d\Omega$$

over a suitable function class, where, for given $\mathbf{v}$, (Φ, ξ) is a solution of the Stokes problem

$$-\nu\Delta\Phi + \nabla\xi = -\nu\Delta\mathbf{v} + \mathbf{v}\cdot\nabla\mathbf{v} - \mathbf{f} \quad \text{in } \Omega\,,$$

$$\nabla\cdot\Phi = 0 \quad \text{in } \Omega\,,$$

and

$$\Phi = \mathbf{0} \quad \text{on } \Gamma\,.$$

Moreover, if $\mathbf{u}$ is a minimizer of $\mathcal{J}$, then

$$\Phi = \mathbf{0}\,, \quad \xi = -p\,, \quad \text{and} \quad \mathcal{J}(\mathbf{u}) = \mathbf{0}\,.$$

The problem of minimizing $\mathcal{J}$ can be solved by a conjugate gradient algorithm having the property that at each iteration, only a sequence of Stokes solves is required. Thus, an efficient CFD algorithm is generated. However, note that no intrinsic property of the flow is being optimized; hence, in this sense, we have *flow control without an objective.*

1.1.4 Flow control and optimization and CFD

In this book, we are concerned with flow control and optimization as a *computational science.* Attempts at combining the use of CFD methodologies with optimization strategies for the control and optimization of flows have been going on for quite some time. The papers in [1] and the references cited therein provide an overview of many early attempts; [46] is an especially noteworthy paper as it set the stage for much of what has been developed and reinvented in the ensuing two and a half decades. Despite these attempts, as of 1990 flow control and optimization as a computational science had not become a routine matter nor a large-scale research subdiscipline. Given the developments of the last decade, many of which are the subject of this book, it is natural to ask the following question:

> *Has flow control and optimization now become a subject ready for* systematic computational resolution *using* sophisticated fluid models *and* sophisticated optimization algorithms*?*

or, equivalently,

> *can one put together* sophisticated CFD codes *and* sophisticated optimization techniques *in order to meet desired objectives?*

The answer to these questions is a qualified *yes*. In many settings, especially in two dimensions, one can indeed use complex CFD codes together with modern optimization algorithms to design and control fluid systems. This is due to the significant advances over the last 30 to 40 years in the development and implementation of efficient and robust *algorithms for CFD* for all types of flow regimes so that one can accurately *simulate* many practical flows.

The answer is only a qualified yes because the state of the art in CFD is not sufficiently advanced to enable the routine solution of flow control and optimization problems, e.g., for three-dimensional turbulent flows. Indeed, the simulation of such flows is not yet a routine endeavor.

Recent years have also seen many advances in the *analysis* of flow control and optimization problems, especially for viscous, incompressible flows governed by the Navier–Stokes system. These advances rely on similar advances in the *theory of partial differential equations* and in the *analysis of algorithms* for such equations. They have also played a significant role in the successful development of computational strategies for solving flow control and optimization problems. However, as is the case for computations, there are many such problems that have not yet yielded to rigorous analyses.

Thus, although there have been many analytic and computational successes in the control and optimization of flows, there is still much to be done.

1.2 The structure of flow control problems

Flow control or optimization problems have the usual three ingredients of such problems. First, one has an

> *objective*, a reason why one wants to control the flow.

There are numerous objectives of interest in applications, e.g., flow matching, drag minimization, lift enhancement, preventing separation, preventing transition to turbulence, deterring temperature variations, enhancing mixing, deterring mixing, etc. Mathematically, such an objective is expressed as a *cost*, or *objective*, or *performance functional*.

Next, one has

> *controls* or *design parameters* at one's disposal in order to meet the objective.

One can have *boundary value controls* such as injection or suction of fluid and heating or cooling or temperature controls; one could have *distributed controls* such as heat sources or magnetic fields; one could have *shape controls* such as leading or trailing edge flaps, movable walls, rudders, propeller pitch, surface roughness, or domain design; or, one could have a combination of a number of controls or design parameters. Mathematically, controls or design parameters are expressed in terms of *unknown data* in the mathematical specification of the problem.

Finally, one has

> *constraints* that determine what type of flow one is interested in and that place direct or indirect limits on candidate optimizers.

One must, for example, decide what type of fluid model is adequate for the flows one is interested in, i.e., decide if one is satisfied with assuming that the flow is a potential flow, an inviscid flow, a viscous flow, an incompressible flow, a compressible flow, a stationary flow, a time-dependent flow, etc. Mathematically, the type of flow is expressed in terms of a specific set of *partial differential equations.* One may also impose constraints motivated by practical necessities, e.g., one may want to minimize the drag on an airfoil subject to the lift and/or the volume being greater than a specified value.

One then puts together the three ingredients in an optimization problem by seeking optimal states and controls that satisfy the constraints and minimize the objective functional.

1.3 Some philosophical observations

Well-posedness is bad for optimization and control.[4] This bold statement may surprise someone who is steeped in the notion that well-posedness is something to be desired. So, let's explain what we mean. A mathematically well-posed problem is one such that a solution exists, the solution is unique, and *the solution depends continuously on the data.* Of course, mathematical well-posedness is closely related to physical reality.

Continuous dependence on data means that if one makes small changes in the data, one effects small changes in the solution, i.e., the solution is largely insensitive to small changes in the data. *This is exactly what one doesn't want for control and optimization problems!* Instead, what one wants is to have at one's disposal control mechanisms or design parameters such that small changes in their values effect large changes in the solution; the solution should be very sensitive to small changes in the data! Achieving the goals of optimization is made much easier when the flow is very sensitive to changes in the controls or design parameters.

Related to this notion of high sensitivity is the notion that *what's bad for CFD is good for optimization and vice versa.* This is another bold statement that requires some explanation. Let us illustrate this point with a simple example optimization problem for stationary, viscous, incompressible flow in a two-dimensional channel. We want the flow, i.e., the velocity field, at the matching location (which we assume to be a vertical line far downstream from the inflow) to match a given profile; see Figure 1.6. We are allowed to specify any inflow velocity we want in order to effect the match. We know that the flow will develop into the parabolic profile of Poiseuille flow regardless of the inflow condition; the lower the value of the Reynolds number (i.e., the larger the viscosity), the quicker (the farther upstream) the parabolic profile will develop. The flow at the matching location is *insensitive* (especially for low values of the Reynolds numbers) to the inflow. Thus, for low values of the Reynolds number, it will take a very wild inflow velocity to effect matching to a profile that is substantially different from that of Poiseuille flow. For higher values of the Reynolds number, it takes longer for the Poiseuille flow to develop so that the flow at the matching location is more sensitive to the inflow than it is at lower values of the Reynolds number. As a result, for high values of the Reynolds number, the inflow velocity need not

[4] In this section, we are, of course, being somewhat facetious. We do not necessarily recommend that one solve ill-posed problems or use bad CFD algorithms just because one is trying to control or optimize a flow. (In fact, ill posedeness can result in nonrobust optimal controls or designs.) We are merely being dramatic in order to emphasize the point that objectives, to be effective, should be highly sensitive to variations in the controls or design parameters.

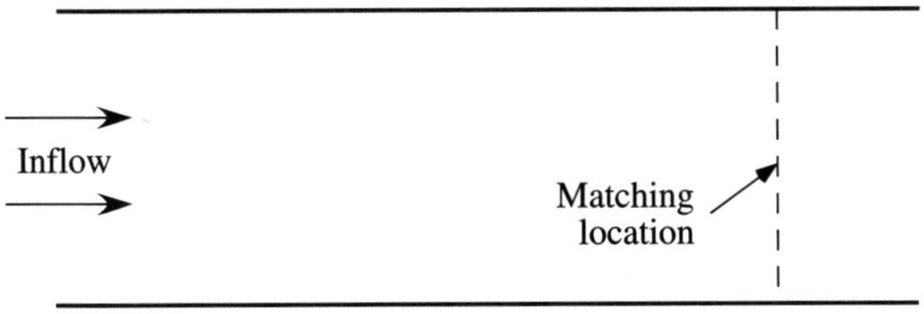

Figure 1.6. *Flow in a channel with the matching location for optimization far downstream from the inflow.*

be so wild in order to achieve a good match downstream. The conclusion is that *low values of the Reynolds number are bad for optimization and high values are good!*

Note that the lower the Reynolds number, the more well-posed the problem is in the sense that the same change in the data results in smaller changes in the solution for low values of the Reynolds number than for high values. Note also that the lower the Reynolds number, the easier it is to simulate the flow in the sense that CFD is "easy" for low values of the Reynolds number and more difficult for high values. We then conclude that, for this example, *what's good for well-posedness and CFD is bad for optimization and control and vice versa* and that in trying to effect control or optimization, *one should look for controls or design parameters that have a major effect on the flow and ignore those that do not.*

Another example is given by compressible, inviscid flows with shock waves. Such flows are considered bad for CFD, at least compared to smooth flows. However, we shall see later (in Section 3.2) that optimizers love shock waves and other flow phenomena that cause the flow to be very sensitive to small changes in a design parameter.

Chapter 2

Three Approaches to Optimal Control and Optimization

2.1 Model control and optimization problems

We consider three approaches to the solution of optimal control and optimization problems:

- one-shot, or adjoint, or co-state, or Lagrange multiplier methods;
- optimization methods based on sensitivity equations; and
- optimization methods based on adjoint equations.

In this chapter, we illustrate the three approaches using an abstract example and several "simple" concrete examples involving finite-dimensional mappings and scalar partial differential equations. In later chapters, the three approaches will be applied to a variety of flow control and optimization problems.

Optimization and control problems are made up of the following ingredients:

- state variables;
- control variables or design parameters;
- an objective, or cost, or performance functional; and
- constraints that candidate state and control variables are required to satisfy.

The optimization problem is then to find state and control variables that minimize (or maximize, as the case may be) the objective functional subject to the requirement that the constraints are satisfied.

The approaches we consider fall into the class of methods that require differentiability of the functionals and constraints. We do not consider other approaches within this class or approaches of different types including those that apply to nondifferentiable problems. In particular, we do not consider global optimization methods (see, e.g., [411]) or genetic or evolutionary algorithms (see group [K] of the bibliography).

In the fluid mechanics setting, the *state variables* could be one or more of the velocity vector, velocity potential, pressure, density, temperature, and internal energy—in short, the mechanical and thermodynamic variables that describe the flow. The *control or design variables* could be the heat flux or temperature at a wall, an inflow mass flow rate, or parameters that determine the shape of the boundary—in short, one or more of the data specified that serve to determine the state variables. There are many possibilities for an *objective functional.* It may measure, e.g., how close the velocity field is to a given field, or the size of the drag or lift, or temperature variations, etc. The constraints are of two types. The *main constraints* are the governing flow equations, e.g., the Navier–Stokes, or Euler, or potential flow equations. In addition, one can encounter *side constraints* such as minimum lift, or minimum volume, or maximum power requirements. In order not to overly complicate the discussion, we postpone consideration of side constraints until Section 2.7.

We now list the three examples that we will first use to illustrate the three methods for dealing with flow control and optimization problems.

Example 1: An abstract optimization problem

Let ϕ denote the state variables, g the control variables or design parameters, $\mathcal{J}(\phi, g)$ the cost or objective functional, and $F(\phi, g) = 0$ the constraints. The optimization problem is then simply[5]

find controls g and states ϕ such that $\mathcal{J}(\phi, g)$ is minimized subject to $F(\phi, g) = 0$.

Many of the objective functionals encountered in practice do not explicitly depend on the controls or design parameters; this may result in unbounded optimal controls or design parameters. In these cases, one must somehow *limit the size of the control.* There are two ways to do this.

- Constrain the size of admissible controls so that they are sought for within a bounded set, e.g., look for optimal controls such that, for some suitable norm $\|\cdot\|$ and constant κ,
$$\|g\| \leq \kappa\,. \tag{2.1}$$
- Penalize the objective functional, i.e., instead of minimizing a given functional $\mathcal{E}(\phi)$ that depends only on the state variables, minimize, for some suitable norm $\|\cdot\|$ and constants σ and β,
$$\mathcal{J}(\phi, g) = \mathcal{E}(\phi) + \sigma\|g\|^{\beta}\,. \tag{2.2}$$

In the first case, one is merely imposing an additional side constraint on the optimization problem; side constraints are discussed in Section 2.7. In the second case, one has changed the problem, i.e., minimizers of $\mathcal{J}$ are not, in general, minimizers of the given functional

[5]At times, the functional $\mathcal{J}(\phi, g)$ will be thought of as a function of both the state ϕ and the control g. At other times, using the solution $\phi(g)$ of the state equation $F(\phi, g) = 0$, it will be thought of as $\mathcal{J}(\phi(g), g)$, i.e., as a function of only the control g. To avoid confusion, this necessitates the adoption of conventions for identifying derivatives of the functional. We will use the following notation. First, we let $\partial\mathcal{J}/\partial\phi$ and $\partial\mathcal{J}/\partial g$ denote the partial derivatives of $\mathcal{J}(\phi, g)$ with respect to ϕ and g, respectively, where ϕ and g are viewed as independent variables, i.e., we have
$$\frac{\partial\mathcal{J}}{\partial\phi} = \left(\frac{\partial\mathcal{J}}{\partial\phi}\right)_g \quad \text{and} \quad \frac{\partial\mathcal{J}}{\partial g} = \left(\frac{\partial\mathcal{J}}{\partial g}\right)_\phi .$$
We then let $D\mathcal{J}/Dg$ denote the derivative of $\mathcal{J}(\phi(g), g)$ with respect to g, where the functional is viewed as being a function only of g. Of course, by the chain rule, we have that
$$\frac{D\mathcal{J}}{Dg} = \frac{\partial\mathcal{J}}{\partial\phi}\frac{d\phi}{dg} + \frac{\partial\mathcal{J}}{\partial g}\,.$$

$\mathcal{E}$. However, in many if not most practical settings, limiting the size of the control via penalization is easier to implement than through placing explicit bounds on the control variables. Since we will not deal with constraints such as (2.1) until Section 2.7, we will, until then, assume that the functional $\mathcal{J}$ depends explicitly on both ϕ and g, i.e., $\mathcal{J} = \mathcal{J}(\phi, g)$.

Example 2: A finite-dimensional optimization problem

Let $\vec{\phi} \in \mathbb{R}^M$ and $\vec{g} \in \mathbb{R}^K$ denote the state variables and the control or design parameters, respectively. Let the functional $\mathcal{J}(\vec{\phi}, \vec{g}) : \mathbb{R}^M \times \mathbb{R}^K \to \mathbb{R}$ be defined by

$$\mathcal{J}(\vec{\phi}, \vec{g}) = \frac{1}{2}\left(\vec{\phi}^T A\vec{\phi} + \vec{g}^T B\vec{g}\right). \tag{2.3}$$

The constraint equation is

$$\vec{F}(\vec{\phi}, \vec{g}) \equiv C\vec{\phi} + (\vec{\phi}^T\vec{\phi})\vec{\phi} - D\vec{g} = \vec{0}, \tag{2.4}$$

where $\vec{F}(\vec{\phi}, \vec{g})$ is a nonlinear mapping from $\mathbb{R}^M \times \mathbb{R}^K$ to $\mathbb{R}^M$. In (2.3)–(2.4), A is a symmetric, positive definite, $M \times M$ matrix; B is a symmetric, positive definite, $K \times K$ matrix; and C and D are full-rank $M \times M$ and $M \times K$ matrices, respectively; $\vec{\phi}^T$ denotes the transpose of the vector $\vec{\phi}$, and $\vec{\xi}^T\vec{\phi}$ denotes the Euclidean inner product of the two vectors $\vec{\xi}$ and $\vec{\phi}$. The constrained finite-dimensional optimization problem we consider is given by

> *find controls $\vec{g}$ and states $\vec{\phi}$ such that the functional given in* (2.3) *is minimized subject to* (2.4).

Example 3: A partial differential equation control problem with a finite number of design parameters

Let Ω be an open domain in $\mathbb{R}^n$, $n = 2, 3$, having boundary Γ.[6] Let ϕ denote the scalar-valued state variable and α_k, $k = 1, \dots, K$, denote the discrete set of control variables or design parameters. The constraints are the second-order, nonlinear, elliptic partial differential equation

$$-\nabla \cdot (a\nabla\phi) + \mathbf{b} \cdot \nabla\phi + \phi^3 = \sum_{k=1}^{K} \alpha_k f_k \quad \text{in } \Omega \tag{2.5}$$

along with the boundary condition

$$\phi = 0 \quad \text{on } \Gamma, \tag{2.6}$$

where a, $\mathbf{b}$, and f_k, $k = 1, \dots, K$, are given functions defined on $\overline{\Omega}$. The objective, or cost, or performance functional is given by

$$\mathcal{J}(\phi, \alpha_1, \dots, \alpha_K) = \frac{1}{2}\int_\Omega (\phi - \Phi)^2 \, d\Omega + \frac{\sigma}{2}\sum_{k=1}^{K}(\alpha_k)^2, \tag{2.7}$$

where Φ is a given function and σ a given penalty parameter. The optimization problem is then given by

[6] The discussions in this chapter will be purely formal; for this reason, we will not be precise about smoothness requirements for Γ or about many other considerations.

find ϕ and α_k, $k = 1, \ldots, K$, such that the functional $\mathcal{J}$ defined in (2.7) *is minimized subject to the requirements that ϕ and α_k, $k = 1, \ldots, K$, satisfy the constraints* (2.5) *and* (2.6).

The functional $\mathcal{J}(\phi, \alpha_1, \ldots, \alpha_K)$ may be viewed as depending on only the design parameters $\{\alpha_1, \ldots, \alpha_K\}$ as follows:

- Through the partial differential equation problem (2.5), (2.6), one may view the state ϕ as a function of the design parameters, i.e., for any set of design parameters $\{\alpha_1, \ldots, \alpha_K\}$, $\phi(\alpha_1, \ldots, \alpha_K)$ may be determined by solving (2.5) and (2.6);
- we then have that $\mathcal{J} = \mathcal{J}(\phi(\alpha_1, \ldots, \alpha_K), \alpha_1, \ldots, \alpha_K)$.

Thus, $\mathcal{J}$ can depend on the design parameters both explicitly and, through the state variables, implicitly.[7] It is important to keep this in mind throughout the discussion of approaches for solving optimization or control problems.

2.2 One-shot methods

The three examples given in Section 2.1 are examples of *constrained* optimization problems. Such problems may be recast as unconstrained optimization problems through the Lagrange-multiplier method. The first-order necessary conditions result in an *optimality system* from which optimal states and controls may be determined.

The optimality system for Example 1

For the abstract setting of Example 1 (see page 12), we introduce the *Lagrange multiplier* or *adjoint variable* or *co-state variable* ξ and the Lagrangian functional

$$\mathcal{L}(\phi, g, \xi) = \mathcal{J}(\phi, g) - \xi^* F(\phi, g),$$

where, in our present formal discussion, $\xi^* F$ can be viewed as an inner product or duality pairing. Then, we pose the following *unconstrained* optimization problem:

Find controls g, states ϕ, and co-states ξ such that $\mathcal{L}(\phi, g, \xi)$ is rendered stationary.

[7] Again, at times, the functional $\mathcal{J}(\phi, \alpha_1, \ldots, \alpha_K)$ will be thought of as a function of both the state ϕ and the design parameters $\{\alpha_1, \ldots, \alpha_K\}$. At other times, using the solution $\phi(\alpha_1, \ldots, \alpha_K)$ of the state equation $F(\phi, \alpha_1, \ldots, \alpha_K) = 0$ (see, e.g., (2.5)–(2.6)), the functional will be thought of as $\mathcal{J}(\phi(\alpha_1, \ldots, \alpha_K), \alpha_1, \ldots, \alpha_K)$, i.e., a function of only the design parameters $\{\alpha_1, \ldots, \alpha_K\}$. To avoid confusion, this again necessitates the adoption of conventions for identifying derivatives of the functional. We will use the following notation. First, we let $\partial\mathcal{J}/\partial\phi$ and $\partial\mathcal{J}/\partial\alpha_k$ for $k = 1, \ldots, K$ denote the partial derivatives of $\mathcal{J}(\phi, \alpha_1, \ldots, \alpha_K)$ with respect to ϕ and $\alpha_1, \ldots, \alpha_K$, respectively, where ϕ and $\alpha_1, \ldots, \alpha_K$ are all viewed as independent variables, i.e., we have that

$$\frac{\partial\mathcal{J}}{\partial\phi} = \left(\frac{\partial\mathcal{J}}{\partial\phi}\right)_{\alpha_1,\ldots,\alpha_K} \quad \text{and} \quad \frac{\partial\mathcal{J}}{\partial\alpha_k} = \left(\frac{\partial\mathcal{J}}{\partial\alpha_k}\right)_{\phi,\alpha_1,\ldots,\alpha_{k-1},\alpha_{k+1},\ldots,\alpha_K} \quad \text{for } k = 1, \ldots, K.$$

We then let $D\mathcal{J}/D\alpha_k$ for $k = 1, \ldots, K$ denote the partial derivative of $\mathcal{J}(\phi(\alpha_1, \ldots, \alpha_K), \alpha_1, \ldots, \alpha_K)$ with respect to α_k, where the functional is viewed as being a function of only $\{\alpha_1, \ldots, \alpha_K\}$. Of course, we have that

$$\frac{D\mathcal{J}}{D\alpha_k} = \frac{\partial\mathcal{J}}{\partial\phi}\frac{\partial\phi}{\partial\alpha_k} + \frac{\partial\mathcal{J}}{\partial\alpha_k}.$$

The first-order necessary conditions then yield an *optimality system* from which optimal states and design parameters may be determined:

$$\frac{\delta\mathcal{L}}{\delta\xi} = 0 \quad \Rightarrow \text{constraint or state equation;}$$
$$\frac{\delta\mathcal{L}}{\delta\phi} = 0 \quad \Rightarrow \text{adjoint or co-state equation;}$$
$$\frac{\delta\mathcal{L}}{\delta g} = 0 \quad \Rightarrow \text{optimality condition.}$$

Note that each argument of $\mathcal{L}(\phi, g, \xi)$ is considered to be independent of the others; this was not true for the original optimization problem involving $\mathcal{J}(\phi, g)$ since there the arguments ϕ and g were constrained to satisfy $F(\phi, g) = 0$ and thus could not be chosen independently.

Setting the first variation of $\mathcal{L}$ with respect to the Lagrange multiplier ξ equal to zero is equivalent to the condition

$$\lim_{\epsilon\to 0}\left(\frac{\mathcal{L}(\phi, g, \xi + \epsilon\widetilde{\xi}) - \mathcal{L}(\phi, g, \xi)}{\epsilon}\right) = 0\,,$$

where the variation $\widetilde{\xi}$ in the Lagrange multiplier ξ is arbitrary. Then, substituting for $\mathcal{L}$, we have

$$\lim_{\epsilon\to 0}\frac{1}{\epsilon}\Big[\big(\mathcal{J}(\phi, g) - (\xi + \epsilon\widetilde{\xi})^* F(\phi, g)\big) - \big(\mathcal{J}(\phi, g) - \xi^* F(\phi, g)\big)\Big] = 0$$

or

$$\widetilde{\xi}^* F(\phi, g) = 0\,.$$

Since the variation $\widetilde{\xi}$ in the Lagrange multiplier ξ is arbitrary, we recover the constraint equation, i.e., the *state equation* $F(\phi, g) = 0$.

Setting the first variation of $\mathcal{L}$ with respect to the state ϕ equal to zero is equivalent to the condition

$$\lim_{\epsilon\to 0}\left(\frac{\mathcal{L}(\phi + \epsilon\widetilde{\phi}, g, \xi) - \mathcal{L}(\phi, g, \xi)}{\epsilon}\right) = 0\,,$$

where the variation $\widetilde{\phi}$ in the state ϕ is arbitrary. Then, substituting for $\mathcal{L}$, we have

$$\lim_{\epsilon\to 0}\frac{1}{\epsilon}\Big[\big(\mathcal{J}(\phi + \epsilon\widetilde{\phi}, g) - \xi^* F(\phi + \epsilon\widetilde{\phi}, g)\big) - \big(\mathcal{J}(\phi, g) - \xi^* F(\phi, g)\big)\Big] = 0$$

or

$$\lim_{\epsilon\to 0}\left[\frac{\mathcal{J}(\phi + \epsilon\widetilde{\phi}, g) - \mathcal{J}(\phi, g)}{\epsilon} - \frac{\xi^*\big(F(\phi + \epsilon\widetilde{\phi}, g) - F(\phi, g)\big)}{\epsilon}\right] = 0\,. \tag{2.8}$$

Introducing Taylor series,[8] we have[9]

$$\mathcal{J}(\phi + \epsilon\widetilde{\phi}, g) = \mathcal{J}(\phi, g) + \epsilon\left(\frac{\partial\mathcal{J}}{\partial\phi}\bigg|_{(\phi,g)}\right)\widetilde{\phi} + O(\epsilon^2)$$

[8]All our steps are purely formal, i.e., we assume that the functionals, variables, etc., are all sufficiently smooth to justify all the steps.

[9]Our notation deserves a little explanation. Following [441], in the finite-dimensional setting, i.e., if ϕ and F are column vectors, $\partial\mathcal{J}/\partial\phi$ is a row vector and $\partial F/\partial\phi$ is a matrix. Thus, for example, $(\partial\mathcal{J}/\partial\phi)\widetilde{\phi}$ denotes the product of a row vector times a column vector. We extend this notation to the abstract example.

and similarly for $F(\phi + \epsilon\widetilde{\phi}, g)$. Then, we have from (2.8) that

$$\lim_{\epsilon\to 0}\left(\left(\frac{\partial \mathcal{J}}{\partial \phi}\Big|_{(\phi,g)}\right)\widetilde{\phi} - \xi^*\left(\frac{\partial F}{\partial \phi}\Big|_{(\phi,g)}\widetilde{\phi}\right) + O(\epsilon)\right) = 0$$

or

$$\left(\frac{\partial \mathcal{J}}{\partial \phi}\Big|_{(\phi,g)}\right)\widetilde{\phi} - \xi^*\left(\frac{\partial F}{\partial \phi}\Big|_{(\phi,g)}\widetilde{\phi}\right) = 0$$

or

$$\widetilde{\phi}^*\left(\left(\frac{\partial \mathcal{J}}{\partial \phi}\Big|_{(\phi,g)}\right)^* - \left(\frac{\partial F}{\partial \phi}\Big|_{(\phi,g)}\right)^*\xi\right) = 0,$$

where $(\partial F/\partial\phi)^*$ denotes the adjoint operator[10] to $(\partial F/\partial\phi)$. Since the variation $\widetilde{\phi}$ in the state ϕ is arbitrary, we obtain the *adjoint* or *co-state equation*

$$\left(\frac{\partial F}{\partial \phi}\Big|_{(\phi,g)}\right)^*\xi = \left(\frac{\partial \mathcal{J}}{\partial \phi}\Big|_{(\phi,g)}\right)^*.$$

Setting the first variation of $\mathcal{L}$ with respect to the control g equal to zero is equivalent to the condition

$$\lim_{\epsilon\to 0}\left(\frac{\mathcal{L}(\phi, g+\epsilon\widetilde{g}, \xi) - \mathcal{L}(\phi, g, \xi)}{\epsilon}\right) = 0,$$

where the variation $\widetilde{g}$ in the control g is arbitrary. Then,

$$\lim_{\epsilon\to 0}\frac{1}{\epsilon}\Big[\big(\mathcal{J}(\phi, g+\epsilon\widetilde{g}) - \xi^*F(\phi, g+\epsilon\widetilde{g})\big) - \big(\mathcal{J}(\phi, g) - \xi^*F(\phi, g)\big)\Big] = 0.$$

Proceeding as we did for for the adjoint equations yields the *optimality condition*

$$\left(\frac{\partial F}{\partial g}\Big|_{(\phi,g)}\right)^*\xi = \left(\frac{\partial \mathcal{J}}{\partial g}\Big|_{(\phi,g)}\right)^*. \tag{2.9}$$

Collecting the results of setting the first variations of the Lagrangian functional to zero yields the *optimality system*

$$\begin{aligned}
\text{state equation} \quad &\Rightarrow \quad F(\phi, g) = 0;\\
\text{adjoint equation} \quad &\Rightarrow \quad \left(\frac{\partial F}{\partial \phi}\Big|_{(\phi,g)}\right)^*\xi = \left(\frac{\partial \mathcal{J}}{\partial \phi}\Big|_{(\phi,g)}\right)^*;\\
\text{optimality condition} \quad &\Rightarrow \quad \left(\frac{\partial F}{\partial g}\Big|_{(\phi,g)}\right)^*\xi = \left(\frac{\partial \mathcal{J}}{\partial g}\Big|_{(\phi,g)}\right)^*.
\end{aligned}$$

If the state system is expensive to solve, then this coupled system is an even more formidable system. However, if it is possible to solve the coupled optimality system through computational methods, then optimal states and controls can be obtained without an optimization iteration.[11] For this reason, such an approach is sometimes called a *one-shot* method for optimization.

[10]In the finite-dimensional case, the adjoint operator is the matrix conjugate transpose.

[11]Of course, if the optimality system is nonlinear, one still has an iteration to contend with to solve the nonlinear equations.

The optimality system for Example 2

For Example 2 given on page 13, we introduce the *Lagrange multiplier* or *adjoint variable* or *co-state variable* $\vec{\xi} \in \mathbb{R}^M$ and the Lagrangian functional[12]

$$\begin{aligned}\mathcal{L}(\vec{\phi}, \vec{g}, \vec{\xi}) &\equiv \mathcal{J}(\vec{\phi}, \vec{g}) - \vec{\xi}^T \vec{F}(\vec{\phi}, \vec{g}) \\ &= \frac{1}{2}\left(\vec{\phi}^T A\vec{\phi} + \vec{g}^T B\vec{g}\right) - \vec{\xi}^T\left(C\vec{\phi} + (\vec{\phi}^T\vec{\phi})\vec{\phi} - D\vec{g}\right).\end{aligned}$$

Then, we pose the following *unconstrained* optimization problem:

> *Find controls $\vec{g}$, states $\vec{\phi}$, and co-states $\vec{\xi}$ such that $\mathcal{L}(\vec{\phi}, \vec{g}, \vec{\xi})$ is rendered stationary.*

Setting the first variation of $\mathcal{L}$ with respect to the Lagrange multiplier $\vec{\xi}$ equal to zero is equivalent to the condition

$$\lim_{\epsilon\to 0}\left(\frac{\mathcal{L}(\vec{\phi}, \vec{g}, \vec{\xi} + \epsilon\widetilde{\vec{\xi}}) - \mathcal{L}(\vec{\phi}, \vec{g}, \vec{\xi})}{\epsilon}\right) = 0,$$

where the variation $\widetilde{\vec{\xi}}$ in the Lagrange multiplier $\vec{\xi}$ is arbitrary in $\mathbb{R}^M$. Then, substituting for $\mathcal{L}$, we have

$$\begin{aligned}\lim_{\epsilon\to 0}\frac{1}{\epsilon}\Bigg[\left(\frac{1}{2}\left(\vec{\phi}^T A\vec{\phi} + \vec{g}^T B\vec{g}\right) - (\vec{\xi} + \epsilon\widetilde{\vec{\xi}})^T\left(C\vec{\phi} + (\vec{\phi}^T\vec{\phi})\vec{\phi} - D\vec{g}\right)\right) \\ -\left(\frac{1}{2}\left(\vec{\phi}^T A\vec{\phi} + \vec{g}^T B\vec{g}\right) - \vec{\xi}^T\left(C\vec{\phi} + (\vec{\phi}^T\vec{\phi})\vec{\phi} - D\vec{g}\right)\right)\Bigg] = 0\end{aligned}$$

or

$$\widetilde{\vec{\xi}}^T\left(C\vec{\phi} + (\vec{\phi}^T\vec{\phi})\vec{\phi} - D\vec{g}\right) = 0.$$

Since the variation $\widetilde{\vec{\xi}} \in \mathbb{R}^M$ in the Lagrange multiplier $\vec{\xi}$ is arbitrary, we recover the constraint equation, i.e., the *state equation*

$$\vec{F}(\vec{\phi}, \vec{g}) = C\vec{\phi} + (\vec{\phi}^T\vec{\phi})\vec{\phi} - D\vec{g} = \vec{0}\ .$$

Setting the first variation of $\mathcal{L}$ with respect to the state $\vec{\phi}$ equal to zero is equivalent to the condition

$$\lim_{\epsilon\to 0}\left(\frac{\mathcal{L}(\vec{\phi} + \epsilon\widetilde{\vec{\phi}}, \vec{g}, \vec{\xi}) - \mathcal{L}(\vec{\phi}, \vec{g}, \vec{\xi})}{\epsilon}\right) = 0,$$

[12] Note that in deriving the optimality systems for this example, we do not merely apply the results obtained for the abstract example, Example 1. We certainly could do this. However, it is more illuminating to instead apply to this example the same steps as were taken in deriving the optimality system for Example 1. This, in the end, is the easiest way to obtain optimality systems in complicated situations.

where the variation $\widetilde{\phi} \in \mathbb{R}^M$ in the state $\vec{\phi}$ is arbitrary. Then, substituting for $\mathcal{L}$, we have

$$\lim_{\epsilon\to 0}\frac{1}{\epsilon}\left[\left\{\frac{1}{2}\left((\vec{\phi}+\epsilon\widetilde{\phi})^T A(\vec{\phi}+\epsilon\widetilde{\phi})+\vec{g}^T B\vec{g}\right)\right.\right.$$
$$\left.-\vec{\xi}^T\left(C(\vec{\phi}+\epsilon\widetilde{\phi})+((\vec{\phi}+\epsilon\widetilde{\phi})^T(\vec{\phi}+\epsilon\widetilde{\phi}))(\vec{\phi}+\epsilon\widetilde{\phi})-D\vec{g}\right)\right\}$$
$$\left.-\left\{\frac{1}{2}\left(\vec{\phi}^T A\vec{\phi}+\vec{g}^T B\vec{g}\right)-\vec{\xi}^T\left(C\vec{\phi}+(\vec{\phi}^T\vec{\phi})\vec{\phi}-D\vec{g}\right)\right\}\right]=0$$

or

$$\lim_{\epsilon\to 0}\left[\widetilde{\phi}^T A\vec{\phi}-\vec{\xi}^T\left(C\widetilde{\phi}+2(\vec{\phi}^T\widetilde{\phi})\vec{\phi}+(\vec{\phi}^T\vec{\phi})\widetilde{\phi}\right)\right.$$
$$\left.+\epsilon\left(\frac{1}{2}\widetilde{\phi}^T A\widetilde{\phi}-(\widetilde{\phi}^T\widetilde{\phi})(\vec{\xi}^T\vec{\phi})-2(\widetilde{\phi}^T\vec{\phi})(\vec{\xi}^T\widetilde{\phi})\right)-\epsilon^2(\widetilde{\phi}^T\widetilde{\phi})(\vec{\xi}^T\widetilde{\phi})\right]=0\,,$$

where we have used the symmetry of A and of the inner product $\vec{\phi}^T\widetilde{\phi}$. As a result, we obtain

$$\widetilde{\phi}^T A\vec{\phi}-\vec{\xi}^T\left(C\widetilde{\phi}+2(\vec{\phi}^T\widetilde{\phi})\vec{\phi}+(\vec{\phi}^T\vec{\phi})\widetilde{\phi}\right)=0$$

or, equivalently,

$$\widetilde{\phi}^T\left(A\vec{\phi}-C^T\vec{\xi}-2(\vec{\phi}^T\vec{\xi})\vec{\phi}-(\vec{\phi}^T\vec{\phi})\vec{\xi}\right)=0\,.$$

Since the variation $\widetilde{\phi} \in \mathbb{R}^M$ in the state $\vec{\phi}$ is arbitrary, we obtain the *adjoint equation*

$$C^T\vec{\xi}+2(\vec{\phi}^T\vec{\xi})\vec{\phi}+(\vec{\phi}^T\vec{\phi})\vec{\xi}=A\vec{\phi}\,.$$

Setting the first variation of $\mathcal{L}$ with respect to the control $\vec{g}$ equal to zero is equivalent to the condition

$$\lim_{\epsilon\to 0}\left(\frac{\mathcal{L}(\vec{\phi},\vec{g}+\epsilon\widetilde{g},\vec{\xi})-\mathcal{L}(\vec{\phi},\vec{g},\vec{\xi})}{\epsilon}\right)=0\,,$$

where the variation $\widetilde{g} \in \mathbb{R}^K$ in the control $\vec{g}$ is arbitrary. Proceeding as we did for the first variation with respect to the state variable ϕ yields the *optimality condition*

$$B\vec{g}+D^T\vec{\xi}=\vec{0}\,. \tag{2.10}$$

Collecting the results of setting the first variations of the Lagrangian functional to zero yields the *optimality system*

$$\begin{aligned}
\text{state equations} &\quad\Rightarrow\quad C\vec{\phi}+(\vec{\phi}^T\vec{\phi})\vec{\phi}-D\vec{g}=\vec{0}\,;\\
\text{adjoint equations} &\quad\Rightarrow\quad C^T\vec{\xi}+2(\vec{\phi}^T\vec{\xi})\vec{\phi}+(\vec{\phi}^T\vec{\phi})\vec{\xi}=A\vec{\phi}\,;\\
\text{optimality conditions} &\quad\Rightarrow\quad B\vec{g}=-D^T\vec{\xi}\,.
\end{aligned}$$

This system may be written in the form

$$\begin{pmatrix} A & -C^T & 0\\ -C & 0 & D\\ 0 & D^T & B\end{pmatrix}\begin{pmatrix}\vec{\phi}\\ \vec{\xi}\\ \vec{g}\end{pmatrix}-\begin{pmatrix}2(\vec{\phi}^T\vec{\xi})\vec{\phi}+(\vec{\phi}^T\vec{\phi})\vec{\xi}\\ (\vec{\phi}^T\vec{\phi})\vec{\phi}\\ \vec{0}\end{pmatrix}=\begin{pmatrix}\vec{0}\\ \vec{0}\\ \vec{0}\end{pmatrix} \tag{2.11}$$

from which it is clear that the optimality system is a coupled system. If the nonlinear optimality system (2.11) can be solved, then one obtains the optimal state $\vec{\phi}$, optimal co-state $\vec{\xi}$, and optimal control $\vec{g}$ in one shot, i.e., without invoking an optimization algorithm.

The optimality system for Example 3

For Example 3 given on page 13, we introduce the Lagrangian functional[13]

$$\mathcal{L}(\phi, \alpha_1, \ldots, \alpha_K, \xi, \eta) = \mathcal{J}(\phi, \alpha_1, \ldots, \alpha_K)$$
$$- \int_\Omega \xi \left(-\nabla \cdot (a\nabla\phi) + \mathbf{b} \cdot \nabla\phi + \phi^3 - \sum_{k=1}^{K} \alpha_k f_k \right) d\Omega - \int_\Gamma \eta\phi \, d\Gamma \,,$$

where ξ and η are the Lagrange multiplier functions introduced to enforce the differential equation (2.5) and boundary condition (2.6), respectively.[14]

Setting the first variation of $\mathcal{L}$ with respect to the Lagrange multiplier ξ equal to zero is equivalent to

$$\lim_{\epsilon \to 0} \left(\frac{\mathcal{L}(\phi, \alpha_1, \ldots, \alpha_K, \xi + \epsilon\widetilde{\xi}, \eta) - \mathcal{L}(\phi, \alpha_1, \ldots, \alpha_K, \xi, \eta)}{\epsilon} \right) = 0$$

or

$$\int_\Omega \widetilde{\xi} \left(-\nabla \cdot (a\nabla\phi) + \mathbf{b} \cdot \nabla\phi + \phi^3 - \sum_{k=1}^{K} \alpha_k f_k \right) d\Omega = 0 \,.$$

Since variations $\widetilde{\xi}$ in the Lagrange multiplier are arbitrary, we recover the differential equation (2.5). Similarly, setting the first variation of $\mathcal{L}$ with respect to the Lagrange multiplier η equal to zero recovers the boundary condition (2.6).

Setting the first variation of $\mathcal{L}$ with respect to the state variable ϕ equal to zero is equivalent to

$$\lim_{\epsilon \to 0} \left(\frac{\mathcal{L}(\phi + \epsilon\widetilde{\phi}, \alpha_1, \ldots, \alpha_K, \xi, \eta) - \mathcal{L}(\phi, \alpha_1, \ldots, \alpha_K, \xi, \eta)}{\epsilon} \right) = 0$$

or, after substituting for $\mathcal{L}$ and some straightforward manipulations,

$$\int_\Omega (\phi - \Phi)\widetilde{\phi} \, d\Omega - \int_\Omega \xi \Big(-\nabla \cdot (a\nabla\widetilde{\phi}) + \mathbf{b} \cdot \nabla\widetilde{\phi} + 3\phi^2\widetilde{\phi} \Big) d\Omega - \int_\Gamma \eta\widetilde{\phi} \, d\Gamma = 0 \,.$$

[13]Again, note that in deriving the optimality systems for this example, we do not merely apply the results obtained for the abstract example, Example 1. We instead again apply the same steps as were taken in deriving the optimality systems for Examples 1 and 2.

[14]It is often the case that no Lagrange multiplier is introduced to enforce boundary conditions such as the Dirichlet condition $\phi = 0$ on Γ. Instead, it may be simpler to require that all candidate functions ϕ used to render $\mathcal{L}$ stationary satisfy the boundary condition. However, there is certainly no harm in enforcing the boundary condition through the introduction of the Lagrange multiplier η and, in this example, the optimality systems obtained by the two approaches are the same.

Integration by parts yields that

$$\begin{aligned}\int_\Omega \widetilde{\phi}\Big((\phi - \Phi) + \nabla\cdot(a\nabla\xi) + \nabla\cdot(\mathbf{b}\xi) - 3\phi^2\xi\Big)\,d\Omega \\ + \int_\Gamma \xi a\frac{\partial\widetilde{\phi}}{\partial n}\,d\Gamma - \int_\Gamma \widetilde{\phi}\Big(a\frac{\partial\xi}{\partial n} + \mathbf{b}\cdot\mathbf{n}\xi + \eta\Big)\,d\Gamma = 0\,.\end{aligned} \tag{2.12}$$

Since variations $\widetilde{\phi}$ in the state are arbitrary, we obtain that

$$-\nabla\cdot(a\nabla\xi) - \nabla\cdot(\mathbf{b}\xi) + 3\phi^2\xi = \phi - \Phi \quad \text{in } \Omega\,, \tag{2.13}$$

$$\xi = 0 \quad \text{on } \Gamma\,, \tag{2.14}$$

and

$$\eta = -a\frac{\partial\xi}{\partial n} - \mathbf{b}\cdot\mathbf{n}\xi \quad \text{on } \Gamma\,. \tag{2.15}$$

To see this, first pick arbitrary variations $\widetilde{\phi}$ that vanish in a neighborhood of the boundary Γ so that $\widetilde{\phi} = \partial\widetilde{\phi}/\partial n = 0$ on Γ. Then, the boundary integral terms in (2.12) vanish. This suffices to show that (2.12) implies the differential equation (2.13). Once we have that the differential equation holds, we have that (2.12) implies that

$$\int_\Gamma \xi a\frac{\partial\widetilde{\phi}}{\partial n}\,d\Gamma - \int_\Gamma \widetilde{\phi}\Big(a\frac{\partial\xi}{\partial n} + \mathbf{b}\cdot\mathbf{n}\xi + \eta\Big)\,d\Gamma = 0 \tag{2.16}$$

for arbitrary $\widetilde{\phi}$. We next choose arbitrary $\widetilde{\phi}$ such that $\widetilde{\phi} = 0$ on Γ; then, the second integral in (2.16) vanishes and, since $\partial\widetilde{\phi}/\partial n$ is arbitrary on Γ, we obtain (2.14). Once we have that (2.13) and (2.14) hold, (2.12) implies that

$$\int_\Gamma \widetilde{\phi}\Big(a\frac{\partial\xi}{\partial n} + \mathbf{b}\cdot\mathbf{n}\xi + \eta\Big)\,d\Gamma = 0$$

for arbitrary $\widetilde{\phi}$. This implies (2.15).

Finally, setting to zero the variations of $\mathcal{L}$ with respect to each of the parameters α_k, $k = 1, \ldots, K$, easily yields that

$$\sigma\alpha_k + \int_\Omega \xi f_k\,d\Omega = 0 \quad \text{for} \quad k = 1, \ldots, K\,. \tag{2.17}$$

Note that the only appearance of the Lagrange multiplier η in the optimality system (2.5), (2.6), (2.13)–(2.15), and (2.17) is in (2.15). Therefore, unless one is interested in η itself, we may ignore it and (2.15). Note that the Lagrange multiplier ξ cannot be ignored—it is coupled to the state ϕ and the design parameters $\{\alpha_1, \ldots, \alpha_K\}$.

Collecting the results of setting the first variations of the Lagrangian function to zero results in the following *optimality system*:

The state equations

$$-\nabla\cdot(a\nabla\phi) + \mathbf{b}\cdot\nabla\phi + \phi^3 = \sum_{k=1}^K \alpha_k f_k \quad \text{in } \Omega \quad \text{and} \quad \phi = 0 \quad \text{on } \Gamma\,; \tag{2.18}$$

the adjoint equations[15]

$$-\nabla \cdot (a\nabla\xi) - \nabla \cdot (\mathbf{b}\xi) + 3\phi^2\xi = \phi - \Phi \quad \text{in } \Omega \quad \text{and} \quad \xi = 0 \quad \text{on } \Gamma\,; \tag{2.19}$$

and the optimality conditions

$$\alpha_k = -\frac{1}{\sigma}\int_\Omega \xi f_k \, d\Omega \quad \text{for} \quad k = 1, \ldots, K\,. \tag{2.20}$$

The necessary conditions (2.18)–(2.20) are a system of *coupled* partial differential equations whose solution yields the optimal parameters $\{\alpha_1, \ldots, \alpha_K\}$, the optimal state ϕ, and the optimal adjoint state ξ. One cannot solve the state system (2.18) for ϕ since the design parameters α_k are not known; one cannot solve the adjoint system (2.19) for ξ since ϕ is not known; one cannot solve the optimality condition (2.20) for the α_k's since the adjoint variable ξ is not known. The optimality system is truly coupled.[16]

Note that the coupled optimality system involves at least twice as many unknowns as the state system. Also, note that the adjoint equations are *linear* in the adjoint variables ξ.

For this example, it is possible to solve the coupled optimality system (especially in two dimensions) so that optimal states and controls can be obtained without an optimization iteration.[17] For problems involving more unknowns, one usually must *iterate* between the equations in the optimality system. For the optimality system (2.18)–(2.20), the simplest form of this iteration is to

- guess values for the design parameters α_k, $k = 1, \ldots, K$;
- solve the state system (2.18) for the state ϕ;
- solve the adjoint system (2.19) for the co-state ξ; and
- find new values of the parameters from the optimality condition (2.20).

This process is repeated until satisfactory convergence is (we hope) achieved.

It can be shown (see, e.g., [81]) that this simple iterative method is equivalent to a steepest descent algorithm (with a fixed step size) for the functional $\mathcal{J}(\phi(\alpha_1, \ldots, \alpha_K), \alpha_1, \ldots, \alpha_K)$, where $\phi(\alpha_1, \ldots, \alpha_K)$ denotes the state corresponding to the parameters α_k, $k = 1, \ldots, K$. Such a method may converge very slowly or may not even converge; obviously, we will need to explore better ways to uncouple the state and adjoint computations.

2.3 Sensitivity- and adjoint-based optimization methods

2.3.1 Optimization-based methods

For most practical flow control and optimization problems, it is not possible to solve the coupled optimality system by the one-shot approach. Instead, optimization algorithms can be used to determine optimal states and controls. Many practical optimization algorithms

[15] The left-hand side of the adjoint equations (2.19) involves the formal adjoint operators of the operators appearing on the left-hand side of the state equations (2.18) linearized about the state.

[16] In this example, one can easily substitute the optimality condition into the state equation to obtain a system involving just ϕ and ξ; this is not always possible.

[17] Of course, one still has to iterate to find a solution since the problem is nonlinear, but not to effect optimization.

require (at least an approximation) to the *gradient of the functional* to be minimized with respect to the controls or design parameters. A typical optimization algorithm (in the notation of Example 1 on page 12) proceeds as follows.

ALGORITHM 2.1.
Start with an initial guess $g^{(0)}$ for the design parameters. For $m = 0, 1, 2, \ldots,$ until "satisfactory convergence is achieved,"

1. solve $F(\phi^{(m)}, g^{(m)}) = 0$ to obtain the corresponding state $\phi^{(m)} = \phi(g^{(m)})$;
2. compute the gradient of the functional $\frac{D\mathcal{J}}{Dg}(\phi^{(m)}, g^{(m)})$;
3. use the results of steps 1 and 2 to compute a step $\delta g^{(m)}$;
4. set $g^{(m+1)} = g^{(m)} + \delta g^{(m)}$.

Step 1 is a state solve; in the flow setting, it requires, for given values of the design parameters or control variables, the solution of the flow equations for the velocity, pressure, etc.; thus, *each iteration of the optimization algorithm* (Algorithm 2.1) *requires at least one solution of the flow equations.*

To complete the description of the optimization algorithm, we need to specify

- how the gradient of the functional in step 2 is determined,
- how the increments in the design parameters in step 3 are determined, and
- what "satisfactory convergence" means.

For step 3 of Algorithm 2.1, one can use one's favorite *optimization method*, e.g., gradient, conjugate gradient, quasi-Newton, etc. For determining convergence, one has to prescribe a *stopping criteria.* Here, we focus on step 2.

For the three examples of Section 2.1, we will first use, in Section 2.3.2, *sensitivity derivatives,* or more simply *sensitivities,* to help determine the gradient of the functional. Sensitivities, of course, are of interest in their own right; they tell what changes are effected in the state, i.e., the flow variables, when the control variables or design parameters are changed; here, however, we are interested in the use of sensitivities for optimization. Alternatively, one can use the solution of an *adjoint system* to determine the gradient of the functional; this approach is discussed for the three examples in Section 2.3.3.

2.3.2 The gradient of the functional through sensitivities

The gradient of the functional through sensitivities for Example 1

We consider the abstract setting of Example 1 on page 12. To determine the gradient of the functional for step 2 of the optimization algorithm (Algorithm 2.1), one can apply the chain rule[18] to $\mathcal{J}(\phi(g), g)$, where $\phi(g)$ denotes the solution of the state equation $F(\phi, g) = 0$, to obtain

$$\left.\frac{D\mathcal{J}}{Dg}\right|_{g^{(m)}} = \left.\frac{\partial\mathcal{J}}{\partial\phi}\right|_{g^{(m)}} \left.\frac{d\phi}{dg}\right|_{g^{(m)}} + \left.\frac{\partial\mathcal{J}}{\partial g}\right|_{g^{(m)}}, \tag{2.21}$$

[18]Note that we need to determine the *total* derivative of the functional with respect to the control. Thus, we take the view that the state equation $F(\phi, g) = 0$ may (at least in principle) be solved to determine the state ϕ as a function of the control g so that the functional $\mathcal{J}(\phi, g) = \mathcal{J}(\phi(g), g)$ may be viewed as a function of only the control g.

where $(\cdot)|_{g^{(m)}}$ denotes evaluation at the current iterate for the control, i.e., at $g^{(m)}$. The terms $\partial\mathcal{J}/\partial\phi$ and $\partial\mathcal{J}/\partial g$ are usually "easy" to determine. We still need to specify how the *sensitivities* $d\phi/dg$ are determined. One obvious method is to compute the state at a value $\widetilde{g}$ close to $g^{(m)}$ and then use a difference quotient approximation such as

$$\frac{d\phi}{dg}\bigg|_{g^{(m)}} \approx \frac{\phi(\widetilde{g}) - \phi(g^{(m)})}{\widetilde{g} - g^{(m)}} \tag{2.22}$$

to determine an approximation for the sensitivity. Having in hand a state simulation code, the finite difference quotient approach is trivial to program; one merely runs the state simulation code another time with a different input $\widetilde{g}$. However, it is expensive; a new state solve, i.e., a *nonlinear* flow solve, is needed for each parameter so that if we have K parameters, at each iteration of the optimization algorithm we need $K+1$ state solves: one in step 1 of Algorithm 2.1, and one more for each parameter to determine an approximation to the gradient of the functional.[19]

Another approach for determining sensitivities is to differentiate the state equation $F(\phi, g) = 0$ to obtain, again by the chain rule, the *sensitivity equations*

$$\left(\frac{\partial F}{\partial\phi}\bigg|_{g^{(m)}}\right)\frac{d\phi}{dg}\bigg|_{g^{(m)}} = -\frac{\partial F}{\partial g}\bigg|_{g^{(m)}}. \tag{2.24}$$

Note that $(\partial F/\partial\phi)|_{g^{(m)}}$ and $(\partial F/\partial g)|_{g^{(m)}}$ depend only on $g^{(m)}$ and $\phi(g^{(m)})$ so that they may be evaluated after step 1 of Algorithm 2.1; then, one may solve (2.24) for the sensitivity $(d\phi/dg)|_{g^{(m)}}$.

The system (2.24) is *linear* in the sensitivities $d\phi/dg$. If we have multiple design parameters, then we would have an equation such as (2.24) for each corresponding sensitivity. However, the left-hand-side operator $(\partial F/\partial\phi)|_{g^{(m)}}$ is independent of the particular design parameter.

Once the sensitivity $(d\phi/dg)|_{g^{(m)}}$ evaluated at the current values of the design parameters is determined by either the finite difference quotient or sensitivity equation approach, it may be used in (2.21) to determine the gradient of the functional evaluated at the current values of the design parameters.

The gradient of the functional through sensitivities for Example 2

We next consider the finite-dimensional optimization problem of Example 2 on page 13. First, let us derive the sensitivity equations. If we make a change $\delta\vec{g}$ in the control[20] $\vec{g}$, then,

[19] Additional flow solves can also be used to directly determine, through the formula

$$\frac{D\mathcal{J}}{Dg}\bigg|_{g^{(m)}} \approx \frac{\mathcal{J}\big(\phi(\widetilde{g}), \widetilde{g}\big) - \mathcal{J}\big(\phi(g^{(m)}), g^{(m)}\big)}{\widetilde{g} - g^{(m)}}, \tag{2.23}$$

an approximation to the derivative of the functional with respect to the control variables or design parameters. In fact, in practice, (2.23) is usually used instead of (2.21) and (2.22) when one uses finite difference quotients to determine approximations to the gradient of the functional with respect to the design parameters. Note that both (2.23) and the pair (2.21), (2.22) require an additional flow solve for each design parameter.

[20] For the sake of notational simplicity, in the rest of Section 2.3.2 we suppress the dependence of the control, etc., on the iteration counter in Algorithm 2.1, e.g., we simply write $\vec{g}$ instead of $\vec{g}^{(m)}$.

from the state equation (2.4), we have that the corresponding change $\delta\vec{\phi}$ in the state $\vec{\phi}$ must satisfy

$$C(\vec{\phi} + \delta\vec{\phi}) + \big((\vec{\phi} + \delta\vec{\phi})^T(\vec{\phi} + \delta\vec{\phi})\big)(\vec{\phi} + \delta\vec{\phi}) = D(\vec{g} + \delta\vec{g})\,.$$

Since the pair $(\vec{\phi}, \vec{g})$ also satisfies the state equations, we have that

$$C\delta\vec{\phi} + (\vec{\phi}^T\vec{\phi})\delta\vec{\phi} + 2(\vec{\phi}^T\delta\vec{\phi})\vec{\phi} + 2(\vec{\phi}^T\delta\vec{\phi})\delta\vec{\phi} + (\delta\vec{\phi}^T\delta\vec{\phi})\vec{\phi} + (\delta\vec{\phi}^T\delta\vec{\phi})\delta\vec{\phi} = D\delta\vec{g}\,. \quad (2.25)$$

If we let $|\delta\vec{g}| \to 0$ and assume that $|\delta\vec{\phi}| = O(|\delta\vec{g}|)$, where $|\cdot|$ denotes the Euclidean norm, we obtain

$$C\delta\vec{\phi} + (\vec{\phi}^T\vec{\phi})\delta\vec{\phi} + 2(\vec{\phi}^T\delta\vec{\phi})\vec{\phi} = D\delta\vec{g}\,. \quad (2.26)$$

This is a sensitivity equation in the sense that, given an infinitesimal change $\delta\vec{g}$ in the control $\vec{g}$, the solution $\delta\vec{\phi}$ of (2.26) gives the corresponding infinitesimal change in the state $\vec{\phi}$.

Since the control $\vec{g} \in \mathbb{R}^K$ is finite dimensional, we can go further. Choose $\delta\vec{g} = \delta g_k \vec{e}_k$, where $\vec{e}_k$ denotes the kth unit vector in $\mathbb{R}^K$. Then, we can define the sensitivity $\vec{\phi}_k$ of the state vector $\vec{\phi}$ with respect to the kth component g_k of the control vector $\vec{g}$, i.e.,

$$\vec{\phi}_k = \frac{\partial\vec{\phi}}{\partial g_k}\,, \qquad k = 1, \ldots, K.$$

Letting $\delta\vec{g} = \delta g_k \vec{e}_k$ in (2.25), dividing by δg_k, and then letting $\delta g_k \to 0$, we obtain

$$C\vec{\phi}_k + (\vec{\phi}^T\vec{\phi})\vec{\phi}_k + 2(\vec{\phi}^T\vec{\phi}_k)\vec{\phi} = D\vec{e}_k\,, \qquad k = 1, \ldots, K. \quad (2.27)$$

These are the K sensitivity equations from which the K sensitivities $\vec{\phi}_k$, $k = 1, \ldots, K$, of the state $\vec{\phi}$ with respect to the components of the control vector $\vec{g}$ may be determined.

Note that the sensitivity equations (2.26) and (2.27) are linear in the sensitivities and that the operator on the left-hand side is the same for all sensitivities. However, the right-hand sides depend on which sensitivity one is seeking. For example, in (2.27), the only dependence on the index k is in the unit vector $\vec{e}_k$ appearing on the right-hand side.

To summarize, the solution of (2.27) provides $\vec{\phi}_k$, $k = 1, \ldots, K$, the sensitivity of the state vector $\vec{\phi}$ with respect to each of the components g_k, $k = 1, \ldots, K$, of the control vector $\vec{g}$. The solution of (2.26) provides the change $\delta\vec{\phi}$ in the state due to a change in the control $\delta\vec{g}$.

We can proceed in a similar manner for the functional given in (2.3). Given an (infinitesimal) change $\delta\vec{g}$ in the control, the corresponding change in the functional is given by

$$\delta\mathcal{J} = \vec{\phi}^T A\delta\vec{\phi} + \vec{g}^T B\delta\vec{g}\,,$$

where $\delta\vec{\phi}$ is the solution of the sensitivity equation (2.26). Also, we have that

$$\frac{D\mathcal{J}}{Dg_k} = \vec{\phi}^T A\vec{\phi}_k + \vec{g}^T B\vec{e}_k\,, \qquad k = 1, \ldots, K, \quad (2.28)$$

where $\vec{\phi}_k$ is the solution of the sensitivity equation (2.27). Note that to determine the gradient of the functional with respect to all K components g_k, $k = 1, \ldots, K$, of the control vector $\vec{g}$, one must solve the K sensitivity equations (2.27).

Now that we have derived formulas for the sensitivities and the gradient of the functional, we recognize that we could have derived them by formal differentiation. For example, (2.27) and (2.28) can be derived directly by differentiating (2.4) and (2.3), respectively, with respect to the kth component of $\vec{g}$. This approach usually works; we will apply it in the next example.

We note that the ability to obtain an explicit formula such as (2.28) for the sensitivity-based gradient of the functional with respect to the design parameters is a feature of problems with a finite-dimensional set of design parameters. As the next example will illustrate, it does not require that the state or constraint set be finite dimensional. The situation for infinite-dimensional control sets is somewhat different; we will examine that case in Section 2.4.2.

The gradient of the functional through sensitivities for Example 3

For Example 3 on page 13, the sensitivities are defined by

$$\phi_k = \frac{\partial \phi}{\partial \alpha_k}, \quad k = 1, \ldots, K .$$

By direct (formal) differentiation with respect to the design parameters α_k, the derivative of the functional (2.7) with respect to a parameter α_k is given by

$$\frac{D\mathcal{J}}{D\alpha_k} = \sigma\alpha_k + \int_\Omega (\phi - \Phi)\phi_k \, d\Omega\,, \quad k = 1, \ldots, K . \tag{2.29}$$

Note the appearance of the sensitivity ϕ_k in (2.29). By direct differentiation of the state equations (2.5) and (2.6) with respect to the design parameters α_k, the sensitivity equations are determined to be

$$-\nabla \cdot (a\nabla\phi_k) + \mathbf{b} \cdot \nabla\phi_k + 3\phi^2\phi_k = f_k \quad \text{in } \Omega\,, \quad k = 1, \ldots, K, \tag{2.30}$$

and

$$\phi_k = 0 \quad \text{on } \Gamma\,, \quad k = 1, \ldots, K . \tag{2.31}$$

Note that for each k, the right-hand side (the data) of the sensitivity system (2.30)–(2.31) is different. Thus, given a candidate state ϕ and K design parameters $\alpha_1, \ldots, \alpha_K$, in order to evaluate the gradient of the functional one must solve K linear sensitivity systems. However, note that the operator

$$-\nabla \cdot (a\nabla) + \mathbf{b} \cdot \nabla + 3\phi^2$$

appearing on the left-hand side of the sensitivity equations is independent of k.

We see that, even when the constraint is a partial differential equation, if one has a finite number of control or design parameters, then one is able to obtain an explicit formula in terms of the sensitivities for the gradient of the functional with respect to the control parameters. For example, from (2.29), we have that

$$\begin{pmatrix} \dfrac{D\mathcal{J}}{D\alpha_1} \\ \vdots \\ \dfrac{D\mathcal{J}}{D\alpha_K} \end{pmatrix} = \begin{pmatrix} \sigma\alpha_1 + \displaystyle\int_\Omega (\phi - \Phi)\phi_1 \\ \vdots \\ \sigma\alpha_K + \displaystyle\int_\Omega (\phi - \Phi)\phi_K \end{pmatrix}, \tag{2.32}$$

where the sensitivities $\phi_1, \ldots, \phi_K$ are determined from (2.30)–(2.31).

2.3.3 The gradient of the functional through adjoint equations

One can also use solutions of *adjoint equations* to determine the gradient of the functional for step 2 of the optimization algorithm (Algorithm 2.1).

The gradient of the functional through adjoints for Example 1

We consider the abstract setting of Example 1 on page 12. Let $\lambda^{(m)}$ be the solution of the adjoint equation

$$\left(\frac{\partial F}{\partial \phi}\bigg|_{g^{(m)}} \right)^* \lambda^{(m)} = \left(\frac{\partial \mathcal{J}}{\partial \phi}\bigg|_{g^{(m)}} \right)^* . \tag{2.33}$$

The left-hand side of the adjoint equation (2.33) involves the adjoint operator $(\partial F/\partial \phi)^*$ evaluated at the current guess $g^{(m)}$ for the control. The right-hand side of the adjoint equation (2.33) involves $\partial \mathcal{J}/\partial \phi$, the derivative of the functional with respect to the state, also evaluated at the current guess $g^{(m)}$ for the control. Note that $(\partial F/\partial \phi|_{g^{(m)}})^*$ and $(\partial \mathcal{J}/\partial \phi|_{g^{(m)}})^*$ depend only on $g^{(m)}$ and $\phi(g^{(m)})$ so that they may be evaluated after step 1 of Algorithm 2.1. Then, one may solve (2.33) for the adjoint variable $\lambda^{(m)}$.

The system (2.33) is *linear* in the adjoint variable $\lambda^{(m)}$. Note that if one has multiple design parameters, then one has a *single* equation (2.33) for the adjoint variable. The last observation will become transparent when we consider Examples 2 and 3.

Recall the expression (2.21),

$$\frac{D\mathcal{J}}{Dg}\bigg|_{g^{(m)}} = \frac{\partial \mathcal{J}}{\partial \phi}\bigg|_{g^{(m)}} \frac{d\phi}{dg}\bigg|_{g^{(m)}} + \frac{\partial \mathcal{J}}{\partial g}\bigg|_{g^{(m)}},$$

for the gradient of the functional in terms of the sensitivity $d\phi/dg$. Using (2.33) to eliminate $\partial \mathcal{J}/\partial \phi$ yields

$$\frac{D\mathcal{J}}{Dg}\bigg|_{g^{(m)}} = (\lambda^{(m)})^* \left(\frac{\partial F}{\partial \phi}\bigg|_{g^{(m)}} \right) \frac{d\phi}{dg}\bigg|_{g^{(m)}} + \frac{\partial \mathcal{J}}{\partial g}\bigg|_{g^{(n)}} .$$

Then, substituting the sensitivity equation (2.24) yields

$$\frac{D\mathcal{J}}{Dg}\bigg|_{g^{(m)}} = -(\lambda^{(m)})^* \frac{\partial F}{\partial g}\bigg|_{g^{(m)}} + \frac{\partial \mathcal{J}}{\partial g}\bigg|_{g^{(m)}} . \tag{2.34}$$

This is an expression for the gradient of the functional in terms of the solution of the adjoint equation (2.33).

Once the adjoint variable $\lambda^{(m)}$ evaluated at the current values of the design parameters is determined by solving the adjoint equations (2.33), it may be used in (2.34) to determine the gradient of the functional evaluated at the current values of the design parameters.

From (2.34), it is easy to see that the optimality condition (2.9) simply states that the gradient of the functional is zero at the optimal values of the state and control variables. This observation remains valid for our other examples.

The gradient of the functional through adjoints for Example 2

We next consider the finite-dimensional optimization problem of Example 2 on page 13. Let[21] $\vec{\lambda}$ denote the solution of the adjoint equation[22]

$$C^T\vec{\lambda} + 2(\vec{\phi}^T\vec{\lambda})\vec{\phi} + (\vec{\phi}^T\vec{\phi})\vec{\lambda} = A\vec{\phi}\,, \tag{2.35}$$

where $\vec{\phi}$ denotes the state determined in step 1 of Algorithm 2.1. Recall the formula (2.28) for the derivative of the functional $\mathcal{J}$ given in (2.3) in terms of the sensitivities $\vec{\phi}_k$ with respect to the components g_k of the control vector $\vec{g}$:

$$\frac{D\mathcal{J}}{Dg_k} = \vec{\phi}_k^T A\vec{\phi} + \vec{e}_k^T B\vec{g}\,, \qquad k = 1, \dots, K.$$

We can substitute the adjoint equation (2.35) for the term $A\vec{\phi}$ to obtain

$$\frac{D\mathcal{J}}{Dg_k} = \vec{\phi}_k^T\left(C^T\vec{\lambda} + 2(\vec{\phi}^T\vec{\lambda})\vec{\phi} + (\vec{\phi}^T\vec{\phi})\vec{\lambda}\right) + \vec{e}_k^T B\vec{g}$$

or, equivalently,

$$\frac{D\mathcal{J}}{Dg_k} = \left(C\vec{\phi}_k + 2(\vec{\phi}^T\vec{\phi}_k)\vec{\phi} + (\vec{\phi}^T\vec{\phi})\vec{\phi}_k\right)^T\vec{\lambda} + \vec{e}_k^T B\vec{g}\,.$$

Combining with the sensitivity equation (2.27) yields

$$\frac{D\mathcal{J}}{Dg_k} = \vec{e}_k^T(D^T\vec{\lambda} + B\vec{g})\,. \tag{2.36}$$

This formula[23] may be used to compute the gradient of the functional in step 2 of Algorithm 2.1 in the following way. At the end of step 1 of the algorithm, we know the current values for control and state, i.e., we know $\vec{g}$ and $\vec{\phi}$. We use the latter in (2.35) to solve for $\vec{\lambda}$. Then, we can substitute in (2.36) to determine all components of the gradient of the functional.

One very important observation is that, using (2.36), all components of the gradient of the functional may be determined with only a *single* solution of the adjoint equation (2.35). This should be contrasted with the use of (2.28), for which one must solve the K sensitivity equations (2.27) to compute the K components of the gradient of the functional.

The gradient of the functional through adjoints for Example 3

We now determine the formula for the gradient of the functional for Example 3 on page 13. Let λ be a solution of the *adjoint equations*

$$-\nabla\cdot(a\nabla\lambda) - \nabla\cdot(\mathbf{b}\lambda) + 3(\phi)^2\lambda = \phi - \Phi \quad \text{in } \Omega \tag{2.37}$$

[21]Again, for the sake of notational simplicity, in the rest of Section 2.3.3 we suppress the dependence of the adjoint variable, etc., on the iteration counter of Algorithm 2.1, e.g., we simply write $\vec{\lambda}$ instead of $\vec{\lambda}^{(m)}$.

[22]For this example and for Example 3, we take advantage of the fact that we have already determined the adjoint equations in Section 2.2. In some cases, the definition of the adjoint system is nearly obvious but, if it is not, one can always determine it by deriving the optimality system as was done in Section 2.2.

[23]Note that (2.36) may also be deduced from the optimality condition (2.10) for this example; this is not surprising since that condition is merely a statement that the gradient of the functional (2.3) vanishes at optimum points.

and

$$\lambda = 0 \quad \text{on } \Gamma\,, \tag{2.38}$$

where ϕ is the solution of the state equations (2.5) and (2.6) for the current values of the design parameters. Recall that, for this example, the gradient of the functional (2.7) in terms of sensitivities is given by (2.29), i.e.,

$$\frac{D\mathcal{J}}{D\alpha_k} = \sigma\alpha_k + \int_\Omega (\phi - \Phi)\phi_k \, d\Omega\,, \quad k = 1, \ldots, K\,, \tag{2.39}$$

where the sensitivities ϕ_k are solutions of the sensitivity equations (2.30)–(2.31). We use the adjoint equation (2.37) to substitute for $(\phi - \Phi)$ in (2.39) to obtain

$$\frac{D\mathcal{J}}{D\alpha_k} = \sigma\alpha_k + \int_\Omega \Big(- \nabla\cdot(a\nabla\lambda) - \nabla\cdot(\mathbf{b}\lambda) + 3\phi^2\lambda\Big)\phi_k \, d\Omega\,, \quad k = 1, \ldots, K\,.$$

Integrating by parts and using the boundary conditions $\phi_k = \lambda = 0$ on Γ results in

$$\frac{D\mathcal{J}}{D\alpha_k} = \sigma\alpha_k + \int_\Omega \Big(- \nabla\cdot(a\nabla\phi_k) + \mathbf{b}\cdot\nabla\phi_k + 3\phi^2\phi_k\Big)\lambda \, d\Omega\,, \quad k = 1, \ldots, K.$$

Substituting the sensitivity equation (2.30) yields[24]

$$\frac{D\mathcal{J}}{D\alpha_k} = \sigma\alpha_k + \int_\Omega \lambda f_k \, d\Omega\,, \quad k = 1, \ldots, K. \tag{2.40}$$

This expresses the components of the gradient of the functional with respect to the design parameters α_k in terms of the solution λ of the adjoint equations.

To compute *all K components* of the gradient of the functional using the adjoint equation approach requires the solution of a *single linear system*, the adjoint equations (2.37)–(2.38); in contrast, determining the K components of the gradient of the functional using the sensitivity equation approach requires the solution of K linear systems, the K sensitivity systems (2.30)–(2.31).

2.4 Other types of controls

In Example 3 defined on page 13, the control acts in the forcing function in the partial differential equation. Moreover, the control is determined by a finite number of parameters. In the chapters that follow, we will encounter several other types of controls, including

- *boundary controls:* controls that act on the boundary Γ of the flow domain Ω;
- *infinite-dimensional controls:* controls that are functions and not merely determined by a finite number of parameters;
- *partial controls:* controls that are applied on part of the flow domain, i.e., on some subset $\Omega_c \subset \Omega$, or on part of the boundary of the flow domain, i.e., on some subset $\Gamma_c \subset \Gamma$;

[24] Note again that (2.40) may also be deduced from the optimality condition (2.17) for this example; this is not surprising since that condition is merely a statement that the gradient of the functional (2.7) vanishes at optimum points.

- *shape controls:* controls that determine the shape of the boundary of the flow domain, or part of that boundary.

Partial controls are considered in Section 2.5 and, in the chapters that follow, we will see a number of additional examples. Likewise, we will see a number of examples of shape controls and we defer discussion of that case until we consider those examples. In this section, we briefly discuss boundary and infinite-dimensional controls. Additional examples of these two cases will also be encountered in the chapters that follow.

2.4.1 Boundary controls

We consider the simple optimization problem of minimizing the functional

$$\begin{aligned}\mathcal{J}(\phi,\alpha_1,\dots,\alpha_K,\beta_1,\dots,\beta_J) &= \frac{1}{2}\int_\Omega (\phi-\Phi)^2\,d\Omega \\ &+\frac{\sigma}{2}\left(\sum_{k=1}^{K}(\alpha_k)^2+\sum_{j=1}^{J}(\beta_j)^2\right)\end{aligned} \tag{2.41}$$

subject to

$$\begin{cases} -\Delta\phi = f \quad \text{in } \Omega\,, \\ \phi = \displaystyle\sum_{k=1}^{K}\alpha_k g_k(x,y) \quad \text{on } \Gamma_d\,, \quad \text{and} \quad \dfrac{\partial\phi}{\partial n} = \displaystyle\sum_{j=1}^{J}\beta_j h_j(x,y) \quad \text{on } \Gamma_n\,, \end{cases} \tag{2.42}$$

where f, g_k, and h_j are given functions and Γ_d and Γ_n are disjoint and together make up the boundary Γ of Ω, i.e., $\Gamma_d \cap \Gamma_n = \emptyset$ and $\overline{\Gamma}_d \cup \overline{\Gamma}_n = \Gamma$. The α_k's and β_j's are the design parameters; note the appearance of the controls in the boundary conditions instead of the right-hand side of the partial differential equation.

The simple problem we consider here suffices to show the effect that the appearance of boundary controls in the state system has on the data in the optimality, sensitivity, and adjoint systems.

The optimality system

The optimality system that is used in a one-shot approach to solving the optimization problem[25] is based on finding stationary points of the Lagrangian functional

$$\begin{aligned}\mathcal{L}(\phi,\alpha_1,\dots,\alpha_K,\beta_1,\dots,\beta_J,\xi,\mu,\nu) &= \mathcal{J}(\phi,\alpha_1,\dots,\alpha_K,\beta_1,\dots,\beta_J) \\ -\int_\Omega \xi(-\Delta\phi - f)\,d\Omega &- \int_{\Gamma_D}\mu\left(\phi-\sum_{k=1}^{K}\alpha_k g_k\right)d\Gamma \\ &-\int_{\Gamma_N}\nu\left(\frac{\partial\phi}{\partial n}-\sum_{j=1}^{J}\beta_j h_j\right)d\Gamma\,,\end{aligned}$$

[25] As we have previously seen, the optimality system can also be used to help define the adjoint system and functional gradients in the adjoint equation approach to determining the gradient of the functional.

where ξ, μ, and ν are Lagrange multiplier functions respectively introduced to enforce the partial differential equation and the Dirichlet and Neumann boundary conditions of the state system (2.42).

Setting the first variation of the Lagrangian functional with respect to the Lagrange multipliers ξ, μ, and ν equal to zero recovers the state system (2.42).

Setting the first variation of the Lagrangian functional with respect to the state ϕ equal to zero yields the adjoint system

$$\begin{cases} -\Delta\xi = \phi - \Phi \quad \text{in } \Omega\,, \\ \xi = 0 \quad \text{on } \Gamma_d\,, \qquad \text{and} \qquad \dfrac{\partial \xi}{\partial n} = 0 \quad \text{on } \Gamma_n \end{cases} \tag{2.43}$$

along with

$$\mu = -\frac{\partial \xi}{\partial n}\bigg|_{\Gamma_D} \qquad \text{and} \qquad \nu = \xi|_{\Gamma_N}\,. \tag{2.44}$$

The derivation of (2.43) and (2.44) proceeds as follows. Setting the first variation of the Lagrangian functional with respect to the state ϕ equal to zero is equivalent to

$$\int_\Omega \big((\phi - \Phi)\widetilde{\phi} + \xi\Delta\widetilde{\phi}\big)\,d\Omega - \int_{\Gamma_D} \mu\widetilde{\phi}\,d\Gamma - \int_{\Gamma_N} \nu\frac{\partial\widetilde{\phi}}{\partial n}\,d\Gamma = 0$$

or, after integrating by parts to remove, as much as possible, derivatives from $\widetilde{\phi}$,

$$\begin{aligned} \int_\Omega \widetilde{\phi}(\phi - \Phi + \Delta\xi)\,d\Omega - \int_{\Gamma_D} \widetilde{\phi}\left(\mu + \frac{\partial\xi}{\partial n}\right)d\Gamma + \int_{\Gamma_D} \frac{\partial\widetilde{\phi}}{\partial n}\xi\,d\Gamma \\ + \int_{\Gamma_N} \frac{\partial\widetilde{\phi}}{\partial n}(\xi - \nu)\,d\Gamma - \int_{\Gamma_N} \widetilde{\phi}\frac{\partial\xi}{\partial n}\,d\Gamma = 0\,. \end{aligned} \tag{2.45}$$

Since we may pick $\widetilde{\phi}$ in Ω and $\widetilde{\phi}$ and $\partial\widetilde{\phi}/\partial n$ on Γ_D and Γ_N arbitrarily, we recover the partial differential equation and the Dirichlet and Neumann boundary conditions in the adjoint system (2.43) by respectively setting the first, third, and fifth terms in the left-hand side of (2.45) separately to zero. Setting the second and fourth terms equal to zero recovers (2.44).

Setting the first variation of the Lagrangian functional with respect to each of the design parameters $\{\alpha_k\}_{k=1}^K$ and $\{\beta_j\}_{j=1}^J$ equal to zero easily yields that

$$\sigma\alpha_k + \int_{\Gamma_D} \mu g_k\,d\Gamma = 0 \qquad \text{for } k = 1, \dots, K$$

and

$$\sigma\beta_j + \int_{\Gamma_N} \nu h_j\,d\Gamma = 0 \qquad \text{for } j = 1, \dots, J$$

or, using (2.44),

$$\sigma\alpha_k - \int_{\Gamma_D} \left(\frac{\partial\xi}{\partial n}\right) g_k\,d\Gamma = 0 \qquad \text{for } k = 1, \dots, K \tag{2.46}$$

and

$$\sigma\beta_j + \int_{\Gamma_N} \xi h_j\,d\Gamma = 0 \qquad \text{for } j = 1, \dots, J. \tag{2.47}$$

To summarize, the *optimality system* from which optimal states, design parameters, and adjoint states may be determined is given by the state system (2.42), the adjoint system (2.43), and the optimality conditions (2.46) and (2.47).

The gradient of the functional in terms of sensitivities

The set of $J + K$ *sensitivity systems* is easy to determine by differentiating the state system with respect to each design parameter. Thus, if

$$\phi_k = \frac{\partial \phi}{\partial \alpha_k} \quad \text{for } k = 1, \ldots, K \qquad \text{and} \qquad \psi_j = \frac{\partial \phi}{\partial \beta_j} \quad \text{for } j = 1, \ldots, J$$

denote the sensitivities, then these are determined from the sensitivity equations

$$\text{for } k = 1, \ldots, K, \qquad \begin{cases} -\Delta \phi_k = 0 & \text{in } \Omega\,, \\ \phi_k = g_k \quad \text{on } \Gamma_d\,, \quad \text{and} & \dfrac{\partial \phi_k}{\partial n} = 0 \quad \text{on } \Gamma_n, \end{cases} \tag{2.48}$$

and

$$\text{for } j = 1, \ldots, J, \qquad \begin{cases} -\Delta \psi_j = 0 & \text{in } \Omega\,, \\ \psi_j = 0 \quad \text{on } \Gamma_d\,, \quad \text{and} & \dfrac{\partial \psi_j}{\partial n} = h_j \quad \text{on } \Gamma_n\,. \end{cases} \tag{2.49}$$

Notice the homogeneous right-hand sides in the partial differential equations and the inhomogeneous boundary conditions. Also note that the inhomogeneity in the boundary conditions for ϕ_k, the sensitivity with respect to a parameter in the Dirichlet boundary condition of the state system, appears in the Dirichlet boundary condition of the sensitivity system for that sensitivity. The analogous relation holds for ψ_j and the Neumann boundary condition.

The components of the gradient of the functional (2.41) in terms of sensitivity variables are determined by the chain rule to be

$$\frac{D\mathcal{J}}{D\alpha_k} = \sigma\alpha_k + \int_\Omega (\phi - \Phi)\phi_k \, d\Omega \qquad \text{for } k = 1, \ldots, K \tag{2.50}$$

and

$$\frac{D\mathcal{J}}{D\beta_j} = \sigma\beta_j + \int_\Omega (\phi - \Phi)\psi_j \, d\Omega \qquad \text{for } j = 1, \ldots, J. \tag{2.51}$$

Thus, given the current guess for the design parameters $\{\alpha_k\}_{k=1}^K$ and $\{\beta_j\}_{j=1}^J$, one can determine the $K + J$ components of the gradient of the functional (2.41) by first solving the state system (2.42) for the state ϕ, then solving the $K + J$ sensitivity systems (2.48) and (2.49) for the sensitivities $\{\phi_k\}_{k=1}^K$ and $\{\psi_j\}_{j=1}^J$, and then substituting the results into (2.50) and (2.51). Once determined, the components of the gradient of the functional may be used in an optimization method such as Algorithm 2.1 to determine a new guess for the design parameters.

The gradient of the functional in terms of adjoint variables

From (2.43), the adjoint system for the adjoint variable λ is given by

$$\begin{cases} -\Delta\lambda = \phi - \Phi \quad \text{in } \Omega\,, \\ \lambda = 0 \quad \text{on } \Gamma_d\,, \qquad \text{and} \qquad \dfrac{\partial\lambda}{\partial n} = 0 \quad \text{on } \Gamma_n\,. \end{cases} \tag{2.52}$$

Notice the homogeneous boundary conditions and that the adjoint system does not depend in any way on the design parameters, i.e., in this case, it does not depend on j or k.

By a process similar to that shown on page 28, the components of the gradient of the functional in terms of the adjoint variable are found to be[26]

$$\frac{D\mathcal{J}}{D\alpha_k} = \sigma\alpha_k - \int_{\Gamma_D} \frac{\partial\lambda}{\partial n} g_k \, d\Gamma \qquad \text{for } k = 1, \dots, K \tag{2.53}$$

and

$$\frac{D\mathcal{J}}{D\beta_j} = \sigma\beta_j + \int_{\Gamma_N} \lambda h_j \, d\Gamma \qquad \text{for } j = 1, \dots, J\,. \tag{2.54}$$

Now, given the current guess for the design parameters $\{\alpha_k\}_{k=1}^K$ and $\{\beta_j\}_{j=1}^J$, one can determine the $K + J$ components of the gradient of the functional (2.41) by first solving the state system (2.42) for the state ϕ, then solving the *single* adjoint system (2.52) for the adjoint variable λ, and then substituting the results into (2.53) and (2.54). Once determined, the components of the gradient of the functional may be used in an optimization method such as Algorithm 2.1 to determine a new guess for the design parameters. Thus, we see that one can determine the $K + J$ components of the gradient of the functional by solving either the $K + J$ sensitivity systems (2.48) and (2.49) or the single adjoint system (2.52).

2.4.2 Infinite-dimensional controls

In the examples we have looked at so far, the controls have been determined by a finite number of parameters, e.g., $\{\alpha_k\}_{k=1}^K$ for Example 3 on page 13. One can also pose problems in which controls are infinite dimensional. We will again use an optimal control problem involving a scalar partial differential equation to illustrate how infinite-dimensional controls are treated and what effect they have on optimality systems and sensitivity and adjoint systems.

A control problem with infinite-dimensional controls

Let Ω be an open domain in $\mathbb{R}^n$, $n = 2, 3$, having boundary Γ. Let ϕ denote the scalar-valued state variable and f and g denote the (infinite-dimensional) scalar-valued control functions. The constraints are the nonlinear, second-order, elliptic partial differential equation

$$-\Delta\phi + \phi^3 = f \quad \text{in } \Omega \tag{2.55}$$

along with the boundary condition

$$\frac{\partial\phi}{\partial n} = g \quad \text{on } \Gamma\,. \tag{2.56}$$

[26]The formulas (2.53) and (2.54) can also be deduced from the optimality conditions (2.46) and (2.47) since the latter two equations are merely a statement of the fact that the gradient of the functional vanishes at the optimum.

The objective, cost, or performance functional is given by

$$\mathcal{J}(\phi, f, g) = \frac{1}{2}\int_\Omega (\phi - \Phi)^2 \, d\Omega + \frac{\sigma_1}{2}\int_\Omega f^2 \, d\Omega + \frac{\sigma_2}{2}\int_\Gamma g^2 \, d\Gamma \,, \qquad (2.57)$$

where Φ is a given function and σ_1 and σ_2 are given penalty parameters. The optimization problem is as follows:

> *Find ϕ, f, and g such that the functional $\mathcal{J}$ defined in* (2.57) *is minimized subject to the requirements that ϕ, f, and g satisfy the constraints* (2.55)–(2.56).

The functional $\mathcal{J}(\phi, f, g) = \mathcal{J}\big(\phi(f, g), f, g\big)$ may be viewed, through the state system (2.55)–(2.56), as depending on only the control functions f and g in much the same way as the functional in (2.7) could be viewed as a function of only the design parameters $\{\alpha_1, \ldots, \alpha_K\}$.

This example involves controls that are *infinite dimensional.*[27] As we shall see, this deviation from Example 3 has a significant effect on how one solves control problems.

The optimality system

For the example problem just defined, we introduce the Lagrangian functional

$$\mathcal{L}(\phi, f, g, \xi, \eta) = \mathcal{J}(\phi, f, g) - \int_\Omega \xi(-\Delta\phi + \phi^3 - f) \, d\Omega - \int_\Gamma \eta \left(\frac{\partial \phi}{\partial n} - g\right) d\Gamma \,,$$

where ξ and η are Lagrange multipliers introduced to enforce the differential equation (2.55) and the boundary condition (2.56), respectively. Then, as for Example 3 (see page 19), setting the first variations of $\mathcal{L}$ with respect to these Lagrange multipliers equal to zero recovers the state system (2.55)–(2.56).

Setting the first variation of $\mathcal{L}$ with respect to the state ϕ equal to zero results in

$$\int_\Omega (\phi - \Phi)\widetilde{\phi} \, d\Omega - \int_\Omega \xi(-\Delta\widetilde{\phi} + 3\phi^2\widetilde{\phi}) \, d\Omega - \int_\Gamma \eta \frac{\partial \widetilde{\phi}}{\partial n} \, d\Gamma = 0 \,,$$

where $\widetilde{\phi}$ denotes an arbitrary variation in the state variable ϕ. Integrating by parts to remove, where possible, derivatives from $\widetilde{\phi}$, one obtains

$$\int_\Omega \widetilde{\phi}\left(\Delta\xi - 3\phi^2\xi + (\phi - \Phi)\right) d\Omega - \int_\Gamma \widetilde{\phi}\frac{\partial \xi}{\partial n} \, d\Gamma - \int_\Gamma \frac{\partial \widetilde{\phi}}{\partial n}(\eta - \xi) \, d\Gamma = 0 \,.$$

Since $\widetilde{\phi}$ is arbitrary, this relation implies (using arguments similar to those used for Example 3 on page 20) that

$$-\Delta\xi + 3\phi^2\xi = (\phi - \Phi) \quad \text{in } \Omega \qquad \text{and} \qquad \frac{\partial \xi}{\partial n} = 0 \quad \text{on } \Gamma \qquad (2.58)$$

[27]The infinite dimensionality of the controls is easy to discern if one thinks of expressing, say, f in (2.55) in a power series or Fourier series representation; in general, such representations involve an infinite number of coefficients, i.e., parameters. Thus, in this sense, it takes an infinite number of parameters to specify the functions f and g. This should be contrasted with the control of Example 3 on page 13. There the control was also a function appearing on the right-hand side of the partial differential equation. However, that control function was restricted to being a linear combination of the K given functions f_k so that only K parameters $\{\alpha_k\}_{k=1}^K$ were required to specify a specific control.

and

$$\eta = \xi|_\Gamma \quad \text{on } \Gamma\,. \tag{2.59}$$

Setting the first variation of $\mathcal{L}$ with respect to the distributed control f to zero results in

$$\int_\Omega \widetilde{f}(\sigma_1 f + \xi)\, d\Omega = 0$$

which, since $\widetilde{f}$ is arbitrary, implies that

$$f = -\frac{1}{\sigma_1}\xi \quad \text{in } \Omega\,.$$

Setting the first variation of $\mathcal{L}$ with respect to the boundary control g to zero results in

$$\int_\Gamma \widetilde{g}(\sigma_2 g + \eta)\, d\Gamma = 0\,,$$

which, since $\widetilde{g}$ is arbitrary, implies that

$$g = -\frac{1}{\sigma_2}\eta \quad \text{on } \Gamma\,. \tag{2.60}$$

Collecting the results of setting the first variations of the Lagrangian functional to zero results in the following optimality system:

$$\begin{array}{rl}
\text{state equations} \;\Rightarrow & \left\{\begin{array}{l} -\Delta\phi + \phi^3 = f \quad \text{in } \Omega, \\ \dfrac{\partial\phi}{\partial n} = g \quad \text{on } \Gamma\,; \end{array}\right. \\
\text{adjoint equations} \;\Rightarrow & \left\{\begin{array}{l} -\Delta\xi + 3\phi^2\xi = \phi - \Phi \quad \text{in } \Omega, \\ \dfrac{\partial\xi}{\partial n} = 0 \quad \text{on } \Gamma\,; \end{array}\right. \\
\text{optimality conditions} \;\Rightarrow & \left\{\begin{array}{l} f = -\dfrac{1}{\sigma_1}\xi \quad \text{in } \Omega, \\ g = -\dfrac{1}{\sigma_2}\xi|_\Gamma \quad \text{on } \Gamma\,, \end{array}\right.
\end{array} \tag{2.61}$$

where we have combined (2.59) and (2.60) to eliminate the Lagrange multiplier η in one of the optimality conditions.

For this example, it is again possible, especially in two dimensions, to solve the coupled optimality system (2.61) so that optimal states and controls can be obtained without an optimization iteration, i.e., in “one shot.”

As far as the derivation and the form of the optimality system (2.61) are concerned, there seems to be no difficulty in treating the infinite-dimensional controls f and g. For example, in the derivation, we were able to use the same techniques we used for the case of finite-dimensional controls.

We can easily eliminate the controls f and g from the optimality system (2.61) to obtain the simplified system[28]

$$-\Delta\phi + \phi^3 = -\frac{1}{\sigma_1}\xi \quad \text{in } \Omega \qquad \text{and} \qquad \frac{\partial\phi}{\partial n} = -\frac{1}{\sigma_2}\xi \quad \text{on } \Gamma\,;$$

$$-\Delta\xi + 3\phi^2\xi = \phi - \Phi \quad \text{in } \Omega \qquad \text{and} \qquad \frac{\partial\xi}{\partial n} = 0 \quad \text{on } \Gamma,$$

which involves only the state ϕ and the adjoint state ξ. Once these have been determined, the optimal controls f and g can be determined from the optimality conditions $f = -(1/\sigma_1)\xi$ and $g = -(1/\sigma_2)\xi|_\Gamma$.

The gradient of the functional through sensitivities

For the example problem given on page 33, let us consider how sensitivities with respect to changes in the (infinite-dimensional) control function f can be determined.

Suppose we change the distributed control f to $f + \epsilon\widetilde{f}$, where $\widetilde{f}$ is arbitrary within whatever function space[29] we are using for the control f. This change in f induces the state to change from ϕ to $\phi + \epsilon\phi_f$. The change ϕ_f in the state is determined from the state system (2.55)–(2.56), i.e., we have that

$$-\Delta(\phi + \epsilon\phi_f) + (\phi + \epsilon\phi_f)^3 = f + \epsilon\widetilde{f} \quad \text{in } \Omega \qquad \text{and} \qquad \frac{\partial}{\partial n}(\phi + \epsilon\phi_f) = g \quad \text{on } \Gamma\,.$$

Since ϕ and f also satisfy the state system (2.55)–(2.56), we obtain, by letting $\epsilon \to 0$,

$$-\Delta\phi_f + 3\phi^2\phi_f = \widetilde{f} \quad \text{in } \Omega \qquad \text{and} \qquad \frac{\partial\phi_f}{\partial n} = 0 \quad \text{on } \Gamma\,. \tag{2.62}$$

The *sensitivity equations* (2.62) should be interpreted as follows. An infinitesimal change of the control function f in the "direction" of $\widetilde{f}$ induces the infinitesimal change in the "direction" of ϕ_f in the state ϕ; the induced change ϕ_f is determined from a given $\widetilde{f}$ by solving (2.62), where the state ϕ is determined from f by solving (2.55)–(2.56).

In a similar manner, we can deduce that an infinitesimal change in the direction of $\widetilde{g}$ in the boundary control g induces an infinitesimal change in the direction of ϕ_g in the state ϕ determined by solving

$$-\Delta\phi_g + 3\phi^2\phi_g = 0 \quad \text{in } \Omega \qquad \text{and} \qquad \frac{\partial\phi_g}{\partial n} = \widetilde{g} \quad \text{on } \Gamma\,. \tag{2.63}$$

Next, let us derive a formula for the change in the functional $\mathcal{J}$ of (2.57) effected by an infinitesimal change in the direction $\widetilde{f}$ in the control f. As always, we will keep track of both the explicit dependence of $\mathcal{J}$ on f and the implicit dependence through the state ϕ. Starting with

$$\delta_f\mathcal{J} \equiv \lim_{\epsilon\to 0}\left(\frac{\mathcal{J}(\phi + \epsilon\phi_f, f + \epsilon\widetilde{f}, g) - \mathcal{J}(\phi, f, g)}{\epsilon}\right),$$

[28]It is not always possible to effect such simplification, i.e., it is seldom possible to eliminate the controls from the optimality system.

[29]As in the rest of this chapter, we are being completely cavalier about what function spaces are used for f, ϕ, etc.

a simple calculation leads to

$$\delta_f \mathcal{J} = \int_\Omega (\phi - \Phi)\phi_f \, d\Omega + \sigma_1 \int_\Omega f \widetilde{f} \, d\Omega \,, \tag{2.64}$$

where ϕ is obtained from f and ϕ_f is obtained from $\widetilde{f}$ by solving (2.55)–(2.56) and (2.62), respectively.

The *functional sensitivity equation* (2.64) should be interpreted as follows. An infinitesimal change of the control function f in the "direction" of $\widetilde{f}$ induces the infinitesimal change $\delta_f \mathcal{J}$ in the functional $\mathcal{J}$; the induced change $\delta_f \mathcal{J}$ is determined from a given $\widetilde{f}$ by evaluating (2.64), where the state ϕ is determined from f by solving (2.55)–(2.56) and the sensitivity ϕ_f is determined from $\widetilde{f}$ by solving (2.62).

In a similar manner, we can deduce that an infinitesimal change in the direction of $\widetilde{g}$ in the boundary control g induces an infinitesimal change $\delta_g \mathcal{J}$ in the functional $\mathcal{J}$ determined by evaluating

$$\delta_g \mathcal{J} = \int_\Omega (\phi - \Phi)\phi_g \, d\Omega + \sigma_2 \int_\Gamma g \widetilde{g} \, d\Omega \,, \tag{2.65}$$

where ϕ is obtained from g and ϕ_g is obtained from $\widetilde{g}$ by solving (2.55)–(2.56) and (2.63), respectively.

What do we mean by a "change in the direction of $\widetilde{f}$?" To explain this terminology, let us first consider the case for which the distributed control[30] f is *finite dimensional* and of the form $f = \sum_{k=1}^K \alpha_k f_k$, where $\{f_k\}_{k=1}^K$ are given functions and $\{\alpha_k\}_{k=1}^K$ are the design parameters to be determined. We want to show that, in this case,[31] (2.62) and (2.64) reduce to the result we would have found employing the process used on page 25 for Example 3 of page 13. To this end, let us choose an infinitesimal change in f in the direction of α_k for some k, i.e., we change $f = \sum_{j=1}^K \alpha_j f_j$ to $f + \widetilde{f}$, where $\widetilde{f} = \delta\alpha_k f_k$. Then, (2.62) reduces to

$$-\Delta\phi_f + 3\phi^2\phi_f = \delta\alpha_k f_k \quad \text{in } \Omega \qquad \text{and} \qquad \frac{\partial \phi_f}{\partial n} = 0 \quad \text{on } \Gamma \,.$$

Since ϕ_f and $\delta\alpha_k$ are infinitesimal changes, dividing by $\delta\alpha_k$ gives

$$-\Delta\phi_k + 3\phi^2\phi_k = f_k \quad \text{in } \Omega \qquad \text{and} \qquad \frac{\partial \phi_k}{\partial n} = 0 \quad \text{on } \Gamma \,,$$

where $\phi_k = \phi_f/\delta\alpha_k = \partial\phi/\partial\alpha_k$ is the sensitivity of the state ϕ with respect to the design parameter α_k. In a similar manner, we deduce from (2.64) that the kth component of the gradient of the functional is given by

$$\frac{D\mathcal{J}}{D\alpha_k} = \sigma_1 \alpha_k + \int_\Gamma (\phi - \Phi)\phi_k \, d\Omega \,.$$

Thus, we do indeed arrive at the same sensitivity system and gradient of the functional as we would have if we had used the same process as was used for Example 3.

[30]We focus on f, but with the obvious notational changes, everything we are about to say holds just as well for the boundary control g.

[31]In order to render the functionals (2.7) and (2.57) identical when f is finite dimensional, we make the additional assumptions that the set $\{f_k\}_{k=1}^K$ is orthonormal in the sense that $\int_\Omega f_j f_k d\Omega = 0$ if $j \neq k$ and $\int_\Omega f_j^2 d\Omega = 1$.

What about the infinite-dimensional case? It again helps to think of f as being expressed in the form $f = \sum_{k=1}^{\infty} \alpha_k f_k$, where we now have an infinite number of design parameters. Clearly, we can determine the sensitivity of the state ϕ and the derivative of the functional $\mathcal{J}$ with respect to any of the design parameters in the same way we did in the finite-dimensional case, but now we have an infinite number of sensitivities and an infinite number of components of the gradient of the functional. Thus,

> *in any computation, the best we can do is to compute the sensitivities and the components of the gradient of the functional for only a finite subset of the design parameters.*

This is what we meant by saying that we can use (2.62) and (2.64) to compute the change in the control f and the functional $\mathcal{J}$ in any "direction" $\widetilde{f}$, i.e., if we pick a particular direction[32] $\widetilde{f}$ in the function space for the control f, we can determine the corresponding changes ϕ_f and $\delta_f \mathcal{J}$ in the state ϕ and functional $\mathcal{J}$.

The obvious conclusions one can draw from this discussion are that

- given any distributed control function f and any change $\widetilde{f}$ in the control function, we can determine the corresponding changes in the state and objective functional from (2.62), (2.64), and (2.55)–(2.56);
- given the same distributed control function f, if we want to determine the change in the objective functional due to a different change $\check{f}$ in the control, *we have to re-solve the sensitivity equation* (2.62) *to determine the new change in the state*;
- similar observations hold for the boundary control function g.

Now we are ready to examine the central effect of the infinite dimensionality of the controls. Suppose we want to use the following infinite-dimensional version of Algorithm 2.1, where, for illustrative purposes, we use a simple gradient method in step 3, specialize to $\Omega \subset \mathbb{R}^2$, and consider the boundary control g to be fixed.

ALGORITHM 2.2.
Start with an initial guess $f^{(0)}(x, y)$ for the control f. For $m = 0, 1, 2, \dots,$ until "satisfactory convergence is achieved,"

1. solve the state system (2.55)–(2.56) to obtain the corresponding state $\phi^{(m)}(x, y)$;
2. compute the corresponding gradient of the functional $(D_f \mathcal{J})^{(m)}(x, y)$;
3. set $\delta f^{(m)}(x, y) = -\beta_m (D_f \mathcal{J})^{(m)}(x, y)$ for some appropriate step size parameter β_m;
4. set $f^{(m+1)}(x, y) = f^{(m)}(x, y) + \delta f^{(m)}(x, y)$ and return to step 1.

First, what do we mean by $(D_f \mathcal{J})^{(m)}(x, y)$? This denotes the change in the functional at the point $(x, y) \in \Omega$ due to a change in the control f at the same point. Next, how do we compute $(D_f \mathcal{J})^{(m)}(x, y)$? For any specific point $(x', y') \in \Omega$, we choose $\widetilde{f} = \delta(x', y')$,

[32] For example, if $f = \sum_{k=1}^{\infty} \alpha_k f_k$, then a change in f in a particular direction is determined by changing a finite number of the parameters α_k.

where $\delta(x, y)$ denotes the Dirac delta function.[33] Then, we have

$$(D_f \mathcal{J})^{(m)}(x', y') = \sigma_1 f^{(m)}(x', y') + \int_\Omega \Big(\phi^{(m)}(x, y) - \Phi(x, y)\Big) \widetilde{\phi}^{(m)}_{(x',y')}(x, y)\, dxdy\,, \tag{2.66}$$

where $\phi^{(m)}(x, y)$ is the solution of the state system (2.55)–(2.56) for $f = f^{(m)}(x, y)$, and $\widetilde{\phi}^{(m)}_{(x',y')}(x, y)$ is the solution of the sensitivity system (2.62) with $\widetilde{f} = \delta(x', y')$, i.e., $\widetilde{\phi}^{(m)}_{(x',y')}(x, y)$ is the solution of

$$-\Delta \widetilde{\phi}^{(m)}_{(x',y')} + 3(\phi^{(m)})^2 \widetilde{\phi}^{(m)}_{(x',y')} = \delta(x', y') \text{ in } \Omega \quad \text{and} \quad \frac{\partial \widetilde{\phi}^{(m)}_{(x',y')}}{\partial n} = 0 \text{ on } \Gamma\,. \tag{2.67}$$

Note that, of course, $\widetilde{\phi}^{(m)}_{(x',y')}$ depends on the choice of point $(x', y') \in \Omega$. There is an infinite number of such points so that in order to determine $(D_f \mathcal{J})^{(m)}(x', y')$ at *all* points $(x', y') \in \Omega$ (which is what is needed in step 3 of Algorithm 2.2) using (2.66) requires the solution of an *infinite* number of sensitivity systems (2.67), one for each point $(x', y') \in \Omega$.

We can go through the same process for the boundary control g, yielding similar results. For example, in the sensitivity system (2.67), the delta function appears in the right-hand side of the boundary conditions instead of in the differential equation.

Since in the end one only can solve discretized systems, all this does not mean that sensitivity-based optimization strategies are impossible to use for infinite-dimensional controls. We will return to discuss this point at the end of this section; see page 40.

The gradient of the functional through adjoints

For the example problem given on page 33, let us now consider how an adjoint equation can be used to determine the change in the functional due to changes in the control functions. Consider the adjoint system

$$-\Delta\lambda + 3\phi^2\lambda = \phi - \Phi \quad \text{in } \Omega \qquad \text{and} \qquad \frac{\partial \lambda}{\partial n} = 0 \quad \text{on } \Gamma \tag{2.68}$$

for the adjoint variable λ; this system is extracted from the optimality system (2.61) for this example. Using the partial differential equation in (2.68) to substitute for $(\phi - \Phi)$ in (2.64) yields

$$\delta_f \mathcal{J} = \int_\Omega \phi_f (-\Delta\lambda + 3\phi^2\lambda)\, d\Omega + \sigma_1 \int_\Omega f \widetilde{f}\, d\Omega\,.$$

Integrating by parts and substituting the sensitivity system (2.62) and the boundary condition in (2.68) into the result yield

$$\delta_f \mathcal{J} = \int_\Omega \widetilde{f}(\lambda + \sigma_1 f)\, d\Omega\,. \tag{2.69}$$

[33] We are now really treading on dangerous mathematical ground, but let's go on anyway.

This formula[34] gives the change in the functional $\mathcal{J}$ resulting from a change in the distributed control f in the direction of $\widetilde{f}$. In a similar manner, we obtain

$$\delta_g \mathcal{J} = \int_\Gamma \widetilde{g}(\lambda + \sigma_2 g)\, d\Omega, \tag{2.70}$$

which gives the change in the functional $\mathcal{J}$ resulting from a change in the boundary control g in the direction of $\widetilde{g}$.

At this point, we make the following obvious observations:

- The adjoint system (2.68) is independent of the particular choice of changes $\widetilde{f}$ and $\widetilde{g}$ in the controls;
- as a result, no matter how many changes $\widetilde{f}$ and $\widetilde{g}$ in the controls one chooses, one can use (2.69) and (2.70) to determine the corresponding change in the functional with only a single solution of the adjoint system (2.68).

This should be contrasted with the sensitivity equation approach for determining changes in the functional for which, each time one chooses a new $\widetilde{f}$ or $\widetilde{g}$, one must re-solve the corresponding sensitivity system.

Finally, let us return to Algorithm 2.2 and see how we can use adjoint equations to effect step 2. As on page 37, we choose $\widetilde{f} = \delta(x', y')$. Using (2.69), we easily deduce that

$$(D_f \mathcal{J})^{(m)}(x', y') = \sigma_1 f^{(m)}(x', y') + \lambda^{(m)}(x', y'), \tag{2.71}$$

where $\lambda^{(m)}$ is the solution of the adjoint system (2.68) with $f = f^{(m)}$, i.e.,

$$-\Delta\lambda^{(m)} + 3(\phi^{(m)})^2\lambda^{(m)} = \phi^{(m)} - \Phi \quad \text{in } \Omega \qquad \text{and} \qquad \frac{\partial \lambda^{(m)}}{\partial n} = 0 \quad \text{on } \Gamma, \tag{2.72}$$

where $\phi^{(m)}$ is determined from (2.55)–(2.56) with $f = f^{(m)}$. Since (2.72) and therefore $\lambda^{(m)}$ does not depend on the choice of points (x', y'), we see that

> one can use (2.71) to determine the gradient of the functional at any point $(x', y') \in \Omega$ by *solving the single adjoint equation* (2.72).[35]

Moreover,

> we can combine steps 2, 3, and 4 of Algorithm 2.2 to obtain the explicit formula
>
> $$f^{(m+1)}(x, y) = (1 - \sigma_1\beta_m) f^{(m)}(x, y) - \beta_m \lambda^{(m)}(x, y) \qquad \forall\, (x, y) \in \Omega$$
>
> for the new value of the control f at every point in Ω.

For the boundary control g we can go through the same process to determine that

$$g^{(m+1)}(x, y) = (1 - \sigma_2\beta_m) g^{(m)}(x, y) - \beta_m \lambda^{(m)}(x, y) \qquad \forall\, (x, y) \in \Gamma$$

for the new value of the control g at every point on the boundary Γ using the solution $\lambda^{(m)}(x, y)$ of the same adjoint system (2.72) as was used for the distributed control f.

---

[34] It is again instructive to consider the finite-dimensional case $f = \sum_{i=1}^{K} \alpha_k f_k$ with f_k given and, for consistency of the functionals, $\int_\Omega f_j f_k d\Omega = 0$ if $j \neq k$ and $\int_\Omega f_j^2 d\Omega = 1$. Then, by choosing the change in f to be only along a particular "direction" α_k, we easily see that (2.69) reduces to (2.40).

[35] This should be contrasted with the use of sensitivity equations (see page 38) for determining the new value of the control f at every point in Ω, which requires the solution of an *infinite number of sensitivity equations*.

Approximation of infinite-dimensional controls

In this section, we considered the controls f and g to be *infinite dimensional.* Of course, in practice, one can find only *finite-dimensional approximations* to the control functions f and g. For example, we could approximate f, for some positive integer K, by the finite sum

$$f(x,y) \approx \sum_{i=1}^{K} \alpha_k f_k(x,y)\,, \tag{2.73}$$

where the $f_k(x,y)$'s are given functions and the α_k's are parameters that determine the approximation to the control. This seems to return us to the case of finite-dimensional controls of, e.g., Example 3 on page 13.

If K is fixed independent of the size of the grid used in the discretization, then we have indeed returned to the case of finite-dimensional controls. However, if (2.73) is truly thought of as an approximation to an infinite-dimensional control, then K must increase as the grid size decreases and, in general, $K \to \infty$ as the grid size tends to zero. Moreover, for any fixed grid size, if one wants the accuracy of an approximation to an infinite-dimensional control to be similar to what one wants for, say, the state variable,[36] then K will be very large. In fact, K would be proportional to the number of grid points in Ω for a distributed control and proportional to the number of grid points on Γ for a boundary control.

The implications of these observations are as follows. If one uses sensitivity-based methods to compute objective function gradients, one will have to solve K sensitivity equations, where K is large, i.e., K is proportional to the number of grid points. If one refines the grid, one must solve more sensitivity equations. On the other hand, if one uses adjoint-based methods to compute objective function gradients, one always has to solve only a *single* adjoint equation regardless of the size of the grid.

2.5 Other types of functionals and partial controls

Even among quadratic functionals, there are many other possible types of objective functionals that one encounters in practical flow control and optimization problems. In addition, controls are most often applied on only a portion of the domain or of the boundary of the domain. Here, we introduce a few of the many possible variations in the objective functional and examine what effect these variations and the use of partial controls have on the optimality system and on the sensitivity and adjoint systems.

We again treat an optimization problem involving a scalar partial differential equation; we also consider the case of infinite-dimensional controls since the variations in the functional that we consider naturally apply to that case.

A problem with partial controls and with a different functional

Let Ω be an open domain in $\mathbb{R}^2$ having boundary Γ and let $\Omega_1 \subset \Omega$ and $\Gamma_1 \subset \Gamma$ be nonempty subsets. Let ϕ denote the scalar-valued state variable and f and g denote the

[36] In fact, unless one approximates the infinite-dimensional control in a similar way to the state, one will not in any case obtain the aimed for accuracy for the state.

(infinite-dimensional) scalar-valued control functions. The constraints are the nonlinear, second-order, elliptic partial differential equation

$$-\Delta\phi + \phi^3 = \begin{cases} f & \text{in } \Omega_1, \\ 0 & \text{in } \Omega\backslash\Omega_1, \end{cases} \tag{2.74}$$

along with the boundary condition

$$\phi = \begin{cases} g & \text{on } \Gamma_1, \\ 0 & \text{on } \Gamma\backslash\Gamma_1 . \end{cases} \tag{2.75}$$

Thus, the *distributed control* f is applied on only part of the domain Ω, i.e., on the subset Ω_1, and the *boundary control* g is applied on only part of the boundary Γ, i.e., on the subset Γ_1.

The objective, or cost, or performance functional is given by

$$\begin{aligned}\mathcal{J}(\phi, f, g) = \frac{\sigma_3}{2}\int_{\Omega_2} |\nabla\phi|^2\, d\Omega + \frac{\sigma_4}{2}\int_{\Gamma_2}\left(\frac{\partial\phi}{\partial n} - \Psi\right)^2 d\Gamma \\ + \frac{\sigma_1}{2}\int_{\Omega_1}(\sigma_5|\nabla f|^2 + f^2)\, d\Omega + \frac{\sigma_2}{2}\int_{\Gamma_1}(\sigma_6|\nabla_s g|^2 + g^2)\, d\Gamma ,\end{aligned} \tag{2.76}$$

where $\Omega_2 \subset \Omega$ and $\Gamma_2 \subset \Gamma$ are nonempty subsets, ∇_s denotes the gradient in the direction tangential to the boundary Γ, Ψ is a given function defined on Γ_2, and σ_i, $i = 1, \dots, 6$, are given nonnegative constants[37] with $\sigma_1 > 0$ and $\sigma_2 > 0$. For simplicity, we assume that no part of the boundary $\partial\Omega_2$ of Ω_2 coincides with any part of the boundary Γ of Ω. A sketch of the geometry for this problem is given in Figure 2.1. Note that the penalty terms in the functional (2.76), i.e., the third and fourth integrals, involve integrals over Ω_1 and Γ_1 and not over Ω and Γ, respectively. This follows from the fact that the controls f and g act only over the subsets Ω_1 and Γ_1, respectively.

The subsets Ω_2 and Γ_2 need not be the same as the subsets Ω_1 and Γ_1, respectively.[38] We can view Ω_1 and Γ_1 as "actuating sets," i.e., subsets of the domain Ω and boundary Γ, respectively, over which the controls act. Similarly, we can view Ω_2 and Γ_2 as "sensing sets," i.e., subsets of the domain Ω and boundary Γ, respectively, over which the state is monitored.

The optimization problem is given by

> *find ϕ, f, and g such that the functional $\mathcal{J}$ defined in* (2.76) *is minimized subject to the requirements that ϕ, f, and g satisfy the constraints* (2.74)–(2.75).

[37]The constants σ_1 and σ_2 are penalty parameters, which is why they should be positive. The other four parameters, $\sigma_3, \dots, \sigma_6$, are introduced in order to allow for flexibility in the definition of the functional. Thus, if we set $\sigma_3 = 0$, then the functional does not contain the first term, which is an area integral over the subdomain Ω_2. If we set $\sigma_4 = 0$, then the functional does not contain a matching functional along the boundary segment Γ_2. Also, if $\sigma_5 = 0$ or $\sigma_6 = 0$, we do not penalize the derivatives of the controls f or g, respectively. In addition, introducing the parameters $\{\sigma_i\}_{i=1}^6$ allows us to keep track of how the different terms in the functional (2.76) affect the optimality system as well as the sensitivity and adjoint systems.

[38]In fact, we are examining the case of partial controls together with functionals defined over subdomains so that we can highlight the fact that the sets Ω_1 and Ω_2 can be chosen independently as can the sets Γ_1 and Γ_2.

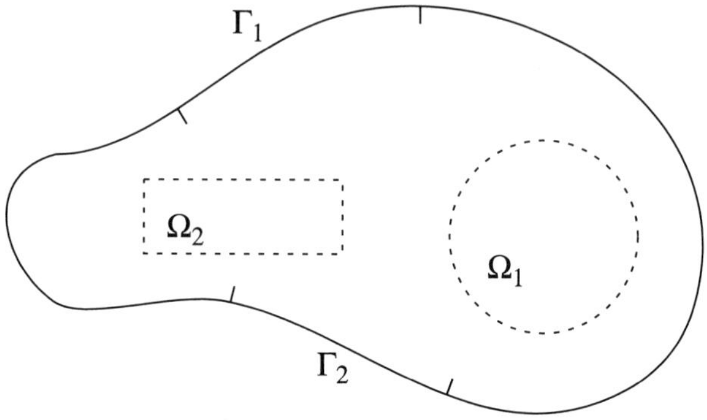

Figure 2.1. *A sketch of the "sensing" subdomains $\Omega_2 \subset \Omega$ and $\Gamma_2 \subset \Gamma$ and the "actuating" subdomains $\Omega_1 \subset \Omega$ and $\Gamma_1 \subset \Gamma$.*

In this example, through the functional (2.76), we have introduced a number of variations in addition to controls that act on only part of the domain or part of the boundary. Specifically, the functional (2.76) involves

- the minimization of the gradient of the state—the first integral in (2.76);
- matching a given function Ψ along the boundary instead of over the domain—the second integral;
- matching the normal derivative of the state along the boundary to a given function instead of matching the state itself—the second integral;
- a functional involving an integral over the subset Ω_2 instead of the whole domain Ω—the first integral;
- a boundary functional involving an integral over the subset Γ_2 instead of the whole boundary Γ—the second integral;
- the penalization of the gradient of the controls—the third and fourth integrals.

The optimality system

For the example problem just defined, we introduce the Lagrangian functional

$$\mathcal{L}(\phi, f, g, \xi, \eta) = \mathcal{J}(\phi, f, g) - \int_\Omega \xi(-\Delta\phi + \phi^3 - \chi_{\Omega_1} f)\, d\Omega - \int_\Gamma \eta\, (\phi - \chi_{\Gamma_1} g)\, d\Gamma\,,$$

where ξ and η are Lagrange multipliers introduced to enforce the differential equation (2.74) and the boundary condition (2.75), respectively. In the Lagrangian functional, χ_{Ω_1} is the characteristic function for the subdomain Ω_1 viewed as a subset of Ω and χ_{Γ_1} is the characteristic function for the subdomain Γ_1 viewed as a subset of Γ. Since the partial differential equation (2.74) holds over all of Ω and the boundary condition (2.75) holds on all of Γ, note that the integrals added to $\mathcal{J}$ in the Lagrangian functional are integrals over Ω and Γ and not just over the subdomains Ω_1 and Γ_1. Also note that, for the same reasons, the Lagrange multiplier functions ξ and η are defined on all of Ω and Γ, respectively.

As usual, setting the first variations of $\mathcal{L}$ with respect to the Lagrange multipliers ξ and η equal to zero recovers the state system (2.74)–(2.75).

Setting the first variation of $\mathcal{L}$ with respect to the state ϕ equal to zero results in

$$\sigma_3 \int_{\Omega_2} \nabla\phi\cdot\nabla\widetilde{\phi}\, d\Omega + \sigma_4 \int_{\Gamma_2} \left(\frac{\partial\phi}{\partial n} - \Psi\right)\frac{\partial\widetilde{\phi}}{\partial n}\, d\Gamma - \int_{\Omega} \xi(-\Delta\widetilde{\phi} + 3\phi^2\widetilde{\phi})\, d\Omega - \int_{\Gamma} \eta\widetilde{\phi}\, d\Gamma = 0\,,$$

where $\widetilde{\phi}$ denotes an arbitrary variation in the state variable ϕ. Integrating by parts to remove, where possible, derivatives from $\widetilde{\phi}$, one obtains

$$\begin{aligned}
&\int_{\Omega} \widetilde{\phi}\Big(\Delta\xi - 3\phi^2\xi - \chi_{\Omega_2}\sigma_3\Delta\phi\Big)\, d\Omega + \int_{\Gamma} \frac{\partial\widetilde{\phi}}{\partial n}\left(\sigma_4\chi_{\Gamma_2}\left(\frac{\partial\phi}{\partial n} - \Psi\right) + \xi\right) d\Gamma \\
&\qquad - \int_{\Gamma} \widetilde{\phi}\left(\eta + \frac{\partial\xi}{\partial n}\right) d\Gamma + \int_{\partial\Omega_2} \left(\left(\xi\frac{\partial\widetilde{\phi}}{\partial n}\right)^- - \left(\xi\frac{\partial\widetilde{\phi}}{\partial n}\right)^+\right) d\Gamma \qquad (2.77)\\
&\qquad + \int_{\partial\Omega_2} \left(\left(\widetilde{\phi}\frac{\partial\phi}{\partial n} - \widetilde{\phi}\frac{\partial\xi}{\partial n}\right)^- + \left(\widetilde{\phi}\frac{\partial\xi}{\partial n}\right)^+\right) d\Gamma = 0\,,
\end{aligned}$$

where χ_{Ω_2} and χ_{Γ_2} denote the characteristic functions of the subdomains $\Omega_2 \subset \Omega$ and $\Gamma_2 \subset \Gamma$, respectively. In the second and third integrals in (2.77), $\partial/\partial n$ denotes the derivative in the direction of the outward normal to Ω along its boundary Γ; in the fourth and fifth integrals, $\partial/\partial n$ denotes the derivative in the direction of the outward normal to Ω_2 along its boundary $\partial\Omega_2$. Also, in the fourth and fifth integrals in (2.77), $(\cdot)^+$ and $(\cdot)^-$ respectively denote the evaluation of a quantity on $\partial\Omega_2$ from the outside and inside with respect to Ω_2.

Since $\widetilde{\phi}$ is arbitrary, (2.77) implies (using arguments similar to those used for Example 3 on page 20) that

$$\begin{aligned}
&-\Delta\xi + 3\phi^2\xi = -\chi_{\Omega_2}\sigma_3\Delta\phi \quad \text{in } \Omega\,, \qquad \xi = -\sigma_4\chi_{\Gamma_2}\left(\frac{\partial\phi}{\partial n}\Big|_\Gamma - \Psi\right) \quad \text{on } \Gamma\,, \\
&\xi^+ = \xi^- \quad \text{on } \partial\Omega_2\,, \qquad \left(\frac{\partial\xi}{\partial n}\right)^+ - \left(\frac{\partial\xi}{\partial n}\right)^- = -\left(\frac{\partial\phi}{\partial n}\right)^- \quad \text{on } \partial\Omega_2\,,
\end{aligned} \qquad (2.78)$$

and

$$\eta = -\frac{\partial\xi}{\partial n}\Big|_\Gamma \quad \text{on } \Gamma\,. \qquad (2.79)$$

The third and fourth equations in (2.78) imply that the adjoint state ξ is continuous across the internal boundary $\partial\Omega_2$ but that the normal derivative of the adjoint state experiences a jump across that boundary.[39]

Setting the first variation of $\mathcal{L}$ with respect to the distributed control f to zero results in

$$\int_{\Omega_1} \Big(\sigma_1(\sigma_5\nabla f\cdot\nabla\widetilde{f} + f\widetilde{f}) + \xi\widetilde{f}\Big)\, d\Omega = 0\,,$$

[39] At first glance, the adjoint system (2.78) seems very complicated due to the appearance of the jump conditions along $\partial\Omega_2$. However, in a finite element discretization of (2.78), the first jump condition can be enforced by using continuous test and trial spaces and the second is a natural interface condition; thus, the jump conditions pose no great difficulty.

where the variation $\widetilde{f}$ is arbitrary. Integrating by parts to remove derivatives from $\widetilde{f}$, one obtains

$$\int_{\Omega_1} \widetilde{f}\Big(\sigma_1(-\sigma_5\Delta f + f) + \xi\Big)\, d\Omega + \sigma_1\sigma_5 \int_{\partial\Omega_1} \widetilde{f}\frac{\partial f}{\partial n}\, d\Gamma = 0$$

from which it follows that

$$-\sigma_5\Delta f + f = -\frac{1}{\sigma_1}\xi \quad \text{in } \Omega_1 \qquad \text{and} \qquad \sigma_5\frac{\partial f}{\partial n} = 0 \quad \text{on } \partial\Omega_1\,.$$

Setting the first variation of $\mathcal{L}$ with respect to the boundary control g equal to zero results in

$$\int_{\Gamma_1} \Big(\sigma_2(\sigma_6\nabla_s g \cdot \nabla_s\widetilde{g} + g\widetilde{g}) + \eta\widetilde{g}\Big)\, d\Gamma = 0,$$

where the variation $\widetilde{g}$ is arbitrary. Integrating by parts to remove derivatives from $\widetilde{g}$, one obtains

$$\int_{\Gamma_1} \widetilde{g}\Big(\sigma_2(-\sigma_6\Delta_s g + g) + \eta\Big)\, d\Omega + \sigma_2\sigma_6 \int_{\partial\Gamma_1} \widetilde{g}\frac{\partial g}{\partial n}\, d\Gamma = 0\,,$$

where Δ_s denotes the Laplace operator along the boundary and $\partial\Gamma_1$ denotes the boundary of Γ_1 viewed as a subset of Γ. It then follows that

$$-\sigma_6\Delta_s g + g = -\frac{1}{\sigma_2}\eta \quad \text{in } \Gamma_1 \qquad \text{and} \qquad \sigma_6\frac{\partial g}{\partial n} = 0 \quad \text{on } \partial\Gamma_1\,. \tag{2.80}$$

Collecting the results of setting the first variations of the Lagrangian functional to zero results in the following optimality system:

$$\text{state equations} \quad \Rightarrow \quad \begin{cases} -\Delta\phi + \phi^3 = \chi_{\Omega_1} f & \text{in } \Omega, \\ \phi = \chi_{\Gamma_1} g & \text{on } \Gamma\,; \end{cases}$$

$$\text{adjoint equations} \quad \Rightarrow \quad \begin{cases} -\Delta\xi + 3\phi^2\xi = \chi_{\Omega_2}\sigma_3(\chi_{\Omega_1} f - \phi^3) & \text{in } \Omega, \\ \xi = -\sigma_4\chi_{\Gamma_2}\left(\dfrac{\partial\phi}{\partial n}\Big|_\Gamma - \Psi\right) & \text{on } \Gamma\,; \end{cases}$$

$$\text{optimality conditions} \quad \Rightarrow \quad \begin{cases} -\sigma_5\Delta f + f = -\dfrac{1}{\sigma_1}\xi & \text{in } \Omega_1, \\ \sigma_5\dfrac{\partial f}{\partial n} = 0 & \text{on } \partial\Omega_1, \\ -\sigma_6\Delta_s g + g = \dfrac{1}{\sigma_2}\dfrac{\partial\xi}{\partial n}\Big|_{\Gamma_1} & \text{on } \Gamma_1, \\ \sigma_6\dfrac{\partial g}{\partial n} = 0 & \text{on } \partial\Gamma_1, \end{cases} \tag{2.81}$$

where we have combined (2.79) and (2.80) to eliminate the Lagrange multiplier η in one of the optimality conditions for the boundary control g, and we have simplified the right-hand sides of the adjoint partial differential equation using the state equations, i.e., we have substituted the state partial differential equations (2.74) into the right-hand side in (2.78) to eliminate the $\Delta\phi$ term.[40] We have also not included the third and fourth equations in the system (2.78), i.e., the jump conditions along $\partial\Omega_2$, since any conforming finite element discretization of the adjoint system in (2.81), i.e., one employing continuous finite element spaces, will automatically take care of them.

We are now in a position to discuss how the fact that we are applying controls on only part of the domain Ω and part of the boundary Γ, as well as how the new terms we introduced into the objective functional (2.76), affect the optimality system.

We first discuss the effects of the fact that in the functional (2.76) the domains of integration of the first two integrals are the *subdomains* $\Omega_2 \subset \Omega$ and $\Gamma_2 \subset \Gamma$ and not all of Ω and Γ, respectively. Note that the right-hand side of the partial differential equation in the adjoint system acts only in the domain Ω_2; this is, of course, due to the fact that, for the first integral in the functional (2.76), the domain of integration is that subdomain. Likewise, in the boundary condition along Γ in the adjoint system, the right-hand side acts only on Γ_2; this is, of course, due to the fact that, for the second integral in the functional (2.76), the domain of integration is that subdomain.

We next see how some other changes we have made in the functional (2.76) have affected the adjoint system. First, note that the matching term involving $(\partial\phi/\partial n - \Psi)$ appears in the right-hand side of the boundary condition in the adjoint system instead of the differential equation as it did in (2.19); this is due to the fact that, in the functional (2.76), the matching term is an integral along the boundary Γ while, in the functional (2.7), the matching term is an integral over the domain Ω. Next, note that, if $\sigma_3 = 0$ so that the first term involving an area integral does not appear in the functional (2.76), then the partial differential equation on Ω for the adjoint variable ξ becomes homogeneous. Likewise, if $\sigma_4 = 0$ so that the boundary-matching term along Γ_2 does not appear in the functional (2.76), then the boundary condition along Γ for the adjoint variable ξ becomes homogeneous.

We next note that the optimality conditions involving the controls f and g are defined on the subdomains $\Omega_1 \subset \Omega$ and $\Gamma_1 \subset \Gamma$, respectively. This is, of course, due to the fact that we are using partial controls, i.e., controls f and g that act only on the sudomains Ω_1 and Γ_1, respectively.

We next see how the fact that we have introduced penalization of the derivatives of the controls f and g in the definition of the functional (2.76) affects the optimality conditions. So long as $\sigma_5 \neq 0$ and $\sigma_6 \neq 0$, the optimality conditions in (2.81) are partial differential equation problems. This is due to the fact that the penalization terms for f and g in the functional (2.76) involve the gradient of f and the surface gradient of g. Note that the partial differential equation for the boundary control g is posed along the boundary and involves the surface Laplacian operator Δ_s.

On the other hand, if there is no penalization of the derivatives of one or both of the controls f and g, i.e., if $\sigma_5 = 0$ and/or $\sigma_6 = 0$ hold, then the optimality conditions in (2.81) reduce to the simple relations

$$f = -\frac{1}{\sigma_1}\xi \quad \text{in } \Omega_1 \quad \text{if } \sigma_5 = 0 \qquad \text{and} \qquad g = \frac{1}{\sigma_2}\frac{\partial\xi}{\partial n} \quad \text{on } \Gamma_1 \quad \text{if } \sigma_6 = 0 \tag{2.82}$$

that directly relate the controls to the adjoint variable. As a result, if $\sigma_5 = 0$, i.e., if derivatives of the distributed control f are not penalized in the functional (2.76), we can easily eliminate

[40]It is not always possible to effect such simplifications, i.e., it is not always possible to eliminate Lagrange multipliers or to simplify the right-hand sides of the adjoint equations.

f from the optimality system (2.81). A similar situation holds for the boundary control g if $\sigma_6 = 0$. For example, if both $\sigma_5 = 0$ and $\sigma_6 = 0$ hold, we can use (2.82) in the state and adjoint equations in (2.81) to arrive at the simplified optimality system

$$
\begin{aligned}
-\Delta\phi + \phi^3 &= -\frac{1}{\sigma_1}\chi_{\Omega_1}\xi \quad \text{in } \Omega\,, \quad \phi = \frac{1}{\sigma_2}\chi_{\Gamma_1}\frac{\partial \xi}{\partial n} \quad \text{on } \Gamma\,,\\
-\Delta\xi + 3\phi^2\xi &= -\chi_{\Omega_2}\sigma_3\left(\frac{1}{\sigma_1}\chi_{\Omega_1}\xi + \phi^3\right) \quad \text{in } \Omega\,, \quad \xi = -\sigma_4\chi_{\Gamma_2}\left(\frac{\partial\phi}{\partial n}\bigg|_{\Gamma} - \Psi\right) \quad \text{on } \Gamma,
\end{aligned}
$$

which involves only the state ϕ and the adjoint state ξ. Once these have been determined, the optimal controls f and g can be determined from the optimality conditions (2.82).

In the general case of $\sigma_5 \neq 0$ and $\sigma_6 \neq 0$, the optimality system (2.81) is a coupled system of *four* partial differential equations, two posed over Ω, one over the subdomain $\Omega_1 \subset \Omega$, and one over the boundary subdomain $\Gamma_1 \subset \Gamma$. For this example, it may again be possible, especially in two dimensions, to solve the coupled optimality system (2.81) so that optimal states and controls can be obtained without an optimization iteration, i.e., in one shot.

The gradient of the functional through sensitivities

Without going into detail, we provide the sensitivity equations and the corresponding formulas for computing changes in the objective function for the example on page 40. We let $\widetilde{f}$ denote an arbitrary change in the distributed control f; we assume that this change is confined to the same domain Ω_1 on which f itself is defined.[41] Then, the corresponding change ϕ_f in the state ϕ is determined from the sensitivity system

$$
-\Delta\phi_f + 3\phi^2\phi_f = \begin{cases} \widetilde{f} & \text{in } \Omega_1, \\ 0 & \text{in } \Omega\backslash\Omega_1, \end{cases} \qquad \text{and} \qquad \phi_f = 0 \quad \text{on } \Gamma\,. \tag{2.83}
$$

Similarly, if $\widetilde{g}$ denotes an arbitrary change in the boundary control g, then the corresponding change ϕ_g in the state ϕ is determined from

$$
-\Delta\phi_g + 3\phi^2\phi_g = 0 \quad \text{in } \Omega \qquad \text{and} \qquad \phi_g = \begin{cases} \widetilde{g} & \text{on } \Gamma_1, \\ 0 & \text{on } \Gamma\backslash\Gamma_1. \end{cases} \tag{2.84}
$$

The change in the objective functional $\mathcal{J}$ defined in (2.76) due to the change $\widetilde{f}$ in the distributed control f is given by

$$
\begin{aligned}
\delta_f \mathcal{J} = \sigma_3 \int_{\Omega_2} \nabla\phi\cdot\nabla\phi_f\, d\Omega + \sigma_4 \int_{\Gamma_2}\left(\frac{\partial\phi}{\partial n} - \Psi\right)\frac{\partial\phi_f}{\partial n}\, d\Gamma \\
+ \sigma_1 \int_{\Omega_1} (\sigma_5 \nabla f\cdot\nabla\widetilde{f} + f\widetilde{f})\, d\Omega\,,
\end{aligned} \tag{2.85}
$$

where ϕ_f is determined from $\widetilde{f}$ (and ϕ) through (2.83). Likewise, the change in the objective

[41] This assumption does not lose us any generality since, if the domains of allowable f and $\widetilde{f}$ are different, we just use the larger of the two.

functional $\mathcal{J}$ due to the change $\widetilde{g}$ in the boundary control g is given by

$$\begin{aligned}\delta_g \mathcal{J} = \sigma_3 \int_{\Omega_2} \nabla\phi \cdot \nabla\phi_g \, d\Omega + \sigma_4 \int_{\Gamma_2} \left(\frac{\partial \phi}{\partial n} - \Psi \right) \frac{\partial \phi_g}{\partial n} \, d\Gamma \\ + \sigma_2 \int_{\Gamma_1} (\sigma_6 \nabla_s g \cdot \nabla_s \widetilde{g} + g\widetilde{g}) \, d\Gamma \,,\end{aligned} \tag{2.86}$$

where ϕ_g is determined from $\widetilde{g}$ (and ϕ) through (2.84).

If the controls f and g are finite dimensional, then (2.83) and (2.84) may be easily specialized to define a finite set of sensitivity equations and (2.85) and (2.86) may be used to easily determine the components of the gradient of the functional with respect to the finite number of design parameters. In the case of infinite-dimensional controls, we interpret (2.83)–(2.86) in the same way as in the discussions of Section 2.4.2.

The gradient of the functional through adjoints

Again, without going into detail, we provide the adjoint system and the corresponding formulas for computing the gradient of the objective function for the example on page 40. Let λ denote the solution of the adjoint system

$$\begin{cases} -\Delta\lambda + 3\phi^2\lambda = \chi_{\Omega_2}\sigma_3(\chi_{\Omega_1} f - \phi^3) & \text{in } \Omega, \\ \lambda = -\sigma_4 \chi_{\Gamma_2} \left(\left.\dfrac{\partial \phi}{\partial n}\right|_{\Gamma} - \Psi \right) & \text{on } \Gamma \,, \end{cases} \tag{2.87}$$

where ϕ is the solution of the state system (2.74)–(2.75) for given f and g. The gradient of the functional $\mathcal{J}$ defined in (2.76) with respect to changes $\widetilde{f}$ in the distributed control f is given by

$$D_f \mathcal{J} = \int_{\Omega_1} \Big(\lambda \widetilde{f} + \sigma_1 (\sigma_5 \nabla f \cdot \nabla \widetilde{f} + f\widetilde{f}) \Big) d\Omega \,. \tag{2.88}$$

The gradient of the functional $\mathcal{J}$ with respect to changes $\widetilde{g}$ in the boundary control g is given by

$$D_g \mathcal{J} = \int_{\Gamma_1} \Big(\lambda \widetilde{g} + \sigma_2 (\sigma_6 \nabla_s g \cdot \nabla_s \widetilde{g} + g\widetilde{g}) \Big) d\Gamma \,. \tag{2.89}$$

In (2.88) and (2.89), ϕ and λ are the solutions of the state system (2.74)–(2.75) and the adjoint system (2.87), respectively. The formulas (2.88) and (2.89) may be used in a gradient-based optimization algorithm to update the control functions f and g. Note that the gradient of the functional can be determined for *any* change in $\widetilde{f}$ (for (2.88)) or $\widetilde{g}$ (for (2.89)) using a *single* adjoint system solution. This should be contrasted with (2.85) which requires a separate solution of the sensitivity system (2.83) for each change $\widetilde{f}$ in the distributed control f and with (2.86) which requires a separate solution of the sensitivity system (2.84) for each change $\widetilde{g}$ in the boundary control g.

2.6 Time-dependent problems

We illustrate the use of the three approaches to finding optimal states and controls for unsteady problems through a simple example problem. Consider the time-dependent problem

$$\begin{cases} \dfrac{\partial \phi}{\partial t} - \Delta \phi = \displaystyle\sum_{k=1}^{K} \alpha_k f_k(t,x,y) \quad \text{in } (0,T)\times\Omega\,, \\ \phi = g \quad \text{on } (0,T)\times\Gamma\,, \\ \phi = \phi_0 \quad \text{in } \Omega \text{ at } t=0\,, \end{cases} \tag{2.90}$$

where f_k, $k = 1, \dots, K$, g, and ϕ_0 are given functions and $(0, T)$ is a given time interval over which control is to be effected. We wish to choose the parameters α_k so as to minimize

$$\begin{aligned} \mathcal{J}(\phi,\alpha_1,\dots,\alpha_K) = {} & \frac{\gamma_1}{2}\int_0^T\int_\Omega (\phi-\Phi)^2\, d\Omega dt \\ & + \frac{\gamma_2}{2}\int_\Omega (\phi|_{t=T}-\Psi)^2\, d\Omega + \frac{\sigma}{2}\sum_{k=1}^{K}(\alpha_k)^2\,, \end{aligned} \tag{2.91}$$

where Φ and Ψ are given functions, σ is a given positive number, and γ_1, γ_2 are given nonnegative numbers such that $\gamma_1 + \gamma_2 \neq 0$. The inclusion of the third term in the functional represents an attempt to limit, through penalization, the size of the design parameters α_k. Making the first term small forces ϕ to be "close" to Φ over $(0, T) \times \Omega$, while making the second term small forces ϕ to be "close" to Ψ at the terminal time $t = T$ and over Ω.

Often, in (2.91), one chooses $\Psi = \Phi|_{t=T}$. It may then seem that the inclusion of the second term in the functional (2.91) is redundant in view of the presence of the first term. However, in practice, it has been found that including the second term improves the quality of both the optimal controls and states, especially near the terminal time $t = T$. In any case, we shall see that the inclusion of the second term does not cause any difficulties for the optimization problem being considered.

Another case of interest is $\gamma_1 = 0$. In this case, the functional (2.91) may be viewed as a "controllability functional," i.e., the object of optimization is then to make the state ϕ at $t = T$ as "close" as possible to a given state Ψ. In (2.91), the measure of closeness is the $L^2(\Omega)$-norm.

One-shot methods

One-shot methods for the minimization of the functional (2.91) subject to (2.90) are based on enforcing the constraints by introducing Lagrange multipliers. To this end, we define the Lagrangian functional

$$\begin{aligned} \mathcal{L}(\phi,\alpha_1,\dots,\alpha_K,\xi,\eta,\chi) = \mathcal{J}(\phi,\alpha_1,\dots,\alpha_K) & - \int_0^T\int_\Omega \xi\left(\frac{\partial\phi}{\partial t} - \Delta\phi - \sum_{k=1}^{K}\alpha_k f_k\right) d\Omega dt \\ & - \int_0^T\int_\Gamma \eta(\phi-g)\, d\Gamma dt - \int_\Omega \chi(\phi|_{t=0}-\phi_0)\, d\Omega\,, \end{aligned}$$

where ξ, η, and χ, respectively, are the Lagrange multipliers introduced to enforce the partial differential equation, the boundary condition, and the initial condition in (2.90).

Setting the first variations of $\mathcal{L}$ with respect to the Lagrange multipliers ξ, η, and χ equal to zero recovers the given state system (2.90). Setting the first variation of $\mathcal{L}$ with respect to a design parameter α_k equal to zero yields the optimality condition

$$\sigma\alpha_k = -\int_0^T \int_\Omega \xi f_k \, d\Omega dt \qquad \text{for } i = 1, \dots, K\,.$$

Next, let us set the first variation of $\mathcal{L}$ with respect to the state variable ϕ equal to zero; by the techniques illustrated in Section 2.2, we obtain that

$$\gamma_1 \int_0^T \int_\Omega (\phi - \Phi)\widetilde{\phi} \, d\Omega dt + \gamma_2 \int_\Omega (\phi|_{t=T} - \Psi)\widetilde{\phi}|_{t=T} \, d\Omega$$
$$- \int_0^T \int_\Omega \xi \Big(\frac{\partial \widetilde{\phi}}{\partial t} - \Delta \widetilde{\phi} \Big) \, d\Omega dt - \int_0^T \int_\Gamma \eta \widetilde{\phi} \, d\Gamma dt - \int_\Omega \chi \widetilde{\phi}|_{t=0} \, d\Omega = 0$$

for arbitrary variations $\widetilde{\phi}$ in the state ϕ. Integrating by parts with respect to time and space so that derivatives are removed from $\widetilde{\phi}$, we obtain

$$\begin{aligned} &\int_0^T \int_\Omega \widetilde{\phi} \Big(\frac{\partial \xi}{\partial t} + \Delta \xi + \gamma_1 (\phi - \Phi) \Big) \, d\Omega dt \\ &\quad + \int_0^T \int_\Gamma \xi \frac{\partial \widetilde{\phi}}{\partial n} \, d\Gamma dt - \int_0^T \int_\Gamma \widetilde{\phi} \Big(\frac{\partial \xi}{\partial n} + \eta \Big) \, d\Gamma dt \\ &\quad - \int_\Omega \widetilde{\phi}|_{t=T} \Big(\xi|_{t=T} - \gamma_2 (\phi|_{t=T} - \Psi) \Big) + \int_\Omega \widetilde{\phi}|_{t=0} \Big(\xi|_{t=0} - \chi \Big) \, d\Omega = 0\,. \end{aligned} \tag{2.92}$$

Since $\widetilde{\phi}$ is arbitrary, each term in (2.92) must vanish separately. Then, we conclude that

$$\begin{cases} -\dfrac{\partial \xi}{\partial t} - \Delta \xi = \gamma_1 (\phi - \Phi) & \text{in } (0, T) \times \Omega\,, \\ \xi = 0 \quad \text{on } (0, T) \times \Gamma\,, \\ \xi = \gamma_2 (\phi|_{t=T} - \Psi) & \text{in } \Omega \text{ at } t = T \end{cases} \tag{2.93}$$

and

$$\begin{cases} \eta = -\Big(\dfrac{\partial \xi}{\partial n} \Big)\Big|_\Gamma & \text{on } (0, T)\,, \\ \chi = \xi|_{t=0} & \text{in } \Omega\,. \end{cases} \tag{2.94}$$

The Lagrange multipliers η and χ appear only in (2.94) and do not couple with the other unknowns, i.e., with ϕ, ξ, and α_k, $k = 1, \dots, K$. Thus, the equations in (2.94) serve solely to determine η and χ and, unless one is interested in these multipliers, (2.94) may be ignored.

To summarize, the *optimality system* from which optimal states, design parameters, and co-states may be determined is given by

$$
\text{state equations} \Rightarrow \begin{cases} \dfrac{\partial \phi}{\partial t} - \Delta \phi = \displaystyle\sum_{k=1}^{K} \alpha_k f_k(t,x,y) & \text{in } (0,T)\times\Omega\,, \\ \phi = g \quad \text{on } (0,T)\times\Gamma\,, \\ \phi = \phi_0 \quad \text{in } \Omega \text{ at } t=0\,; \end{cases}
$$

$$
\text{adjoint equations} \Rightarrow \begin{cases} -\dfrac{\partial \xi}{\partial t} - \Delta \xi = \gamma_1(\phi - \Phi) & \text{in } (0,T)\times\Omega\,, \\ \xi = 0 \quad \text{on } (0,T)\times\Gamma\,, \\ \xi = \gamma_2(\phi|_{t=T} - \Psi) \quad \text{in } \Omega \text{ at } t=T\,; \end{cases}
$$

$$
\text{optimality conditions} \Rightarrow \left\{ \alpha_k = -\frac{1}{\sigma}\int_0^T \int_\Omega \xi f_k \, d\Omega dt \qquad \text{for } i = 1,\ldots,K\,. \right.
$$

This is a coupled system, i.e., the α_k's are not known in the state system, the state ϕ is not known in the adjoint system, and the co-state ξ is not known in the optimality condition. If, however, one solves this coupled system, one can determine, without invoking an optimization algorithm, the optimal state ϕ, co-state ξ, and design parameters α_k, $k = 1,\ldots,K$, such that the functional (2.91) is minimized subject to (2.90).

The most important observation about the optimality system is that the *adjoint equations* (2.93) *are posed backward in time with a terminal condition at* $t = T$ while the *state system* (2.90) *is posed forward in time with an initial condition at* $t = 0$. Note the different signs of the time derivative term in these two systems so that *both systems are well posed.* When one discretizes the optimality system, one cannot march in time since the marching directions for the state and adjoint systems are opposite to each other. This means that in the one-shot approach, one must solve the optimality system as a system in which the solutions at all time levels are coupled to each other. This makes the one-shot method prohibitively expensive, even for the simple, scalar partial differential example we are discussing. Thus, even for the simplest optimization problems in the time-dependent case, one must *introduce iterative methods that uncouple the state and adjoint computations.*

One obvious iterative approach is to guess values for the design parameters α_k, $k = 1,\ldots,K$, then sequentially solve the state system forward in time and the adjoint system backward in time and then use the optimality condition to update the design parameters. There are flow optimization examples for which this simple iteration succeeds, i.e., it converges. However, in general, one should expect slow convergence or, more likely, divergence from this simple uncoupling scheme.

We note that if the second term in the functional (2.91) is not present, i.e., if $\gamma_2 = 0$, then the optimality system involves the homogeneous terminal condition $\xi|_{t=T} = 0$. In general, it is not more difficult to discretize the nonhomogeneous terminal condition $\xi|_{t=T} = \gamma_2(\phi|_{t=T} - \Psi)$ that results whenever $\gamma_2 \neq 0$.

Sensitivity-based optimization methods

We next consider solving for minimizers of (2.91) subject to (2.90) through an optimization algorithm such as Algorithm 2.1. We first consider sensitivity-based methods for determining the gradient of the functional for step 2 of that algorithm.

By direct (formal) differentiation of (2.91), the K components of the gradient of the functional are given by, for $k = 1, \ldots, K$,

$$\frac{D\mathcal{J}}{D\alpha_k} = \sigma\alpha_k + \gamma_1 \int_0^T \int_\Omega (\phi - \Phi)\phi_k \, d\Omega dt + \gamma_2 \int_\Omega (\phi|_{t=T} - \Psi)\phi_k|_{t=T} \, d\Omega \,, \tag{2.95}$$

where $\phi_k = \partial\phi/\partial\alpha_k$, $k = 1, \ldots, K$, denote the sensitivities. By direct (formal) differentiation of (2.90), the sensitivity equations are given by

$$\text{for } k = 1, \ldots, K, \quad \begin{cases} \dfrac{\partial\phi_k}{\partial t} - \Delta\phi_k = f_k & \text{in } (0,T)\times\Omega\,, \\ \phi_k = 0 & \text{on } (0,T)\times\Gamma\,, \\ \phi_k = 0 & \text{in } \Omega \text{ at } t = 0\,. \end{cases} \tag{2.96}$$

Note that both the state system and the sensitivity system are well-posed problems forward in time with initial conditions specified at $t = 0$.

Given a guess for the parameters, one may solve the state system (2.90) and the sensitivity system (2.96) to obtain the information needed to determine the gradient of the functional through (2.95), which in turn may be used to determine a new set of parameters through an optimization algorithm, e.g., Algorithm 2.1. Note that to determine all K components of the gradient of the functional, one must solve the K sensitivity systems in (2.96).

Adjoint-based optimization methods

We next consider adjoint-equation–based methods for determining the gradient of the functional for step 2 of Algorithm 2.1. The adjoint variable λ is defined as the solution of the adjoint system[42]

$$\begin{cases} -\dfrac{\partial\lambda}{\partial t} - \Delta\lambda = \gamma_1(\phi - \Phi) & \text{in } (0,T)\times\Omega, \\ \lambda = 0 & \text{on } (0,T)\times\Gamma, \\ \lambda = \gamma_2(\phi|_{t=T} - \Psi) & \text{in } \Omega \text{ at } t = T\,. \end{cases} \tag{2.97}$$

We substitute these equations into the formula (2.95) for the gradient of the functional in terms of sensitivities, i.e., we substitute for $(\phi - \Phi)$ and $(\phi|_{t=T} - \Psi)$, and then integrate by parts to remove derivatives from the adjoint variable λ. After the substitution of the sensitivity equations (2.96), we arrive at the expression

$$\frac{D\mathcal{J}}{D\alpha_k} = \sigma\alpha_k + \int_0^T \int_\Omega \lambda f_k \, d\Omega dt\,, \quad k = 1, \ldots, K, \tag{2.98}$$

for the components of the gradient of the functional (2.91) in terms of the adjoint variable λ.

[42]The adjoint system has already been derived in our consideration of the one-shot approach; see (2.93).

Given a guess for the parameters, one may solve the state system (2.90) forward in time and, regardless of the number of design parameters, solve the *single* adjoint system (2.97) backward in time to obtain the information needed to determine the gradient of the functional from (2.98); this gradient information may be used in a gradient-based optimization algorithm such as Algorithm 2.1. Thus, as in the steady-state case, for a given set of K design parameters, one may determine the gradient of the functional through a single solve of the state equations and solving either K sensitivity equations or a single adjoint equation.

However, there are serious practical problems with the adjoint equation approach to determining the gradient of the functional in the time-dependent case. These will be discussed in Section 5.1.

2.7 Side constraints

We briefly consider how side constraints may be treated in control and optimization problems. Side constraints are constraints that take the form of equalities or inequalities involving functionals, i.e., scalar-valued mappings, over the state space and the space of controls or design parameters. For the abstract optimal control or optimization problem of Example 1 on page 12 we denote the side constraints by $\Lambda(\phi, g) \leq 0$. Then, for that example, the control or optimization problem with the addition of *side constraints* is given by

> *find controls* g *and states* ϕ *such that* $\mathcal{J}(\phi, g)$ *is minimized subject to* $F(\phi, g) = 0$ *and* $\Lambda(\phi, g) \leq 0$.

Side constraints are divided into two categories, namely those that involve only the control or design parameters and those that involve the state, either on its own or along with the control. This division is made because there is a big difference in how side constraints that involve only the control are handled compared to those that also involve the state.

Theoretical treatments of side constraints result in additional adjoint variables in adjoint systems but no additional equations or variables for sensitivity-based methods. In practice, many textbook-type methods (see, e.g., [403]) for treating side constraints are not used; they are often too cumbersome to implement in the complicated setting of nonlinear partial differential equations.

Some examples of side constraints are provided by minimum lift, minimum thickness-to-chord ratio, minimum volume requirement, or maximum weight requirements in airfoil design problems, e.g., in a drag minimization problem for an airfoil, one usually wants to make sure that the volume of the airfoil is bounded from below or that the lift is bounded from below. As noted on page 12, an explicit bound on the control can be viewed as a side constraint; note that, in this case, the constraint involves only the control.

Side constraints involving only the control

A common method for treating side constraints that involve only the control is to *retreat or project back to the feasible set.* Let $\Lambda(g) \leq 0$ be a constraint on the control g and assume that it is satisfied by the current iterate g_{old} for the control. Next, assume that the new iterate for the control determined by

$$g_{new} = g_{old} + \delta g ,$$

where δg is a step supplied by the optimizer, does not satisfy the constraint, i.e., $\Lambda(g_{new}) > 0$. One obvious thing to do is shorten the step, i.e., redefine g_{new} to be

$$g_{new} = g_{old} + \gamma \delta g \,,$$

where γ is small enough so that now g_{new} satisfies the constraint $\Lambda(g_{new}) \leq 0$; such a γ exists and, moreover, the value of the functional is still lower than it was at g_{old}. Another approach is to project back to the constraint set, e.g., find a g that satisfies the constraint and that is nearest g_{new} and use that point instead as the new iterate. Figure 2.2 provides a pictorial description of these processes.

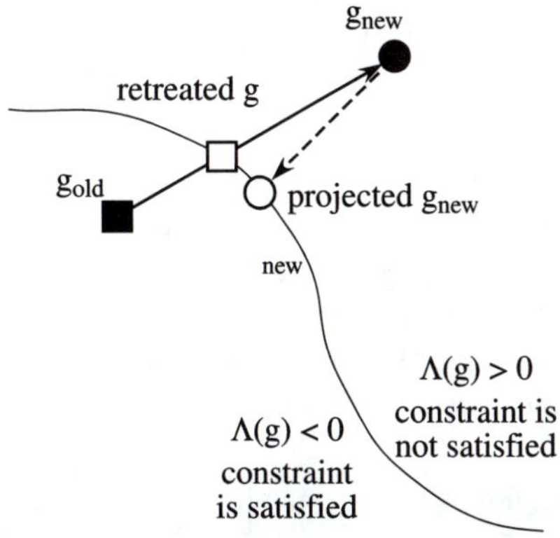

Figure 2.2. *The old iterate (filled square) and the new iterate (filled circle). The new iterate leaves the constraint set. It can be brought back to the constraint set by either retreating along the path from the old iterate (open square) or by projecting it back so that it satisfies the constraint (open circle).*

There are more sophisticated yet efficient algorithms available for handling constraints on the design parameters. These become especially important when there is a large number of design parameters, as is the case when one approximates infinite-dimensional controls; see, e.g., [395].

Side constraints involving the state

Reformulation. Side constraints that involve the state are more difficult to handle. Sometimes an optimization problem can be reformulated so that the side constraint can actually be made into one of the controls. For example, suppose we need to solve the following optimization problem: find a state ϕ and design parameters α_1 and α_2 such that the functional $\mathcal{J}(\phi, \alpha_1, \alpha_2)$ is minimized subject to the constraint $F(\phi, \alpha_1, \alpha_2) = 0$ and the side constraint

$$\Lambda(\phi, \alpha_1, \alpha_2) \leq 0 \,.$$

Suppose $\Lambda = \Lambda(\phi, \alpha_1, \alpha_2)$ can be solved for α_1 so that we have a relation of the form

$$\alpha_1 = f(\phi, \Lambda, \alpha_2) \,. \tag{2.99}$$

We then define a new functional

$$\mathcal{K}(\phi, \Lambda, \alpha_2) \equiv \mathcal{J}(\phi, f(\phi, \Lambda, \alpha_2), \alpha_2)$$

and a new state equation $G(\phi, \Lambda, \alpha_2) = 0$, where

$$G(\phi, \Lambda, \alpha_2) \equiv F(\phi, f(\phi, \Lambda, \alpha_2), \alpha_2)\,.$$

We can then reformulate the optimization problem so that now we seek a state ϕ and "design parameters" Λ and α_2 such that the functional $\mathcal{K}(\phi, \Lambda, \alpha_2)$ is minimized subject to the reformulated state equation $G(\phi, \Lambda, \alpha_2) = 0$ and the side constraint $\Lambda \le 0$. In this new optimization problem, the side constraint involves only the "design parameter" Λ. Once the reformulated optimization problem is solved, the optimal value of the actual design parameter α_1 may be determined from (2.99).

Penalization. Another popular means of enforcing side constraints is through *penalization.* We have already seen that penalization is the popular means of limiting the size of the control. For example, (2.2) may be viewed as an attempt to enforce a constraint of the type (2.1). Penalization can also be used to enforce side constraints involving the state. For example, if an optimization problem involves the minimization of a functional $\mathcal{J}(\phi, g)$ subject to the side constraint $\Lambda(\phi, g) \le 0$, one can instead minimize

$$\mathcal{J}(\phi, g) + \sigma\Big(\max\big[0, \Lambda(\phi, g)\big]\Big)^2. \tag{2.100}$$

Note that the penalty is active only when the constraint is violated, i.e., when $\Lambda(\phi, g) > 0$. It can be shown that as $\sigma \to \infty$, the solution of the minimization problem involving the functional (2.100) converges to the solution of the minimization problem involving the functional $\mathcal{J}(\phi, g)$ and the side constraint $\Lambda(\phi, g) \le 0$. Of course, if one solves a minimization problem having a finite value for σ, then there is no guarantee that the constraint $\Lambda(\phi, g) \le 0$ is satisfied. In practice, one would increase σ during the iterative process to make sure the constraint is satisfied. See, e.g., [403], for details about penalty methods.

Barriers. Another way to enforce a side constraint such as $\Lambda(\phi, g) \le 0$ is through the introduction of *barrier functions.* For example, if the minimization problem involves the minimization of the functional $\mathcal{J}(\phi, g)$ subject to the side constraint $\Lambda(\phi, g) \le 0$, one can instead minimize the barrier functional

$$\mathcal{J}(\phi, g) - \frac{1}{\sigma}\left(\frac{1}{\Lambda(\phi, g)}\right) \tag{2.101}$$

or the functional with a *logarithmic barrier*

$$\mathcal{J}(\phi, g) - \frac{1}{\sigma}\ln\big(-\Lambda(\phi, g)\big)\,. \tag{2.102}$$

The purpose of the barrier terms in the functionals (2.101) and (2.102) is to prevent the iterates in an optimization algorithm from escaping to values such that the constraint is violated. It

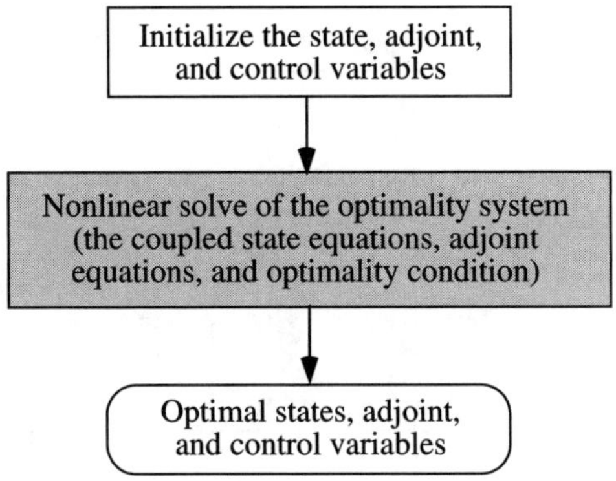

Figure 2.3. *Schematic diagram of the one-shot approach.*

can be shown that as $\sigma \to \infty$, the solutions of the minimization problems involving either of the functionals (2.101) or (2.102) converge to the solution of the minimization problem involving the functional $\mathcal{J}(\phi, g)$ and the side constraint $\Lambda(\phi, g) \leq 0$. In practice, one would again increase σ during the iterative process. Again, see, e.g., [403], for details about barrier methods.

2.8 One-shot vs. sensitivity vs. adjoint approaches

We now discuss some of the pros and cons of the different approaches for optimization and control that have been considered in Sections 2.2–2.6. Many of these comments have already been made in those sections.

One-shot vs. optimization-based approaches

At first glance, the one-shot approach seems very attractive. For example, compare Figures 2.3 and 2.4 in which schematic diagrams of the one-shot and gradient-based optimization approaches are respectively given. Note that the one-shot approach does not involve an optimization iteration. A single solve of the optimality system, i.e., the coupled state and adjoint systems and optimality condition, suffices to determine the optimal states, adjoint states, and control. If the state equations are nonlinear, then one still has to perform an iteration to solve the nonlinear optimality system (the shaded box in Figure 2.3); however, no optimization iteration is needed. Thus, if the nonlinear solve of the fully coupled optimality system is a practical possibility, then the one-shot approach is certainly worth considering. Unfortunately, in many flow control and optimization settings, it is not feasible to solve the coupled optimality system. For example, for three-dimensional, incompressible, viscous flows at "high" values of the Reynolds number in the primitive variable formulation, the unknowns are eight scalar fields (the pressure, the three components of the velocity, the adjoint pressure, and the three components of the adjoint velocity) in addition to any control or design variables. Just solving a simulation problem for the state equations, i.e., the Navier–Stokes system, is a formidable task!

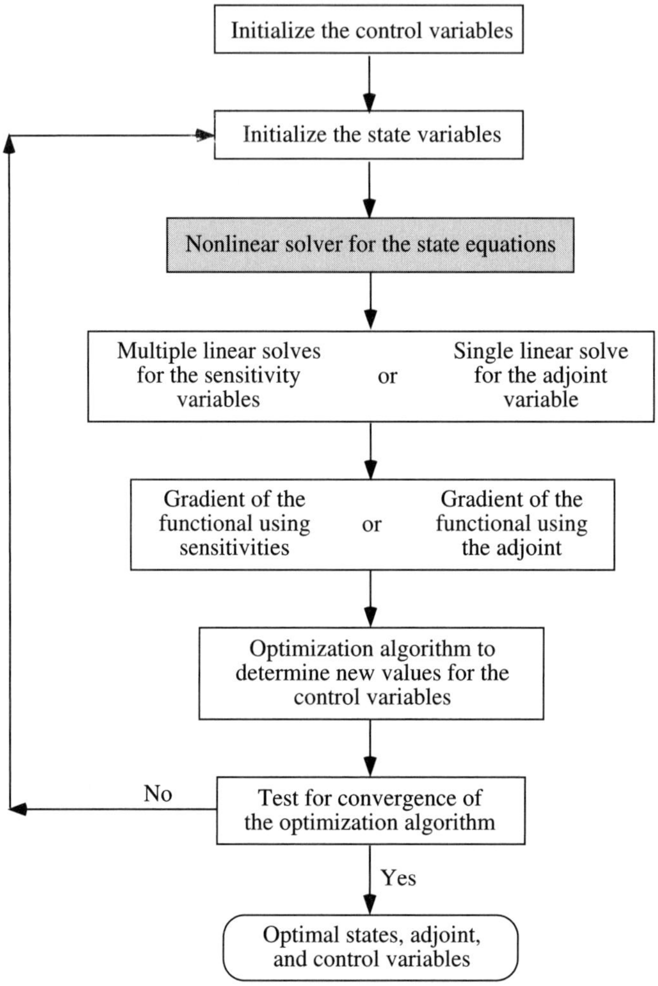

Figure 2.4. *Schematic diagram of the use of gradient-based optimization algorithm approaches.*

The schematic diagram in Figure 2.4 for gradient-based optimization algorithms points out the iteration that is engendered by the optimization algorithm. In addition, if the state system is nonlinear, one needs an inner iteration to solve the state equation (the shaded box in Figure 2.4). The nonlinear solves required for gradient-based optimization algorithms involve only the state system as opposed to the one-shot method, for which the nonlinear solve involves the coupled state and adjoint systems and the optimality condition. On the other hand, for gradient-based optimization algorithms, one must perform multiple nonlinear solves, e.g., for sensitivity-equation–based or adjoint-equation–based methods, one per iteration of the optimization algorithm.[43] Thus, in the one-shot approach, one has

[43]If finite difference quotient approximations to the gradient of the functional or to the sensitivities are used, then the number of nonlinear state solves is at least the number of design parameters plus one for each iteration of the optimization algorithm.

a single nonlinear solve of the larger coupled optimality system and in a gradient-based optimization approach one has multiple nonlinear solves of the smaller state system. The trade-offs are clear. For most practical flow control and optimization problems, the winner is the gradient-based optimization approach, i.e., it is more practical to solve the nonlinear state system several times than it is to solve the nonlinear optimality system once.

Sensitivities vs. adjoints

Having decided to use a gradient-based optimization method for the solution of a control or optimization problem, one still must decide between using sensitivities or adjoints for determining the gradient of the functional. Using Algorithm 2.1 and Example 3 on page 13 as prototypes, a summary of the two approaches is given in Figure 2.5.

For stationary problems such as Example 3, it is clear that the use of adjoint equations is preferable to the use of sensitivity equations. Regardless of the number of design parameters, one only has to solve a single adjoint system to determine the gradient of the functional, as opposed to multiple sensitivity systems. This advantage is even more apparent for infinite-dimensional controls. Furthermore, this advantage remains in effect even if one imposes side constraints by introducing additional Lagrange multipliers, i.e., additional adjoint variables. This is because side constraints are invariable scalar-valued functionals of the design parameters and the state so that the number of additional unknowns, i.e., additional Lagrange multipliers, is merely equal to the number of side constraints. This number is miniscule compared to the number of discrete state variables, sensitivity variables, and adjoint variables, which is proportional to the number of grid points.

For unsteady problems, the situation is quite different. Although one still has to solve a single adjoint system as opposed to multiple sensitivity systems, the adjoint systems must be solved backward in time, which is opposite to the sense of the state and sensitivity systems. Thus, a naive approach to solving the adjoint system results in considerably greater storage costs and perhaps CPU costs than that necessary for using sensitivity equations. However, the fact that only a single adjoint equation needs to be solved is so attractive that it is worth pursuing ways to mitigate the additional storage costs incurred. We will examine this issue in Section 5.1.

2.9 Differentiate-then-discretize vs. discretize-then-differentiate

So far, we have hardly discussed the discretization of optimization and control problems involving partial differential equation constraints. It is obvious that, even for the relatively simple Example 3 on page 13, one cannot solve such optimization problems exactly; of course, this is even more true for the considerably more complicated flow control and optimization setting. Thus one must be content with obtaining approximate solutions.

There are many issues that arise when one considers the discretization of flow control and optimization problems. In later chapters, we will encounter these issues as we examine a number of specific flow control problems. However, there is one philosophical issue, namely the choice of the differentiate-then-discretize vs. discretize-then-differentiate avenues for effecting the discretization of adjoint and sensitivity variables, that is important enough and

Choose initial values $\{\alpha_k^{(0)}\}_{k=1}^K$ for the design parameters. Then, for $m = 0, 1, \ldots,$

1. determine $\phi^{(m)}$ by solving the nonlinear state system

$$-\nabla \cdot (a\nabla\phi^{(m)}) + \mathbf{b} \cdot \nabla\phi^{(m)} + \left(\phi^{(m)}\right)^3 = \sum_{k=1}^K \alpha_k^{(m)} f_k \quad \text{in } \Omega\,,$$
$$\phi^{(m)} = 0 \quad \text{on } \Gamma\,;$$

2a. determine either

the K sensitivities $\{\phi_k^{(m)}\}_{k=1}^K$ by solving, for $k = 1, \ldots, K$, the K linear sensitivity systems

$$-\nabla \cdot (a\nabla\phi_k^{(m)}) + \mathbf{b} \cdot \nabla\phi_k^{(m)} + 3\left(\phi^{(m)}\right)^2 \phi_k^{(m)} = f_k \quad \text{in } \Omega,$$
$$\phi_k^{(m)} = 0 \quad \text{on } \Gamma$$

or

the adjoint variable $\lambda^{(m)}$ by solving the single linear adjoint system

$$-\nabla \cdot (a\nabla\lambda^{(m)}) - \nabla \cdot (\mathbf{b}\lambda^{(m)}) + 3\left(\phi^{(m)}\right)^2 \lambda^{(m)} = \phi^{(m)} - \Phi \quad \text{in } \Omega,$$
$$\lambda^{(m)} = 0 \quad \text{on } \Gamma\,.$$

2b. For $k = 1, \ldots, K$, determine the components of the gradient of the functional from either

$$\frac{D\mathcal{J}}{D\alpha_k} = \sigma\alpha_k^{(m)} + \int_\Omega (\phi^{(m)} - \Phi)\phi_k^{(m)}\, d\Omega \quad \text{or} \quad \frac{D\mathcal{J}}{D\alpha_k} = \sigma\alpha_k^{(m)} + \int_\Omega \lambda^{(m)} f_k\, d\Omega.$$

3. Use the result of step 2b to determine an increment in the design parameters; for example, in a simple gradient method, one would set

$$\delta\alpha_k^{(m)} = -\beta_m \frac{D\mathcal{J}}{D\alpha_k} \quad \text{for } k = 1, \ldots, K,$$

where β_m is an appropriately chosen step size; alternatively, one could use a more sophisticated method, e.g., BFGS, to determine $\delta\alpha_k^{(m)}$.

4. Determine the new values of the design parameters from

$$\alpha_k^{(m+1)} = \alpha_k^{(m)} + \delta\alpha_k^{(m)} \quad \text{for } k = 1, \ldots, K.$$

Then, either return to step 1, or, if satisfactory convergence has been achieved, exit.

Figure 2.5. *A summary of gradient-based optimization algorithms using either sensitivity or adjoint equations for Example* 3 *on page* 13.

probably controversial enough to merit discussion at this point.

In a *differentiate-then-discretize* approach, one obtains the adjoint or sensitivity equations at the partial differential equation level and then discretizes the result. Thus, one first differentiates the partial differential equations and the boundary and initial conditions with respect to the design or control parameters. Then, one discretizes the continuous sensitivity or adjoint systems using, e.g., algorithms similar to those used for discretizing the flow equations. Solving the discretized sensitivity or adjoint equations yields discrete approximations to the sensitivities or adjoint variables, respectively.

In a *discretize-then-differentiate* approach, one first discretizes the continuous flow equations to obtain a set of discrete flow equations. One then differentiates the discrete approximate flow equations to obtain a discrete sensitivity system or a discrete adjoint system. One popular way to obtain the discrete sensitivity or adjoint equations resulting from differentiating the discretized flow equations is to use *automatic differentiation software* [194, 197, 198, 199, 200, 375, 376, 377, 381, 384, 385, 386, 407]. Solving the sensitivity or adjoint equations corresponding to the discretized flow equations yields discrete approximations to the sensitivities or adjoint variables, respectively.

Thus, we have two paths for arriving at discrete sensitivity or discrete adjoint equations whose solutions are approximations to the exact sensitivity or adjoint variables, respectively. In general, the two paths lead to different approximations because *the differentiation and discretization steps do not commute.* Again, we emphasize that the approximate sensitivities or adjoint variables obtained by differentiating-then-discretizing and by discretizing-then-differentiating are not the same.

For both the differentiate-then-discretize and discretize-then-differentiate approaches, as grid sizes go to zero, discrete sensitivities and adjoint variables and therefore gradients of discretized objective functionals all converge (at least for smooth flows) to the same thing, i.e., the corresponding entities of the partial differential equation optimization problem. However, for finite (and practical) values of the grid size, there can be substantial differences in these entities as determined by each of the two approaches. Thus, we are led to the issue of deciding which approach is better.

Advantages of the discretize-then-differentiate approach

Consistency of functional gradients. Using discrete adjoints or sensitivity variables obtained by discretizing the continuous adjoint or sensitivity equations, i.e., the differentiate-then-discretize approach, can yield inconsistent gradients of objective functionals. The approximate gradient obtained in this way is not a true gradient of anything—not of the continuous functional nor of a discretized functional. This can result in serious difficulties; e.g., the downhill direction as determined by an inconsistent gradient may actually be an uphill direction for the functional; we will examine this point in more detail in Section 4.1.

On the other hand, discrete adjoints or sensitivity systems obtained through differentiation of the discretized flow equations can be used to obtain the exact gradient of a discretized functional and therefore inconsistencies between the computed gradient and the discretized functional do not arise.

Use of automatic differentiation software. In the discretize-then-differentiate approach, automatic differentiation software can be used to generate codes that solve for the exact sensitivities or adjoint variables for the discretized flow system. This can greatly simplify the writing of adjoint or sensitivity codes, especially for complex flows. On the other hand, although substantial savings in labor can be effected, sensitivity or adjoint codes created by automatic differentiation software usually require more storage and often require more CPU time than that for corresponding handwritten codes.

Advantages of the differentiate-then-discretize approach

Treatment of computational facilitators. Using adjoint variables or sensitivities determined through differentiating the discretized flow equations, e.g., by using automatic differentiation software, will generally involve differentiating computational facilitators such as turbulence models, shock-capturing devices, or outflow boundary treatments. Computational facilitators often complicate the approximate discrete flow equations, and therefore also complicate the derivation of the adjoint or sensitivity systems; this is one reason researchers who take the discretize-then-differentiate approach often turn to automatic differentiation software to determine discrete sensitivity or adjoint systems. However, when determining the adjoint or sensitivity systems for the continuous flow equations, one could (and probably should) take the attitude that computational facilitators are merely computational artifacts that are introduced to obtain better or cheaper approximate discrete flows; the basic flow models do not contain the computational facilitators. Thus, one should not have to deal with computational facilitators when determining adjoint or sensitivity equations. Thus, for example, for turbulent flows, the basic model is the Navier–Stokes equations without any turbulence model and one should merely find the sensitivity or adjoint systems using the standard Navier–Stokes system. Likewise, for compressible, inviscid flows, the basic model is the Euler equations without any flux limiters, artificial viscosity, or other shock-capturing devices. For flows posed on infinite domains, another type of computational facilitator is introduced, namely outflow boundary conditions; any special treatment of outflows in the state equations will get differentiated at the discrete level by an automatic differentiation code when perhaps what one needs to do, since the fluid inflow is the adjoint outflow and vice versa, is give the adjoint system special treatment at its outflow boundary.

On the other hand, once having obtained the continuous adjoint or sensitivity equations by differentiating a basic flow model, i.e., one not containing any computational facilitators, one may then introduce, if necessary, computational facilitators to help obtain good or cheap approximate sensitivity or adjoint variables. For the sensitivity or adjoint equations, one may in fact need to use different computational facilitators than those used for the flow equations.

Use of different grids for the flow, sensitivity, and adjoint systems. Using adjoint variables or sensitivities determined through differentiating the discretized flow equations usually leads to using the same grid for the sensitivity or adjoint systems as that used for the flow; this is certainly the case if one uses automatic differentiation software. However, it is clear that adjoint or sensitivity systems are often best discretized using a different grid than that used for the flow; see, e.g., [277, 285, 292]. For example, consider Figure 2.6 which provides velocity plots of an incompressible, viscous flow in a channel with a bump and of the sensitivity of that flow with respect to a parameter that determines the shape of the bump. Clearly, any self-respecting adaptive grid strategy will end up with substantially

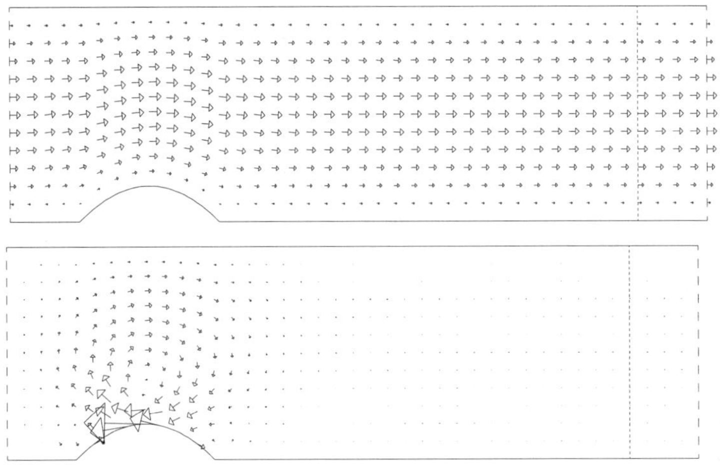

Figure 2.6. *A flow and its sensitivity with respect to a parameter determining the shape of the bump.*

different grids for the flow and sensitivity.

Using the differentiate-then-discretize approach, one can design grids that are specifically well suited to the sensitivity or adjoint systems and that may be different from a grid that is well suited to the flow equations.

Accounting for grid motion in shape optimization problems. Using adjoint variables or sensitivities determined through differentiating the discretized flow equations, i.e., a discretize-then-differentiate approach, may bring grid movements into the optimization problem if the grid changes from one iteration to another as it does in most algorithms for shape optimization problems. This results in the need to determine derivatives of the grid with respect to the design parameters. On the other hand, for the differentiate-then-discretize approach, one starts each iteration of the optimizer with partial differential equation problems, i.e., the state system and the sensitivity or adjoint system, posed on the latest domain shape. One then discretizes these problems on the new domain without needing to account for how the previous grid deformed into the new grid, e.g., without having to account for grid movements.

Conclusions

The choice between the differentiate-then-discretize and discretize-then-differentiate strategies has been, and remains to this day, more a matter of taste or of which method one first used than resulting from a decision made on the basis of perceived clear advantages possessed by one approach over the other. Certainly both strategies have been used with great success and it is likely that both strategies will continue to be used for some time to come.

In the end, which approach one uses probably depends on which one is easier to amend so that one can overcome its disadvantages. In this regard, it is probably true that the differentiate-then-discretize approach is preferable. Its main disadvantage, namely the possiblity of encountering inconsistent gradients, is relatively easy to avoid by, e.g., using regularization; see, e.g., Section 4.1. On the other hand, the disadvantages of the discretize-then-differentiate approach, because it is most often implemented using automatic differentiation software, are relatively more difficult to avoid. Although automatic differentiation is attractive because it obviates the need for hand coding of discrete adjoint or senstivity systems, once one commits to its use one gives up a lot of the flexibility needed to avoid complications that arise due to added costs, computational facilitators, grid motion, etc.

Chapter 3

Illustrations of the Three Approaches

In this chapter, we present, in the context of fluid flows, some concrete problems that serve to illustrate each of the three approaches discussed in Chapter 2 for solving optimization and control problems.

3.1 One-shot method for a temperature matching problem in a coupled solid-fluid heat transfer problem

We illustrate the use of one-shot or Lagrange multiplier methods for optimal control and optimization with a two-dimensional, coupled solid-fluid heat conduction problem. Not all aspects of this example problem are realistic, but it also serves to illustrate some of the types of controls and functionals that we previously encountered in Chapter 2. For details concerning this illustration, see [81]. Several other papers cited in the bibliography deal with similar problems.

The model problem

The objective is to match the temperature along part of a fluid-solid interface to a given temperature by controlling the temperature at the fluid inflow. A sketch of the geometric set-up of the problem we are considering is given in Figure 3.1. The fluid velocity $\mathbf{u}$ is found by solving the following Navier–Stokes system:

$$\begin{aligned}
\text{in the fluid:} \quad & \begin{cases} -\nu\Delta\mathbf{u} + \mathbf{u}\cdot\nabla\mathbf{u} + \nabla p = \mathbf{0}, \\ \nabla\cdot\mathbf{u} = 0; \end{cases} \\
\text{at the fluid inflow:} \quad & \mathbf{u} = (u_0, 0)^T; \\
\text{at the bottom and top walls of the fluid region:} \quad & \mathbf{u} = \mathbf{0}; \\
\text{at the fluid outflow:} \quad & \frac{\partial u}{\partial x} = 0 \quad \text{and} \quad v = 0,
\end{aligned}$$

where u and v denote the components of $\mathbf{u}$ in the x and y directions, respectively, p denotes the pressure, and u_0 denotes a prescribed inflow velocity. For our simple illustration, the fluid momentum and continuity equations are uncoupled from the energy equations so that the fluid velocity and pressure are unaffected by changes in the temperature.

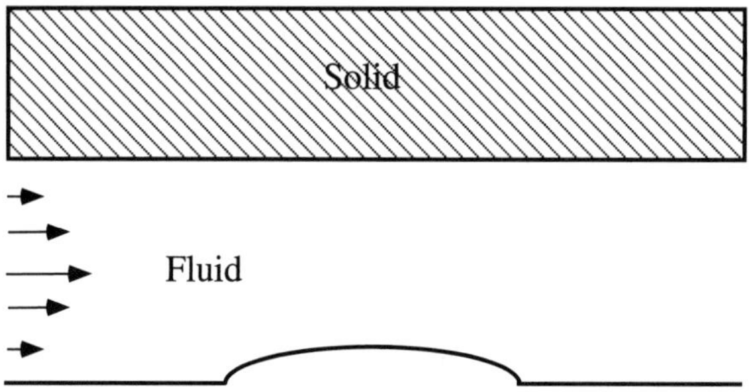

Figure 3.1. *A channel with a bump with a solid object above the channel.*

Let Γ_c denote the fluid inflow boundary along which the temperature control is applied. The *constraint* or *state equations* are the heat conduction equation in the solid and the energy equation for the fluid, both in terms of T, the temperature:

$$\begin{aligned} \text{in the solid:} \quad & -a_1 \Delta T = Q; \\ \text{in the fluid:} \quad & -a_2 \Delta T + \mathbf{u} \cdot \nabla T = 0; \\ \text{along the fluid-solid interface:} \quad & \begin{cases} T_{solid} = T_{fluid}, \\ \left(a_1 \dfrac{\partial T}{\partial y}\right)_{solid} = \left(a_2 \dfrac{\partial T}{\partial y}\right)_{fluid}; \end{cases} \\ \text{at the fluid inflow } \Gamma_c\text{:} \quad & T = g; \\ \text{at other boundaries:} \quad & \frac{\partial T}{\partial n} = 0, \end{aligned}$$

where a_1 and a_2 are constants, Q is a source term, and g *is the control function acting along the fluid inflow* Γ_c. Note the appearance of the fluid velocity $\mathbf{u}$, which is presumed known, as a coefficient in the constraint system.

Let Γ_d denote the portion of the fluid-solid interface along which we want to match the temperature. Let T_d denote the desired temperature distribution along Γ_d and let δ and γ be constants. The *objective functional* is of matching type and is given by

$$\mathcal{T}(T, g) = \frac{1}{2\gamma} \int_{\Gamma_d} |T - T_d|^2 \, dx + \frac{\delta}{2} \int_{\Gamma_c} \left(\left| \frac{dg}{dy} \right|^2 + |g|^2 \right) dy .$$

The first term in the objective functional measures, along Γ_d, the difference between the temperature T and the desired temperature T_d. The second term in the functional is a penalty term introduced in order to limit the size of the control, i.e., the temperature along the fluid inflow Γ_c. Note that, in order to avoid *oscillations* in the control, *the penalty term involves the derivative of the boundary control g along the boundary segment* Γ_c.

The *optimization problem* we consider is to find a control g such that the functional $\mathcal{T}$ is minimized, subject to the constraint equations being satisfied. Note that this is a two-dimensional problem involving scalar partial differential equations; as was alluded to in Chapter 2, this is the type of optimization problem that is amenable to solution by the one-shot approach.

The optimality system

The *adjoint equations* for the adjoint variable θ are as follows:

$$
\begin{aligned}
&\text{in the solid:} && -a_1\Delta\theta = 0;\\
&\text{in the fluid:} && -a_2\Delta\theta - \mathbf{u}\cdot\nabla\theta = 0;\\
&\text{along the fluid-solid interface:} && \begin{cases} \theta_{solid} = \theta_{fluid},\\ \left(a_2\dfrac{\partial\theta}{\partial y}\right)_{fluid} - \left(a_1\dfrac{\partial\theta}{\partial y}\right)_{solid} = \dfrac{1}{\gamma}\chi_d(T-T_d); \end{cases}\\
&\text{at the fluid inflow } \Gamma_c\text{:} && \theta = 0;\\
&\text{at the fluid outflow:} && a_2\frac{\partial\theta}{\partial x} + u\theta = 0;\\
&\text{at other boundaries:} && \frac{\partial\theta}{\partial n} = 0,
\end{aligned}
$$

where χ_d denotes the characteristic function for Γ_d, i.e.,

$$
\chi_d = \begin{cases} 1 & \text{on } \Gamma_d,\\ 0 & \text{on the rest of the interface.} \end{cases}
$$

The *optimality condition* is given by

$$
\begin{cases} -\dfrac{d^2g}{dy^2} + g = -\dfrac{1}{\delta}\left(a_2\dfrac{\partial\theta}{\partial x} + u_0\theta\right) & \text{on } \Gamma_c,\\ \dfrac{dg}{dy} = 0 \quad \text{at the endpoints of } \Gamma_c, \end{cases}
$$

where u denotes the x-component of the fluid velocity.

The constraint equations, the adjoint equations, and the optimality condition form the *optimality system*, whose solution provides the optimal state T, co-state θ, and control function g.

We provide a few details about the derivation of the optimality system, although the process we follow is exactly that already presented in Chapter 2. We introduce Lagrange multipliers θ to enforce the partial differential equations for the state T in the solid and fluid and μ to enforce the Dirichlet boundary condition on the state T at the fluid inflow. We assume that all the other boundary conditions on T are satisfied by all candidate minimizers. The Lagrangian functional is then given by

$$
\begin{aligned}
\mathcal{L}(T,g,\theta,\mu) = {}& \frac{1}{2\gamma}\int_{\Gamma_d} |T-T_d|^2\,dx + \frac{\delta}{2}\int_{\Gamma_c}\left(\left|\frac{dg}{dy}\right|^2 + |g|^2\right)dy - \int_{\Gamma_c}\mu(T-g)\,dy\\
& - \int_{\text{fluid}} \theta(-a_2\Delta T + \mathbf{u}\cdot\nabla T)\,dxdy - \int_{\text{solid}} \theta(-a_1\Delta T - Q)\,dxdy\,.
\end{aligned}
$$

The first two terms on the right-hand side are the given functional $\mathcal{T}(T, g)$; the third term results from enforcing the boundary condition $T = g$ along the inflow Γ_c; the last two terms result from enforcing the partial differential equations in the fluid and solid.

Setting the first variations of $\mathcal{L}$ with respect to the Lagrange multipliers θ and μ equal to zero yields the corresponding constraint equations.

Setting the first variations of $\mathcal{L}$ with respect to the state T equal to zero results in

$$\frac{1}{\gamma}\int_{\Gamma_d}(T - T_d)\widetilde{T}\,dx - \int_{\Gamma_c}\mu\widetilde{T}\,dy$$
$$- \int_{\text{fluid}}\theta(-a_2\Delta\widetilde{T} + \mathbf{u}\cdot\nabla\widetilde{T})\,dxdy - \int_{\text{solid}}\theta(-a_1\Delta\widetilde{T})\,dxdy = 0$$

for all variations $\widetilde{T}$ in the state T that are arbitrary, except that $\widetilde{T}$ is constrained so that all candidate minimizers satisfy the boundary and interface conditions that have not been enforced through the Lagrangian functional. We next integrate by parts to remove derivatives from $\widetilde{T}$ to obtain

$$- \int_{\text{fluid}}\widetilde{T}(-a_2\Delta\theta - \mathbf{u}\cdot\nabla\theta)\,dxdy - \int_{\text{solid}}\widetilde{T}(-a_1\Delta\theta)\,dxdy$$
$$+ \int_{\text{fluid boundary}}\left(a_2\left(\theta\frac{\partial\widetilde{T}}{\partial n} - \widetilde{T}\frac{\partial\theta}{\partial n}\right) - \mathbf{u}\cdot\mathbf{n}\theta\widetilde{T}\right)d\Gamma$$
$$+ \int_{\text{solid boundary}} a_1\left(\theta\frac{\partial\widetilde{T}}{\partial n} - \widetilde{T}\frac{\partial\theta}{\partial n}\right)d\Gamma$$
$$+ \frac{1}{\gamma}\int_{\Gamma_d}(T - T_d)\widetilde{T}\,dx - \int_{\Gamma_c}\mu\widetilde{T}\,dy = 0\,,$$

where the fluid and solid boundaries include the interface. We first choose variations $\widetilde{T}$ that vanish in the neighborhood of all boundaries and in the solid region; this yields

$$- \int_{\text{fluid}}\widetilde{T}(-a_2\Delta\theta - \mathbf{u}\cdot\nabla\theta)\,dxdy = 0,$$

which, since $\widetilde{T}$ is arbitrary in the fluid, implies that

$$-a_2\Delta\theta - \mathbf{u}\cdot\nabla\theta = 0 \qquad \text{in the fluid.}$$

In an analogous manner, we arrive at

$$-a_1\Delta\theta = 0 \qquad \text{in the solid.}$$

We are now left with

$$\int_{\text{fluid boundary}}\left(a_2\left(\theta\frac{\partial\widetilde{T}}{\partial n} - \widetilde{T}\frac{\partial\theta}{\partial n}\right) - \mathbf{u}\cdot\mathbf{n}\theta\widetilde{T}\right)d\Gamma$$
$$+ \int_{\text{solid boundary}} a_1\left(\theta\frac{\partial\widetilde{T}}{\partial n} - \widetilde{T}\frac{\partial\theta}{\partial n}\right)d\Gamma$$
$$+ \frac{1}{\gamma}\int_{\Gamma_d}(T - T_d)\widetilde{T}\,dx - \int_{\Gamma_c}\mu\widetilde{T}\,dy = 0\,.$$

We have enforced only the boundary condition $T = g$ on Γ_c through the Lagrangian functional; this means that allowable variations $\widetilde{T}$ in the state T are required to satisfy[44]

$$\text{along the solid fluid interface:} \quad \begin{cases} \widetilde{T}_{solid} = \widetilde{T}_{fluid}, \\ \left(a_1 \dfrac{\partial \widetilde{T}}{\partial y}\right)_{solid} = \left(a_2 \dfrac{\partial \widetilde{T}}{\partial y}\right)_{fluid} \end{cases};$$

$$\text{along boundaries other than } \Gamma_c\text{:} \quad \frac{\partial \widetilde{T}}{\partial n} = 0\,.$$

Also, recall that the fluid velocity components u and v vanish along the channel bottom and the fluid-solid interface. Putting all this together and keeping in mind that $\mathbf{n}$ denotes the outward unit normal vector leaves us with

$$\begin{aligned} &\int_{\text{fluid inflow}} \left(-a_2\theta \frac{\partial \widetilde{T}}{\partial x} + \left(a_2 \frac{\partial \theta}{\partial x} + u_0\theta - \mu \right) \widetilde{T} \right) dy \\ &- \int_{\text{fluid outflow}} \widetilde{T} \left(a_2 \frac{\partial \theta}{\partial x} + u\theta \right) dy + \int_{\text{channel bottom}} \widetilde{T} \left(a_2 \frac{\partial \theta}{\partial y} \right) dx \\ &- \int_{\text{outer solid boundary}} a_1 \widetilde{T} \frac{\partial \theta}{\partial n} d\Gamma + \int_{\text{interface}} (\theta_{fluid} - \theta_{solid}) \left(a_2 \frac{\partial \widetilde{T}}{\partial y} \right) dx \\ &- \int_{\text{interface}} \widetilde{T} \left(\left(a_2 \frac{\partial \theta}{\partial y} \right)_{fluid} - \left(a_1 \frac{\partial \theta}{\partial y} \right)_{solid} - \frac{1}{\gamma} \chi_d (T - T_d) \right) dx = 0\,. \end{aligned}$$

We have integrals over four different portions of the boundary and the interface; one at a time, we choose variations $\widetilde{T}$ that vanish in the neighborhood of four of these; for example, after choosing variations that vanish in a neighborhood of all boundaries and interfaces except the fluid inflow, all that remains is

$$-\int_{\text{fluid inflow}} a_2\theta \frac{\partial \widetilde{T}}{\partial x} dy + \int_{\text{fluid inflow}} \left(a_2 \frac{\partial \theta}{\partial x} + u_0\theta - \mu \right) \widetilde{T}\, dy = 0\,.$$

Next, we choose variations $\widetilde{T}$ that are arbitrary in a neighborhood of the fluid inflow but such that $\partial \widetilde{T}/\partial x = 0$ at the inflow. Then we are left with

$$\int_{\text{fluid inflow}} \left(a_2 \frac{\partial \theta}{\partial x} + u_0\theta - \mu \right) \widetilde{T}\, dy = 0$$

from which we conclude that

$$\mu = a_2 \frac{\partial \theta}{\partial x} + u_0\theta \quad \text{along the fluid inflow.}$$

If instead we choose variations $\widetilde{T}$ that are arbitrary in a neighborhood of the fluid inflow but such that $\widetilde{T} = 0$ at the inflow, then we are left with

$$\int_{\text{fluid inflow}} a_2\theta \frac{\partial \widetilde{T}}{\partial x} dy = 0$$

[44] For example, we have not enforced the boundary condition $T_{solid} = T_{fluid}$ through the Lagrangian functional. Therefore, any variations $\widetilde{T}_{solid}$ and $\widetilde{T}_{fluid}$ must be such that $T_{solid} + \widetilde{T}_{solid} = T_{fluid} + \widetilde{T}_{fluid}$, which, since $T_{solid} = T_{fluid}$, implies that $\widetilde{T}_{fluid} = \widetilde{T}_{solid}$.

from which we conclude that

$$\theta = 0 \quad \text{along the fluid inflow.}$$

Proceeding in a similar manner for the other boundary segments and the interface, we determine all the boundary conditions and interface conditions for the adjoint variable θ.

We now turn to the derivation of the optimality condition; setting the first variation in the Lagrangian functional with respect to the control function g equal to zero results in

$$\int_{\Gamma_c} \left(\delta \Big(\frac{dg}{dy}\frac{d\widetilde{g}}{dy} + \widetilde{g}g \Big) + \mu\widetilde{g} \right) dy = 0$$

for arbitrary variations $\widetilde{g}$ in the control function g. Integrating by parts, we obtain

$$\int_{\Gamma_c} \widetilde{g} \left(\delta\Big(-\frac{d^2g}{dy^2} + g \Big) + \mu \right) dy + \delta \left(\frac{dg}{dy}\widetilde{g} \right) \Bigg|_{\text{bottom of } \Gamma_c}^{\text{top of } \Gamma_c} = 0$$

from which we obtain

$$\delta\Big(-\frac{d^2g}{dy^2} + g \Big) = -\mu \quad \text{along } \Gamma_c$$

along with $dg/dy = 0$ at the top and bottom of Γ_c. Since $\mu = a_2 \partial\theta/\partial x + u_0\theta$ along the fluid inflow Γ_c, we obtain

$$-\frac{d^2g}{dy^2} + g = -\frac{1}{\delta}\left(a_2\frac{\partial\theta}{\partial x} + u_0\theta \right)$$

which completes the derivation of the optimality system.

This example has provided us with the opportunity to once again explore the derivation of an optimality system for a problem for which the control acts along a boundary segment, the control is an infinite-dimensional function, the functional contains integrals along the boundary, and the functional contains derivatives of the control function.

A computational example

We choose $a_1 = 1$, $a_2 = 2$, and $Q = 6$. The corners of the fluid region are located at (0, 0), (0, 0.75), (1, 0.75), and (1, 0); the corners of the solid region are located at (0, 0.75), (0, 1), (1, 1), and (1, 0.75). Thus, Γ_c is the line segment $\{0\} \times (0, 0.75)$. We also choose $T_d = 1.2$ and $\Gamma_d = (0.075, 1) \times \{0.75\}$; thus, Γ_d is part of the fluid-solid boundary $(0, 1) \times \{0.75\}$. For the "uncontrolled case," we set $T = g = 1$ along Γ_c. We choose two sets of values for the parameters in the functional:

- $\gamma = \delta = 1$, which allows for relatively small controls;
- $\gamma = 0.002$ and $\delta = 0.01$, which allows for relatively large controls.

Finite element discretizations are used for the state and control functions.

Figures 3.2 and 3.3, respectively, show the optimal control, i.e., the temperature along the fluid inflow, and the temperature along the fluid-surface interface, part of which is the matching surface. For both sets of parameters in the functional, the match of the optimal temperature to the desired temperature along Γ_d is much better than that for the uncontrolled case. For relatively larger controls, the match of the optimal temperature to the desired temperature along Γ_d is better than when controls are smaller.

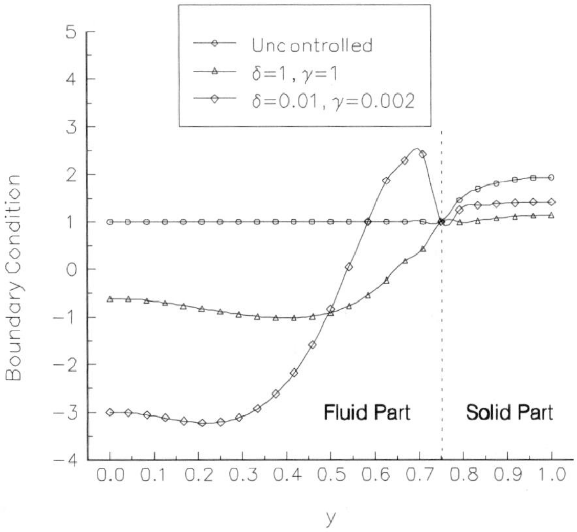

Figure 3.2. *Temperature at $x = 0$; recall that Γ_c extends from $y = 0$ to $y = 0.75$ so that the temperature along that segment is the optimal control.*

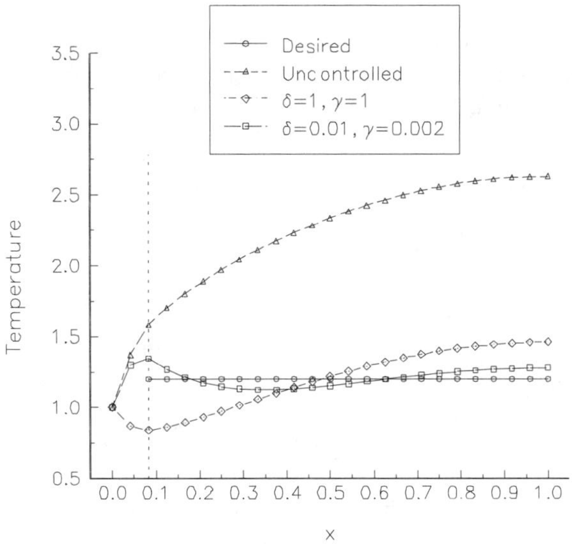

Figure 3.3. *Temperature at the fluid-solid interface $y = 0.75$; recall that Γ_d extends from $x = 0.075$ to $x = 1$ so that we can expect the match to the desired temperature to be good only along that segment. The horizontal line is the desired temperature.*

Remarks about the uncoupling of the mechanical-thermal fluid calculation

In the example we have just studied, the fluid velocity and pressure determination was uncoupled from that of the temperature. This was possible for two reasons: the fluid momentum and continuity equations did not involve the temperature, and the control was chosen to act in the energy equation for the temperature. If either of these were not true, then the fluid motion and temperature calculations become coupled. If the fluid equations depend on the temperature, then obviously we have coupling between the temperature and the fluid velocity. This happens, e.g., for temperature-dependent viscosities or for the Boussinesq equations. Of course, this also happens for compressible flows since the density and pressure are related to the temperature through an equation of state.

Suppose we change the problem so that everything remains the same, including the temperature matching term in the functional, except the control is the x-component of the inflow velocity instead of the temperature. Then, we cannot solve for the velocity by itself because the control is unknown and we cannot solve for the temperature by itself because the velocity appears as a coefficient in the energy equation. Thus, the fluid velocity, the temperature, and the control are determined from a fully coupled optimality system that includes the momentum, continuity, and energy equations.

Suppose instead we change the problem so that everything remains the same, including the use of the temperature at the fluid inflow as the control, except the functional involves the x-component of the velocity instead of the temperature. Then, since the velocity is unaffected by the temperature, we cannot effect control!

Thus, we see that if part of the fluid system is uncoupled from the rest in a simulation setting, they can remain uncoupled in an optimization setting (the example considered in detail in this section), become coupled in an optimization setting (the example of two paragraphs above), or even make optimization impossible (the example of the previous paragraph). Which situation is in effect depends on the specific choices made for the control and the objective functional.

3.2 Sensitivity equation method for an optimization problem for Euler flows

We illustrate the use of sensitivity-based methods for the solution of optimal flow control and optimization problems by considering an optimization problem for Euler flows. This example also serves to highlight some issues that arise for flows with discontinuities and for shape design problems. We use the *sensitivity equation approach*, in which we first differentiate the Euler equations to obtain the continuous sensitivity equations, and then discretize the result to obtain a set of discrete sensitivity equations, i.e., we use a differentiate-then-discretize approach. For details concerning this illustration, see [181, 210, 217, 272]. Several other papers cited in the bibliography deal with similar problems.

The model problem

Consider the two-dimensional Euler equations

$$\frac{\partial \mathbf{F}}{\partial x} + \frac{\partial \mathbf{G}}{\partial y} = \mathbf{0} \qquad \text{in } \Omega\,, \tag{3.1}$$

where Ω denotes a two-dimensional flow domain, $\mathbf{F}(\mathbf{Q})$ and $\mathbf{G}(\mathbf{Q})$ are flux functions, and $\mathbf{Q}$ is the vector of conservative variables, e.g., $\mathbf{Q} = (\rho, m_1, m_2, e)^T$, where ρ denotes the density, m_1 and m_2 denote the x- and y-components of the momentum, and e denotes the internal energy per unit mass. The partial differential equation (3.1) is supplemented by the inflow condition

$$\mathbf{Q} = (\beta, \alpha_0, 0, \sigma)^T \qquad \text{on } \Gamma_0 \tag{3.2}$$

and the no-penetration condition

$$m_1 n_1 + m_2 n_2 = 0 \qquad \text{on } \Gamma\,. \tag{3.3}$$

We assume that the obstacle Γ is determined by a finite number of parameters $\alpha_1, \ldots, \alpha_K$ through a given function Ψ, i.e.,

$$y = \Psi(x; \alpha_1, \ldots, \alpha_K)\,. \tag{3.4}$$

Then, we can rewrite the no-penetration condition (3.3) in the form

$$\begin{aligned} -m_1\Big(x, y = \Psi(x;\alpha_1,\ldots,\alpha_K); \alpha_1,\ldots,\alpha_K\Big)\frac{\partial \Psi}{\partial x}(x;\alpha_1,\ldots,\alpha_K) \\ +m_2\Big(x, y = \Psi(x;\alpha_1,\ldots,\alpha_K); \alpha_1,\ldots,\alpha_K\Big) = 0\,. \end{aligned} \tag{3.5}$$

The design parameters enter into (3.5) in three ways:

- the components m_1 and m_2 of the momentum depend explicitly on the design parameters;
- the function Ψ that determines the shape of the obstacle depends on the design parameters;
- the condition is applied on the obstacle Γ whose shape and normal vector depend on the design parameters.

It is important to keep track of all three dependencies when one is interested in sensitivities.

The objective functional is defined to be

$$\mathcal{J}(\mathbf{Q}) = \frac{1}{2}\int_{\Gamma_1} \left|\mathbf{Q} - \widehat{\mathbf{Q}}\right|^2 d\Gamma\,, \tag{3.6}$$

where Γ_1 denotes a part of the boundary of the flow domain, e.g., the outflow boundary, and $\widehat{\mathbf{Q}}$ is a given flow field. Thus, (3.6) is a matching-type functional.

The optimization problem we consider is

> *find parameters α_k, $k = 0, \ldots, K$, such that the functional $\mathcal{J}(\mathbf{Q})$ given by* (3.6) *is minimized subject to the state system, i.e., the Euler system* (3.1)–(3.3), *being satisfied.*

For this problem one value parameter α_0, the inflow x momentum, and K shape parameters α_k, $k = 1, \ldots, K$, are at our disposal to effect the optimization.

Sensitivity systems

Let $\mathbf{Q}_0 = (\rho_0, (m_1)_0, (m_2)_0, e_0)^T = \partial\mathbf{Q}/\partial\alpha_0$, i.e., $\mathbf{Q}_0$ is the sensitivity with respect to the inflow x momentum component α_0. Then, by formal[45] differentiation of the state system (3.1)–(3.3), we have that $\mathbf{Q}_0$ satisfies the *sensitivity system with respect to the value parameter* α_0

$$\begin{cases} \dfrac{\partial}{\partial x}\left(\dfrac{\partial \mathbf{F}}{\partial \mathbf{Q}}\cdot \mathbf{Q}_0\right) + \dfrac{\partial}{\partial y}\left(\dfrac{\partial \mathbf{G}}{\partial \mathbf{Q}}\cdot \mathbf{Q}_0\right) = \mathbf{0} & \text{in } \Omega\,, \\ \mathbf{Q}_0 = \begin{pmatrix} 0 \\ 1 \\ 0 \\ 0 \end{pmatrix} & \text{at the inflow } \Gamma_0\,, \\ (m_1)_0 n_1 + (m_2)_0 n_2 = 0 & \text{on the obstacle } \Gamma\,. \end{cases} \tag{3.7}$$

Let $\mathbf{Q}_k = (\rho_k, (m_1)_k, (m_2)_k, e_k)^T = \partial\mathbf{Q}/\partial\alpha_k$, $k = 1, \dots, K$, i.e., $\mathbf{Q}_k$ is the sensitivity with respect to the kth shape parameter α_k. Then, by formal differentiation of the state system (3.1)–(3.3), we have that, for $k = 1, \dots, K$, $\mathbf{Q}_k$ satisfies the *sensitivity system with respect to the shape parameter* α_k

$$\begin{cases} \dfrac{\partial}{\partial x}\left(\dfrac{\partial \mathbf{F}}{\partial \mathbf{Q}}\cdot \mathbf{Q}_k\right) + \dfrac{\partial}{\partial y}\left(\dfrac{\partial \mathbf{G}}{\partial \mathbf{Q}}\cdot \mathbf{Q}_k\right) = \mathbf{0} & \text{in } \Omega\,, \\ \mathbf{Q}_k = \mathbf{0} & \text{at the inflow } \Gamma_0\,, \\ -(m_1)_k \dfrac{\partial \Psi}{\partial x} + (m_2)_k & \\ \quad = \left(\dfrac{\partial m_1}{\partial y}\dfrac{\partial \Psi}{\partial x} - \dfrac{\partial m_2}{\partial y}\right)\Psi_k + m_1 \dfrac{\partial \Psi_k}{\partial x} & \text{on the obstacle } \Gamma\,, \end{cases} \tag{3.8}$$

where $\Psi_k = \partial\Psi/\partial\alpha_k$, $k = 1, \dots, K$.

Often, the function Ψ describing the shape is given by

$$\Psi(x; \alpha_1, \dots, \alpha_K) = \sum_{k=1}^{K} \alpha_k \psi_k(x)\,,$$

where $\psi_k(x)$, $k = 1, \dots, K$, are given functions, e.g., B-splines or Bezier polynomials. In this case,

$$\Psi_k = \frac{\partial \Psi}{\partial \alpha_k} = \psi_k(x) \qquad \text{and} \qquad \frac{\partial \Psi_k}{\partial x} = \frac{\partial \psi_k}{\partial x}\,.$$

If the state $\mathbf{Q}$ is given, then the sensitivity systems (3.7) and (3.8) are *linear* in the sensitivities $\mathbf{Q}_k$. The left-hand-side operators in the differential equations and boundary conditions for the sensitivities are independent of k. This means that if $\mathbf{Q}$ is given, a discretization of the sensitivity system will yield a linear system for an approximation of $\mathbf{Q}_k$ having a coefficient matrix that is independent of k and is therefore the same for all

[45] Strictly speaking, the discussions of Chapter 2 do not apply to this problem because neither the state system nor the functional we will consider are differentiable; this is due to the discontinuous nature of the flow field. However, we will ignore this fact and proceed to differentiate both the state system and functional to obtain sensitivity systems and functional gradients, respectively. Later, in Section 4.3, we shall discuss some of the consequences of this criminal activity.

sensitivities $\mathbf{Q}_k$, $k = 0, \ldots, K$. However, for each k, the right-hand sides of the differential equations and boundary conditions are different. For the shape-parameter sensitivities, the only dependence on k is through the appearance of Ψ_k on the right-hand side of the differentiated no-penetration condition along Γ.

Note the appearance of the derivatives $\partial m_1/\partial y$ and $\partial m_2/\partial y$ of the flow evaluated at the boundary segment Γ_1 on the right-hand side of the shape sensitivity system. This is generic to sensitivity equations in shape optimization problems and may lead to inaccuracies under discretizations. Note that the right-hand sides of the value sensitivity system do not involve derivatives of the flow field.

The gradient of the matching functional

The derivative of the functional with respect to a parameter α_k, $k = 0, \ldots, K$, is easily found to be

$$\begin{aligned} \frac{D\mathcal{J}}{D\alpha_k} = \int_{\Gamma_1} \Bigg(& (\rho - \widehat{\rho})\frac{\partial \rho}{\partial \alpha_k} + (m_1 - \widehat{m}_1)\frac{\partial m_1}{\partial \alpha_k} \\ & + (m_2 - \widehat{m}_2)\frac{\partial m_2}{\partial \alpha_k} + (e - \widehat{e})\frac{\partial e}{\partial \alpha_k}\Bigg) d\Gamma \qquad \text{for } k = 0, \ldots, K. \end{aligned} \tag{3.9}$$

The flow sensitivities

$$\rho_k = \frac{\partial \rho}{\partial \alpha_k}, \qquad (m_1)_k = \frac{\partial m_1}{\partial \alpha_k}, \qquad (m_2)_k = \frac{\partial m_2}{\partial \alpha_k}, \qquad \text{and} \qquad e_k = \frac{\partial e}{\partial \alpha_k}$$

are determined by solving the appropriate sensitivity systems (3.7) and (3.8).

An optimization algorithm

Start with an initial guess $\alpha_0^{(0)}, \alpha_1^{(0)}, \ldots, \alpha_K^{(0)}$ for the design parameters. Then, for $m = 0, 1, 2, \ldots,$

1. solve the Euler system (3.1)–(3.3) to obtain the corresponding flow field $\mathbf{Q}^{(m)}$;

2. solve the $K+1$ sensitivity systems (3.7) and (3.8) to obtain the sensitivities $\mathbf{Q}_k^{(m)}$, $k = 0, \ldots, K$, and then use (3.9) to determine the components $\frac{D\mathcal{J}}{D\alpha_k}(\mathbf{Q}^{(m)}, \mathbf{Q}_0^{(m)}, \ldots, \mathbf{Q}_K^{(m)})$, $k = 0, \ldots, K$, of the gradient of the functional;

3. use the results of steps 1 and 2 to compute steps $\delta\alpha_k^{(m)}$, $k = 0, \ldots, K$, by, e.g., the BFGS method;

4. set $\alpha_k^{(m+1)} = \alpha_k^{(m)} + \delta\alpha_k^{(m)}$, $k = 0, \ldots, K$.

Other optimization methods may be chosen in step 3; however, once one chooses an optimization method for determining the increments in step 3, the optimization algorithm is completely defined.

A computational example

We present the results for a simple computational example. The *target flow* $\widehat{\mathbf{Q}}$ in the functional (3.6) is chosen to be the flow about a given body shape with a given free-stream Mach number. The *matching boundary* Γ_1 in that functional is chosen to be the exit plane. The *goal of optimization* is to try to match the target flow along the exit plane, in the sense of minimizing the functional (3.6), with a flow around a body half as long. The *design parameters* are the free-stream Mach number and two Bezier parameters that determine the shape of the body. A sketch of the target flow body and a candidate body for the optimizer is given in Figure 3.4.

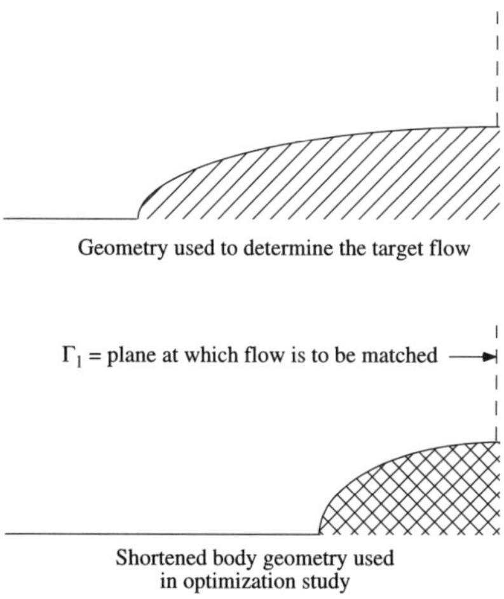

Figure 3.4. *Body for the target flow* $\widehat{\mathbf{Q}}$ *(top) and a candidate body for the optimal flow (bottom).*

The initial body for the optimization is simply the squeezed (in the x direction) body of the target flow; the initial free-stream Mach number is that of the target flow. Specifically, we have the following initial data: Bezier parameters = 0.2 and 0.3 and Mach number = 2.

An unstructured grid, finite volume method is used to solve the flow equations. The same grid and discretization method is used to solve the sensitivity equations; we have also used finite difference quotients requiring multiple flow solves to determine the approximate sensitivities. A BFGS-trust region, quasi-Newton method [382] is used to effect the optimization. Roughly five or six optimization steps produce an optimal solution that is accurate to within discretization error. The optimal values of the three design parameters are found to be Bezier parameters = 0.415 and 0.279 and the Mach number = 1.98. Pressure contours for the target flow, the initial flow fed to the optimizer, and the flow for the optimal values of the parameters are given in Figure 3.5.

The object of optimization was to match the flow at the exit plane Γ_1 and not to match throughout the flow field. As a result, the matching at the exit plane is superior to that at other portions of the flow. Note that the *optimal body shape is substantially more blunt* than the reference or initial body shape; this is predictable since the most effective way the shock

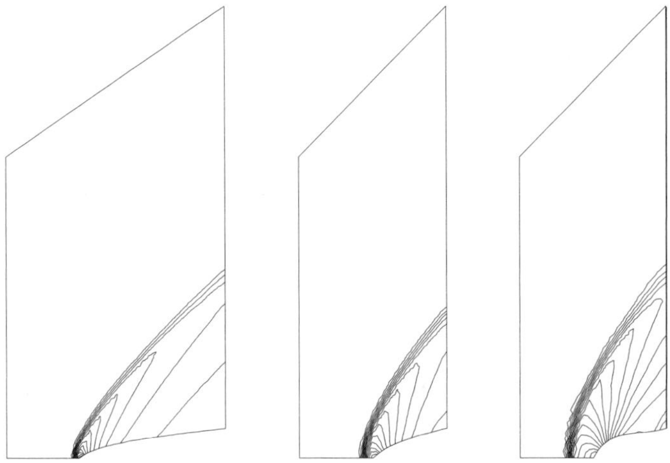

Figure 3.5. *Pressure contours for the target flow* (*left*), *the initial flow in the optimization* (*middle*), *and the optimal flow* (*right*).

position can be moved forward to match the shock position of the target flow is to make the body more blunt. In fact, the optimizer very quickly (in one or two iterations) makes the body blunt enough so as to get the correct shock position. The remaining iterations are used to match, as well as possible with the few parameters at our disposal, the continuous portions of the flow at the exit plane. Note that the *optimal body shape is not convex*; this is not so predictable.

Figure 3.6 provides a plot of one of the sensitivities along the exit plane and illustrates a potential problem, i.e., *the appearance of a large spike in the sensitivities at the shock position.* Note that the spike is present for both a discretize-then-differentiate approach (finite difference quotient approximations for the gradient of the functional) and the differentiate-then-discretize approach discussed in this chapter. We will be discussing the meaning and effect of such spikes in detail in Section 4.3.

3.3 Computational experiments for the adjoint equation approach to optimization using flow matching functionals

In this section, we consider simple academic problems involving flow matching functionals in order to explore the effects of several of the variations possible in defining optimal control problems. The variations we consider include different types of controls, different means of limiting the size of the control, and different specifications of the objective functional. For details concerning these illustrations, see [82, 83, 84]. Several other papers cited in the bibliography deal with similar problems.

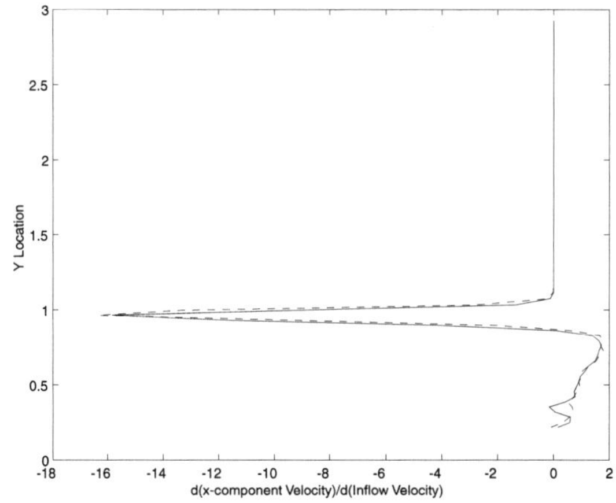

Figure 3.6. *Sensitivity of the x-component of the velocity at the exit plane computed by finite difference quotients* (—) *and by the sensitivity equation method* (- - -).

The basic objective functional we use is the *matching functional*

$$\mathcal{J}(\mathbf{u}) = \frac{1}{2}\int_0^T \int_\Omega |\mathbf{u} - \mathbf{U}|^2 d\Omega dt\,, \tag{3.10}$$

where $\mathbf{U}$ is a given velocity field, Ω is the given flow domain, and $T > 0$ is a given time. Thus, the primary goal of the control problems we consider in this section is to match as well as possible, in a root mean square sense over a given time interval and over the flow domain, the velocity field $\mathbf{u}$ to a given velocity field $\mathbf{U}$.

We will also consider the objective functional

$$\mathcal{J}_\beta(\mathbf{u}) = \mathcal{J}(\mathbf{u}) + \frac{\beta}{2}\int_\Omega |(\mathbf{u} - \mathbf{U})|_{t=T}|^2 d\Omega\,, \tag{3.11}$$

where $\mathcal{J}(\mathbf{u})$ is defined in (3.10) and $\beta \geq 0$ is a parameter at our disposal. Now we not only match the velocity field $\mathbf{u}$ of the flow to the given velocity field $\mathbf{U}$ over $(0, T) \times \Omega$, but we have added a term that has the specific goal of matching $\mathbf{u}$ to $\mathbf{U}$ at the terminal time T. The new term in (3.11) is useful in order to keep the solution $\mathbf{u}$ close to the target flow $\mathbf{U}$ near the terminal time T. Of course, if $\beta = 0$, we are back to the case of (3.10).

3.3.1 Distributed control with a penalized functional

We first consider the case of distributed controls, i.e., controls that act over the flow domain, for which the size of the control is limited through penalization of the objective functional.

The state system is the Navier–Stokes system

$$\left\{\begin{aligned} \partial_t \mathbf{u} + \mathbf{u}\cdot\nabla\mathbf{u} - \nu\Delta\mathbf{u} + \nabla p &= \mathbf{f} && \text{in } (0,T)\times\Omega\,,\\ \nabla\cdot\mathbf{u} &= 0 && \text{in } (0,T)\times\Omega\,,\\ \mathbf{u} &= \mathbf{0} && \text{on } (0,T)\times\Gamma\,,\\ \mathbf{u}|_{t=0} &= \mathbf{u}_0 && \text{on } \Omega\,,\\ \int_\Omega p\, d\Omega &= 0 && \text{on } (0,T)\,, \end{aligned}\right. \tag{3.12}$$

where $\mathbf{u}$ and p denote the velocity and pressure, ν the constant kinematic viscosity, $\mathbf{u}_0$ a given initial velocity field, and $\mathbf{f}$ the control function acting over $(0, T) \times \Omega$. The penalized objective functional is defined by

$$\begin{aligned} \mathcal{J}_{\beta\gamma}(\mathbf{u},\mathbf{f}) &= \mathcal{J}_\beta(\mathbf{u}) + \frac{\gamma}{2}\int_0^T\int_\Omega |\mathbf{f}|^2\, d\Omega dt\\ &= \frac{1}{2}\int_0^T\int_\Omega |\mathbf{u}-\mathbf{U}|^2\, d\Omega dt + \frac{\beta}{2}\int_0^T\int_\Omega |(\mathbf{u}-\mathbf{U})|_{t=T}|^2\, d\Omega\\ &\quad + \frac{\gamma}{2}\int_0^T\int_\Omega |\mathbf{f}|^2\, d\Omega dt\,, \end{aligned} \tag{3.13}$$

where $\gamma > 0$ is a penalty parameter.

The first optimization problem we consider is to find an optimal state $(\mathbf{u}, p)$ and an optimal control $\mathbf{f}$ such that the functional $\mathcal{J}_{\beta\gamma}(\mathbf{u}, \mathbf{f})$ defined in (3.13) is minimized subject to $\mathbf{u}$, p, and $\mathbf{f}$ satisfying the Navier–Stokes system (3.12).

Using the process introduced in Section 2.3.3, the optimality system for this problem is given by the Navier–Stokes system (3.12), the adjoint system

$$\left\{\begin{aligned} -\partial_t\mathbf{w} + (\nabla\mathbf{u})^T\mathbf{w} - \mathbf{u}\cdot\nabla\mathbf{w} - \nu\Delta\mathbf{w} + \nabla r &= (\mathbf{u}-\mathbf{U}) && \text{in } (0,T)\times\Omega\,,\\ \nabla\cdot\mathbf{w} &= 0 && \text{in } (0,T)\times\Omega\,,\\ \mathbf{w} &= \mathbf{0} && \text{on } (0,T)\times\Gamma\,,\\ \mathbf{w}|_{t=T} &= \beta(\mathbf{u}-\mathbf{U})|_{t=T} && \text{on } \Omega\,,\\ \int_\Omega r\, d\Omega &= 0 && \text{on } (0,T)\,, \end{aligned}\right. \tag{3.14}$$

where $\mathbf{w}$ and r denote the adjoint velocity and pressure, respectively, and the optimality condition

$$\gamma\mathbf{f} + \mathbf{w} = \mathbf{0} \qquad \text{in } (0,T)\times\Omega\,. \tag{3.15}$$

As expected, the adjoint system is posed as a backward-in-time problem with a terminal condition at $t = T$. In this example, the optimality condition (3.15) may be used to easily eliminate the control $\mathbf{f}$ from the state system (3.12).

If $\beta = 0$, i.e., no terminal matching is added to the functional, then the adjoint velocity satisfies $\mathbf{w}|_{t=T} = \mathbf{0}$ and, by (3.15), so does the control function, i.e., $\mathbf{f}|_{t=T} = \mathbf{0}$. This imposition on the control, i.e., that it vanish at the terminal time, may be the desired outcome of the optimization, i.e., it may be something that is dictated by external (to the optimization process) requirements. On the other hand, as we shall soon see, forcing the control to vanish at the terminal time deteriorates the ability of the optimal solution to match the target flow near that time.

A gradient method

Due to the large number of unknowns in the optimality system (3.12), (3.14), and (3.15), and due to the forward-in-time nature of the state equations and the backward-in-time nature of the adjoint equations, any practical algorithm necessarily involves an uncoupling of these two parts of the optimality system. We consider a *gradient method* for the solution of the optimal control problem; the gradient of the functional is determined with the help of the solution to the adjoint system (3.14). The optimality condition (3.15), which is merely a statement of the necessary condition that the gradient of the objective functional with respect to the controls vanish at the optimum, is not satisfied until the algorithm converges, i.e., the individual iterates do not satisfy (3.15).

In the algorithm, τ denotes a prescribed tolerance used to test for the convergence of the functional, s denotes a step-size parameter that is automatically selected by the algorithm, and $\mathbf{f}^{(0)}$ denotes an initial guess for the control function.

Algorithm 3.1. (Gradient algorithm with automatic step-length selection)

(a) initialization:

(i) choose τ and $\mathbf{f}^{(0)}$; set $m = 0$ and $s = 1$;

(ii) solve for the starting velocity field $\mathbf{u}^{(0)}$ from (3.12) with $\mathbf{f} = \mathbf{f}^{(0)}$;

(iii) evaluate $\mathcal{J}_{\beta\gamma}^{(0)} = \mathcal{J}_{\beta\gamma}(\mathbf{u}^{(0)}, \mathbf{f}^{(0)})$ using (3.13);

(b) main loop:

(iv) set $m = m + 1$;

(v) solve for $\mathbf{w}^{(m)}$ from (3.14) with $\mathbf{u} = \mathbf{u}^{(m-1)}$;

(vi) set $\mathbf{f}^{(m)} = \mathbf{f}^{(m-1)} - s(\gamma \mathbf{f}^{(m-1)} + \mathbf{w}^{(m)})$;

(vii) solve for $\mathbf{u}^{(m)}$ from (3.12) with $\mathbf{f} = \mathbf{f}^{(m)}$;

(viii) evaluate $\mathcal{J}_{\beta\gamma}^{(m)} = \mathcal{J}_{\beta\gamma}(\mathbf{u}^{(m)}, \mathbf{f}^{(m)})$ using (3.13);

(ix) if $\mathcal{J}_{\beta\gamma}^{(m)} \geq \mathcal{J}_{\beta\gamma}^{(m-1)}$, set $s = 0.5s$ and go to step (vi); otherwise, continue;

(x) if $|\mathcal{J}_{\beta\gamma}^{(m)} - \mathcal{J}_{\beta\gamma}^{(m-1)}|/|\mathcal{J}_{\beta\gamma}^{(m)}| > \tau$, set $s = 1.5s$ and go to step (iv); otherwise, stop;

The bulk of the computational costs are found in the forward-in-time solution of the discrete state system in step (vii) and, to a lesser extent, the backward-in-time solution of the discrete adjoint system in step (v). Steps (ix) and (x) together define an automatic step length determination subalgorithm.

Of course, in practice, Algorithm 3.1 is applied to the discretized state system, adjoint system, and objective functional. Other gradient-based algorithms, e.g., conjugate gradient or quasi-Newton methods, can also be easily defined.

Computational experiments for a steady flow matching problem

We consider a unit square domain $\Omega = (0, 1) \times (0, 1) \subset \mathbb{R}^2$. We assume that the time interval $[0, 1]$ is divided into equal intervals of duration Δt. Spatial discretization is effected using the Taylor–Hood pair of finite element spaces on a rectangular mesh, i.e., the finite element spaces are chosen to be continuous piecewise biquadratic polynomials for the velocity and continuous piecewise bilinear polynomials for the pressure [428, 430]. Temporal discretization is effected using the backward Euler method. The temporal and spatial mesh sizes are set to $\Delta t = 0.0125$ and $h = 1/16$, respectively. Calculations with varying mesh sizes have been performed; the mesh sizes we use here are sufficient to obtain accurate results. For the sake of clarity, all vector plots are normalized by the maximum magnitudes.

We are interested in the convergence history with respect to all the parameters involved so that a simple stationary target velocity $\mathbf{U} = (U, V)$ is chosen, where

$$U(x, y) = 10\frac{d}{dy}\big(\phi(x)\phi(y)\big) \qquad \text{and} \qquad V(x, y) = -10\frac{d}{dx}\big(\phi(x)\phi(y)\big)$$

with $\phi(z) = \big(1 - \cos(0.8\pi z)\big)(1 - z)^2$. Note that $\mathbf{U}$ is divergence free.

Velocity matching evolution. We first choose the initial velocity for the optimization to be given by $\mathbf{u}_0 = (u_0, v_0)$, where

$$u_0(x, y) = -10U(x, y) \qquad \text{and} \qquad v_0(x, y) = -10V(x, y)\,.$$

Note that the initial flow rotates in a clockwise direction, which is opposite to the counterclockwise rotation of the target flow. Furthermore, the initial flow is much more energetic than the target flow. We set $T = 1$, $\gamma = 0.0001$, and $\beta = 0.5$. The target flow $\mathbf{U}$, the initial flow $\mathbf{u}_0$, and the optimal flow at $t = 0.3$ and $t = T = 1$ are depicted in Figure 3.7. We see that by $t = 0.3$, the optimal flow looks very much like the target flow.

Figure 3.8 shows the norm of the error $\|\mathbf{u} - \mathbf{U}\|$ between the controlled flow $\mathbf{u}$ and the target flow $\mathbf{U}$ and the norm of the control $\|\mathbf{f}\|$ plotted vs. time. We see that the error rapidly goes to zero and that the control works hard, i.e., its norm is relatively large, at the beginning in order to steer the controlled flow to the target flow and then, after a good match is achieved, its norm remains relatively constant and small. Near $t = T = 1$, the control strength slightly increases in order to render small the term in the functional that is evaluated at $t = T$.

Dependence on $\boldsymbol{\beta}$ and $\boldsymbol{\gamma}$. We now examine the effects of changes in the parameters β and γ. The initial velocity is now set to zero.

In Figure 3.9, we show the norm of the error $\|\mathbf{u} - \mathbf{U}\|$ between the controlled flow $\mathbf{u}$ and the target flow $\mathbf{U}$ and the norm of the control $\|\mathbf{f}\|$ plotted vs. time for different values of γ. The value of β for these calculations is held fixed at 0.5. The time step Δt is again 0.0125 and the spatial grid size is $h = 1/16$. We note that the controlled flow matches the target flow very well for values of $\gamma < 0.001$. For smaller values of the penalty parameter γ, larger controls $\mathbf{f}$ are available. This is borne out in Figure 3.9, where we see that the maximum value of $\|\mathbf{f}\|$ increases as the value of γ decreases. We also see that, for smaller values of γ, i.e., for larger controls, the matching to the target flow is effected faster, i.e.,

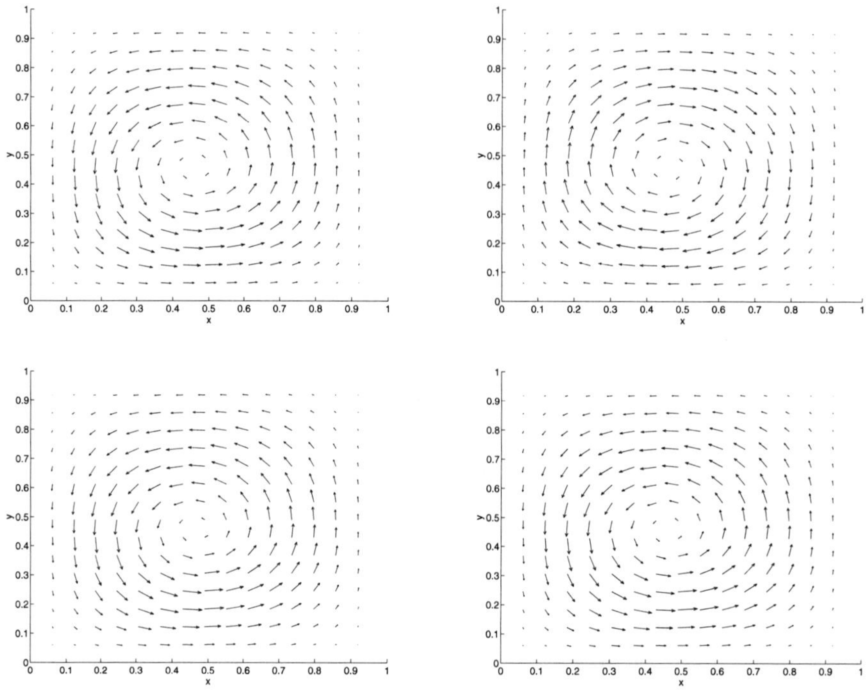

Figure 3.7. *Target flow* **U** (*top left*), *initial flow* $\mathbf{u}_0$ (*top right*), *and optimal flow at* $t = 0.3$ (*bottom left*) *and* $t = T = 1$ (*bottom right*).

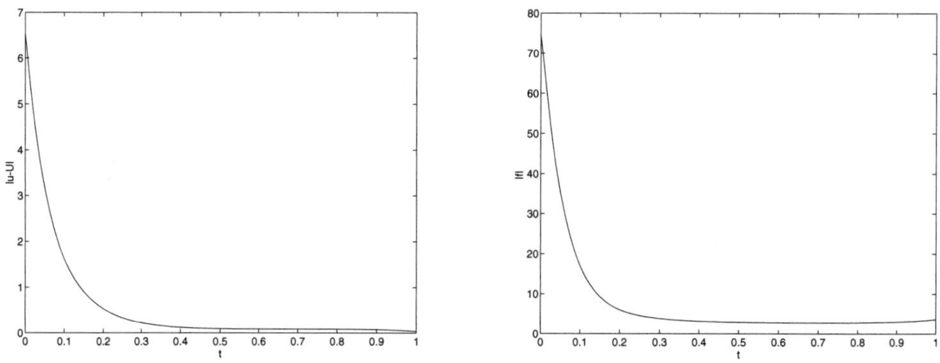

Figure 3.8. *The norm* $\|\mathbf{u} - \mathbf{U}\|$ *of the difference between the optimal flow and the target flow* (*left*) *and the norm* $\|\mathbf{f}\|$ *of the optimal control* (*right*) *vs. time.*

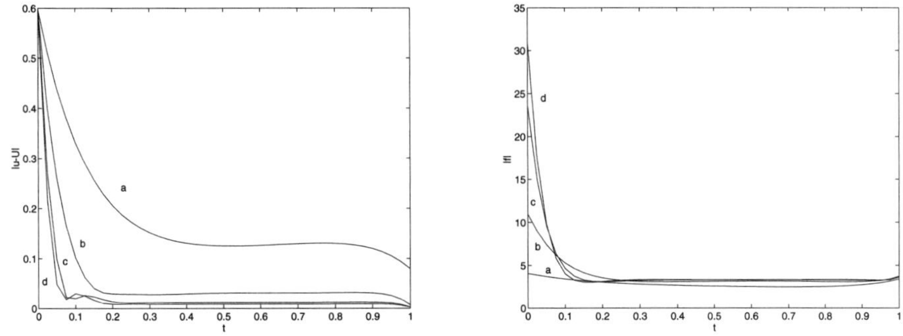

Figure 3.9. *The norm* $\|\mathbf{u} - \mathbf{U}\|$ *of the difference between the optimal flow and the target flow* (*left*) *and the norm* $\|\mathbf{f}\|$ *of the optimal control* (*right*) *vs. time for different values of* γ; $\gamma = 0.01$ *for* (a), 0.001 *for* (b), 0.0001 *for* (c), *and* 0.00001 *for* (d).

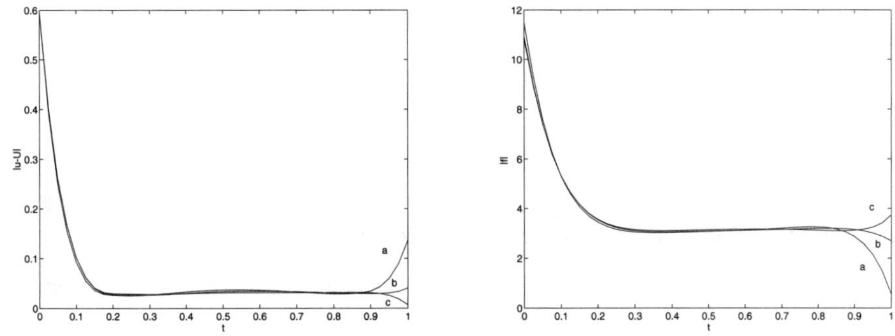

Figure 3.10. *The norm* $\|\mathbf{u} - \mathbf{U}\|$ *of the difference between the optimal flow and the target flow* (*left*) *and the norm* $\|\mathbf{f}\|$ *of the optimal control* (*right*) *vs. time for different values of* β; $\beta = 0.01$ *for* (a), 0.1 *for* (b), *and* 0.5 *for* (c).

it takes less time to appreciably reduce the error $\|\mathbf{u} - \mathbf{U}\|$. Also, the quality of the match improves as one has more control available, i.e., for smaller γ.

In Figure 3.10, we show the norm of the error $\|\mathbf{u} - \mathbf{U}\|$ between the controlled flow $\mathbf{u}$ and the target flow $\mathbf{U}$ and the norm of the control $\|\mathbf{f}\|$ plotted vs. time for different values of β. We can now see the importance of the term in the functional (3.13) evaluated at the terminal time $t = T$. For small values of β, the controlled flow tends to wander far[46] from the target flow near $t = T$. As expected, as β becomes smaller, the control function is forced to become smaller at the terminal time $t = T$ and, as is shown in Figure 3.10, the error $(\mathbf{u} - \mathbf{U})$ becomes less controllable near the terminal time. Thus, the term evaluated at the terminal time $t = T$ in the functional (3.13) is necessary for effecting good control. A good choice for β is about 0.5 for most values of γ.

[46]This effect becomes more pronounced if γ is small.

Dependence on terminal time T. In Figure 3.11, we show the norm of the error $\|\mathbf{u} - \mathbf{U}\|$ between the controlled flow $\mathbf{u}$ and the target flow $\mathbf{U}$ for different choices of the terminal time T. The initial flow is again the zero flow and we choose $\gamma = 0.001$ and $\beta = 0.1$. For these choices for the terminal time T, the time history of the state and control in the overlapping time intervals is not affected.

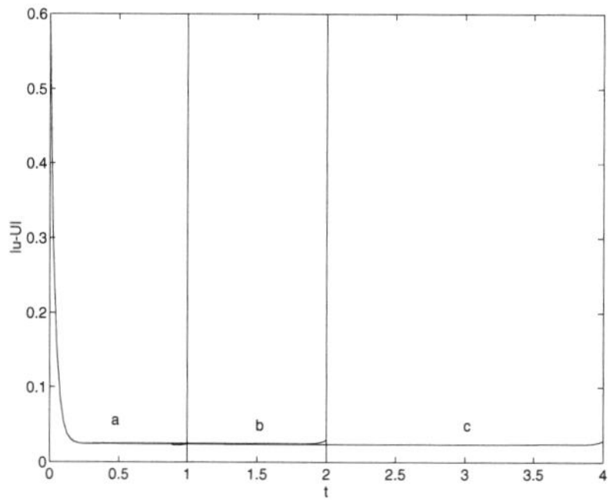

Figure 3.11. *The norm $\|\mathbf{u} - \mathbf{U}\|$ of the difference between the optimal flow and the target flow vs. time for different values of the terminal time T; $T = 1$ for* (a), 2 *for* (b), *and* 4 *for* (c).

Computational experiments for an unsteady flow matching problem

The target velocity $\mathbf{U}$ for this test is a solution of the time-dependent Stokes system with zero initial velocity and a body force $\widetilde{\mathbf{f}} = (\widetilde{f}_1, \widetilde{f}_2)$ given by

$$\widetilde{f}_1 = a(0.4, x, y) - e^{-2t} a(0.6, x, y) \qquad \text{and} \qquad \widetilde{f}_2 = b(0.4, x, y) - e^{-2t} b(0.6, x, y)\,,$$

where

$$a(z, x, y) = 10 \frac{d\psi(z, x, y)}{dy}\,, \qquad b(z, x, y) = -10 \frac{d\psi(z, x, y)}{dx}\,,$$

and

$$\psi(z, x, y) = \big(1 - \cos(4\pi z x)\big)(1 - x)^2 \big(1 - \cos(4\pi z y)\big)(1 - y)^2\,.$$

The target velocity is a combination of two flows: a vortex at the center of the domain with large radius and another vortex with small radius centered near the lower left-hand corner. Each of these flows is prominent at different times of the evolution. The initial velocity $\mathbf{u}_0 = (u_0, v_0)$ for the controlled flow is chosen to be

$$u_0(x, y) = \big(\cos(2\pi x) - 1\big) \sin(2\pi y) \quad \text{and} \quad v_0(x, y) = -\big(\cos(2\pi y) - 1\big) \sin(2\pi x)\,.$$

We set $\gamma = 1/5000$, $\beta = 0.5$, $\Delta t = 0.025$, and $h = 1/16$.

In Figure 3.12, we show the norm of the error $\|\mathbf{u} - \mathbf{U}\|$ between the controlled flow $\mathbf{u}$ and the target flow $\mathbf{U}$ and the norm of the control $\|\mathbf{f}\|$ plotted vs. time. We see that the error reduces quickly. Again, the control works hard at the beginning in order to steer the controlled flow to the target flow and subsequently remains relatively small. Near the terminal time $t = T$, the strength of the control again increases slightly in order to reduce the error at the terminal time. The evolution is illustrated in Figures 3.13 and 3.14. The controlled fluid is on the left and the target flow is on the right. We see that by $t = 0.2$ we achieve a very good match.

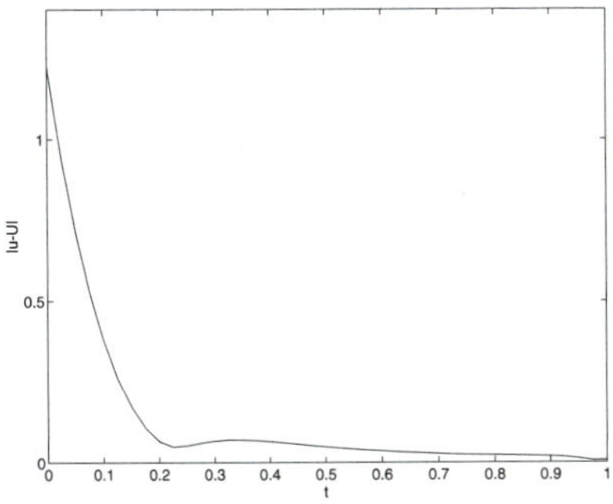

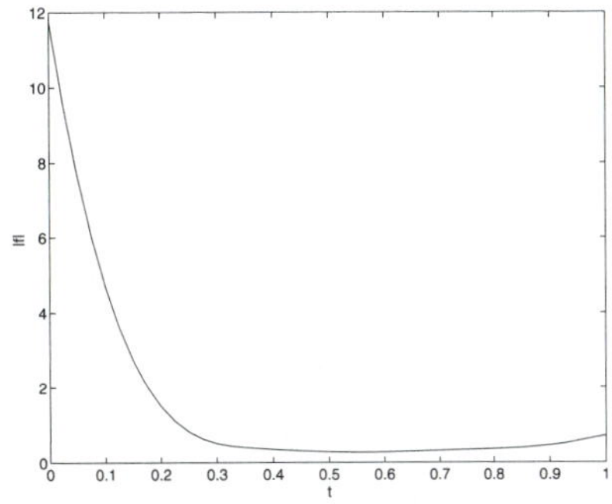

Figure 3.12. *The norm* $\|\mathbf{u} - \mathbf{U}\|$ *of the difference between the optimal flow and the target flow* (*left*) *and the norm* $\|\mathbf{f}\|$ *of the optimal control* (*right*) *vs. time.*

3.3.2 Distributed control with an explicit bound on the control

We again consider the case of distributed controls, but now the size of the control is limited by an explicit bound on the control. The state system remains the Navier–Stokes system (3.12) with a distributed control $\mathbf{f}$. The objective functional is the unpenalized functional (3.11) containing a terminal time matching term. The size of the control is limited by the explicit constraint

$$\left(\int_\Omega |\mathbf{f}\,|^2\,d\Omega\right)^{1/2} \le \kappa \qquad \text{for all } t \in (0,T)\,, \tag{3.16}$$

where $\kappa > 0$ is a prescribed constant. We may view κ as some measure of the maximum power that is available to control the system.

Now, the optimization problem we consider is to find an optimal state $(\mathbf{u}, p)$ and an optimal control $\mathbf{f}$ such that the functional $\mathcal{J}_\beta(\mathbf{u}, \mathbf{f})$ defined in (3.11) is minimized subject to $\mathbf{u}$, p, and $\mathbf{f}$ satisfying the Navier–Stokes system (3.12) and the control $\mathbf{f}$ satisfying the constraint (3.16).

As long as $\mathbf{w} \not\equiv \mathbf{0}$, the optimality system for this problem is given by the Navier–Stokes system (3.12), the adjoint system (3.14), and the optimality condition

$$\mathbf{f} = -\kappa \frac{\mathbf{w}}{\|\mathbf{w}\|} \qquad \text{in } (0,T) \times \Omega\,, \tag{3.17}$$

where

$$\|\mathbf{w}\| \equiv \left(\int_\Omega |\mathbf{w}\,|^2\,d\Omega\right)^{1/2}.$$

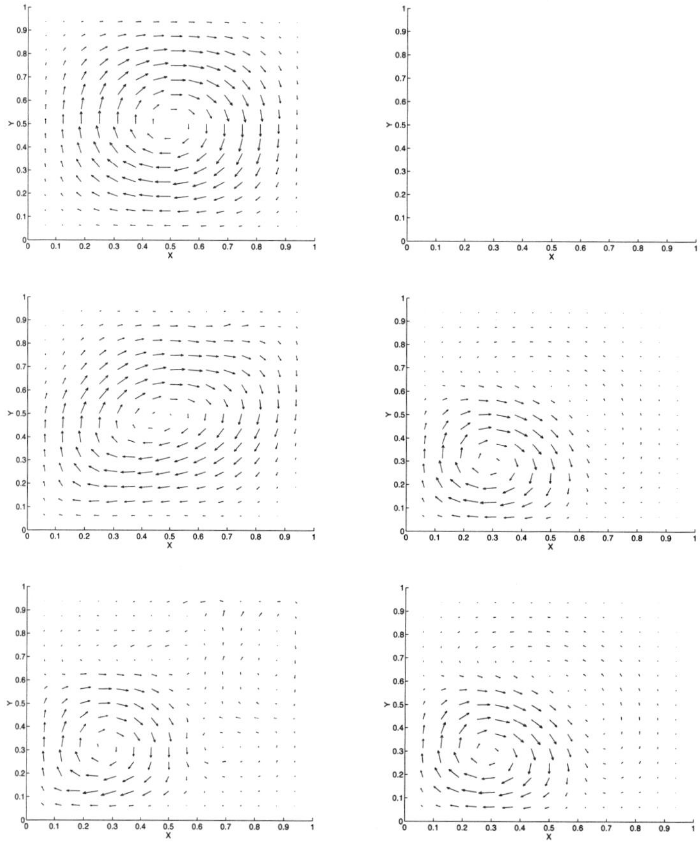

Figure 3.13. *Controlled (left) and target (right) flows at* $t = 0$ *(first row),* $t = 0.05$ *(second row), and* $t = 0.1$ *(third row).*

Again, in this example, the optimality condition (3.17) may be used to easily eliminate the control $\mathbf{f}$ from the state system (3.12).

Note from (3.17) that, for the optimal control, (3.16) holds with equality; this is not surprising. It is usually the case that optimal controls are found at the boundary of their constraint sets.

A projected gradient method

If one applies an algorithm such as Algorithm 3.1 to the current optimization problem, then it is possible that a computed iterate for the control function $\mathbf{f}$ does not satisfy the explicit constraint (3.16). In this case, we add a simple projection step to the algorithm that brings the iterate back to the feasible set. In fact, the projected gradient algorithm is identical to Algorithm 3.1 except for step (vi), the update of the control function.

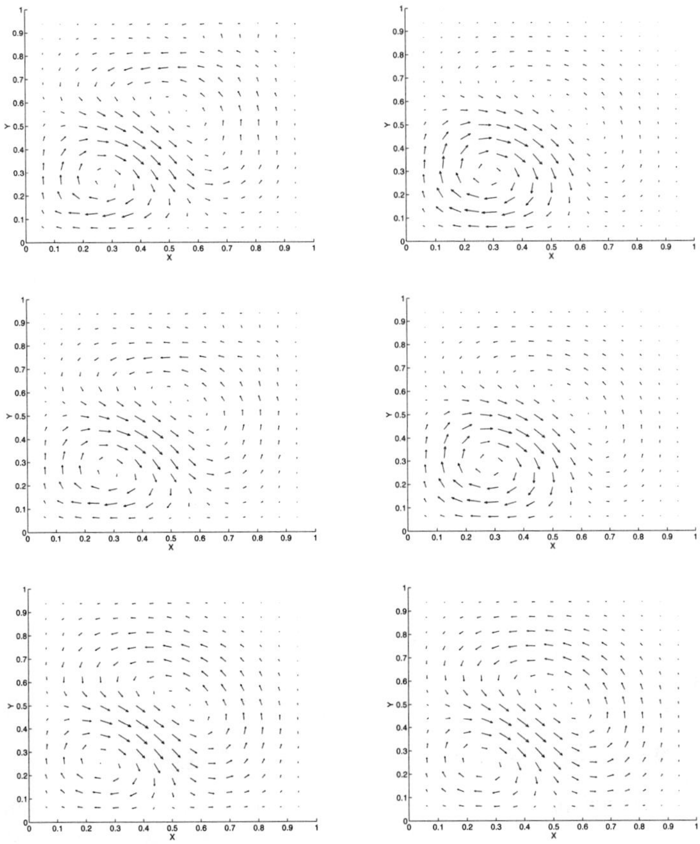

Figure 3.14. *Controlled* (*left*) *and target* (*right*) *flows at* $t = 0.125$ (*first row*), $t = 0.15$ (*second row*), *and* $t = 0.175$ (*third row*).

Algorithm 3.2. (Projected gradient algorithm with automatic step-length selection)

(a) initialization:

(i) choose τ and $\mathbf{f}^{(0)}$; set $m = 0$ and $s = 1$;

(ii) solve for the starting velocity field $\mathbf{u}^{(0)}$ from (3.12) with $\mathbf{f} = \mathbf{f}^{(0)}$;

(iii) evaluate $\mathcal{J}_\beta^{(0)} = \mathcal{J}_\beta(\mathbf{u}^{(0)})$ using (3.11);

(b) main loop:

(iv) set $m = m + 1$;

(v) solve for $\mathbf{w}^{(m)}$ from (3.14) with $\mathbf{u} = \mathbf{u}^{(m-1)}$;

(vi) set $\mathbf{q} = \mathbf{f}^{(m-1)} - s\kappa \dfrac{\mathbf{w}^{(m)}}{\|\mathbf{w}^{(m)}\|}$;

– if $\|\mathbf{q}\| \leq \kappa$, set $\quad \mathbf{f}^{(m)} = \mathbf{q}$;

– if $\|\mathbf{q}\| > \kappa$, set $\quad \mathbf{f}^{(m)} = \kappa \dfrac{\mathbf{q}}{\|\mathbf{q}\|}$;

(vii) solve for $\mathbf{u}^{(m)}$ from (3.12) with $\mathbf{f} = \mathbf{f}^{(m)}$;

(viii) evaluate $\mathcal{J}_\beta^{(m)} = \mathcal{J}_\beta(\mathbf{u}^{(m)})$ using (3.11);

(ix) if $\mathcal{J}_\beta^{(m)} \geq \mathcal{J}_\beta^{(m-1)}$, set $s = 0.5s$ and go to (vi); otherwise, continue;

(x) if $|\mathcal{J}_\beta^{(m)} - \mathcal{J}_\beta^{(m-1)}|/|\mathcal{J}_\beta^{(m)}| > \tau$, set $s = 1.5s$ and go to (iv); otherwise, stop.

The gradient of the functional (3.11) is given by

$$\left.\frac{D\mathcal{J}_\beta}{D\mathbf{f}}\right|_{\mathbf{f}^{(m)}} = \mathbf{w}^{(m)} .$$

Then, it is clear that Algorithm 3.2 is based on the iteration

$$\mathbf{f}^{(m)} = \Pi_\kappa\left(\mathbf{f}^{(m-1)} - \rho_m \left.\frac{d\mathcal{J}_\beta}{d\mathbf{f}}\right|_{\mathbf{f}^{(m-1)}}\right) \tag{3.18}$$

for some sequence of step size factors $\{\rho_m\}$, where Π_κ denotes the $L^2(\Omega)$-projection onto the ball of radius κ.

Computational experiments for a steady flow matching problem

We consider a unit square domain $\Omega = (0,1) \times (0,1) \subset \mathbb{R}^2$. We assume that the time interval [0,1] is divided into equal intervals of duration Δt. Spatial discretization is effected using the Taylor–Hood pair of finite element spaces on a rectangular mesh, i.e., the finite element spaces are chosen to be continuous piecewise biquadratic polynomials for the velocity and continuous piecewise bilinear polynomials for the pressure [428, 430]. Temporal discretization is effected using the backward Euler method. The temporal and spatial mesh sizes are set to $\Delta t = 0.025$ and $h = 1/16$, respectively. Calculations with varying mesh sizes have been performed; the mesh sizes we use here are sufficient to obtain accurate results. For the sake of clarity, all vector plots are normalized by the maximum magnitudes.

We are interested in the convergence history with respect to the bound κ on the control. We choose a simple stationary target velocity $\mathbf{u}(x,y) = (U(x,y), V(x,y))$, where

$$U = 10\frac{d}{dy}\Big(\phi(0.4,x)\phi(0.4,y)\Big) \quad \text{and} \quad V = -10\frac{d}{dx}\Big(\phi(0.4,x)\phi(0.4,y)\Big),$$

with $\phi(t,z) = (1-z)^2\big(1-\cos(2\pi t z)\big)$.

Velocity tracking evolution. We first choose the initial velocity for the optimization to be $\mathbf{u}_0 = (u_0, v_0)$, where

$$u_0(x,y) = -U(x,y) \qquad \text{and} \qquad v_0(x,y) = -V(x,y)$$

so that the initial flow rotates in an opposite sense to the target flow. We set $\beta = 0.5$, $T = 1$, and $\kappa = 3.2$.

The target flow $\mathbf{U}$, the initial flow $\mathbf{u}_0$, and the optimal flow at $t = 0.125$ and $t = 0.2$ are depicted in Figure 3.15. We see that by $t = 0.2$, the optimal flow looks very much like the target flow. Figure 3.16 shows the error $\|\mathbf{u} - \mathbf{U}\|$ between the optimally controlled flow $\mathbf{u}$ and the target flow $\mathbf{U}$. As we can see, the error does not go to zero very rapidly. For this calculation, by $t = 0.2$ we achieve a good match to the shape of the flow, and after $t = 0.6$ we achieve a good match in the magnitude as well.

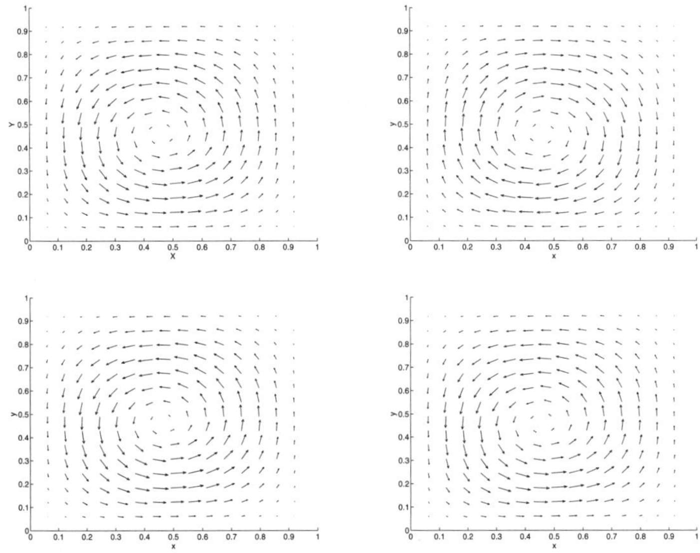

Figure 3.15. *Target flow* $\mathbf{U}$ *(top left), initial flow* $\mathbf{u}_0$ *(top right), and optimal flow at* $t = 0.125$ *(bottom left) and* $t = 0.2$ *(bottom right).*

Effect of the size of the bound κ on the control. We next want to see what effect changing κ, the a priori bound for the norm of control $\mathbf{f}$, has on the quality of the match. The value $\beta = 0.5$ is used in all cases and we now set the initial velocity of the controlled flow to zero.

In Figure 3.17, we provide the value of the objective functional $\mathcal{J}_\beta$ defined by (3.11), evaluated at the optimal solution, vs. the bound on the control κ. As is expected, the functional decreases as κ increases. In Figure 3.18, we give the error $\|\mathbf{u} - \mathbf{U}\|$ between the optimally controlled flow $\mathbf{u}$ and the target flow $\mathbf{U}$ vs. time for different values of κ. We observe that for $\kappa \leq 3.4$, as κ increases, the controlled flow matches better over the whole time interval $(0, T)$. However, for $\kappa > 3.4$, there is a better match only up to some time $t^* < T = 1$; for $t > t^*$, the performance of the controlled flow becomes slightly worse as κ increases. This shows that although the value of the root mean square functional $\mathcal{J}_\beta$ decreases as the bound κ increases, it does not follow that the difference between the optimal flow $\mathbf{u}$ and the target flow $\mathbf{U}$ decreases at *every* instant of time in the interval $(0, T)$.

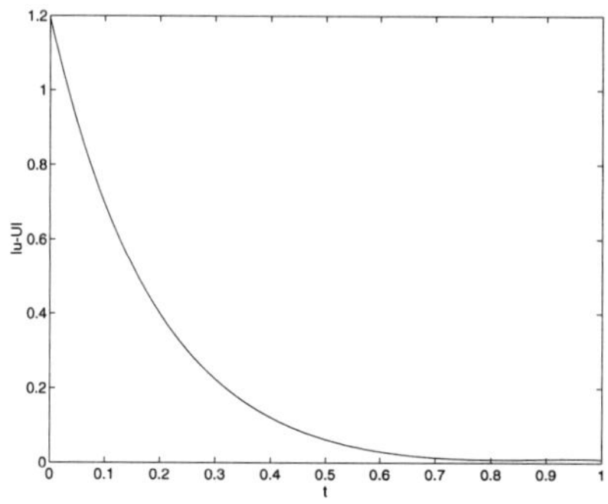

Figure 3.16. *The norm* $\|\mathbf{u} - \mathbf{U}\|$ *of the difference between the optimal flow and the target flow vs. time.*

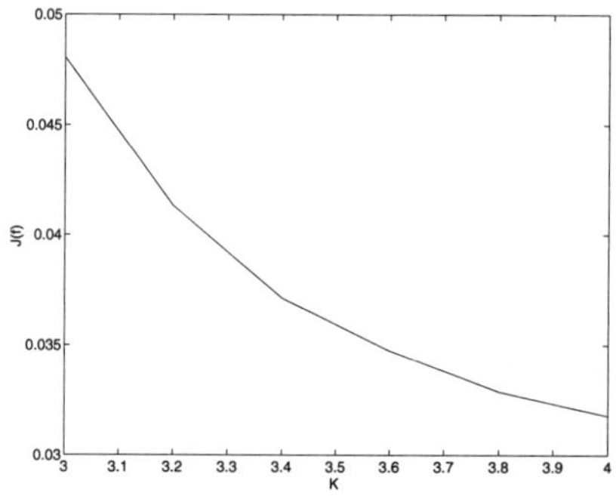

Figure 3.17. *Value of the functional* $\mathcal{J}_\beta$ *vs. the bound* κ *on the control.*

Computational experiments for an unsteady flow matching problem

We now choose the time-dependent target velocity $\mathbf{U}(t, x, y) = (U, V)$ to be

$$U(t,x,y) = a(1, .4, x, y) + a(2, t, x, y)/(4\pi t + 1)\,,$$
$$V(t,x,y) = b(1, .4, x, y) + b(2, t, x, y)/(4\pi t + 1)\,,$$

where

$$a(k,t,x,y) = \frac{d}{dy}\Big(\phi(k,t,x)\phi(k,t,y)\Big) \quad \text{and} \quad b(k,t,x,y) = -\frac{d}{dx}\Big(\phi(k,t,x)\phi(k,t,y)\Big)$$

with $\phi(k,t,z) = (1 - zt)^2\big(1 - \cos(2k\pi tz)\big)$. This velocity field is a combination of two flows, one having a vortex at the center of the domain and another with four vortices. Each of these flows dominates at different times of the evolution. The initial velocity $\mathbf{u}_0 = (u_0, v_0)$ for the controlled flow is chosen to be

$$u_0(x,y) = -8U(1/4, x, y) \qquad \text{and} \qquad v_0(x,y) = -8V(1/4, x, y)$$

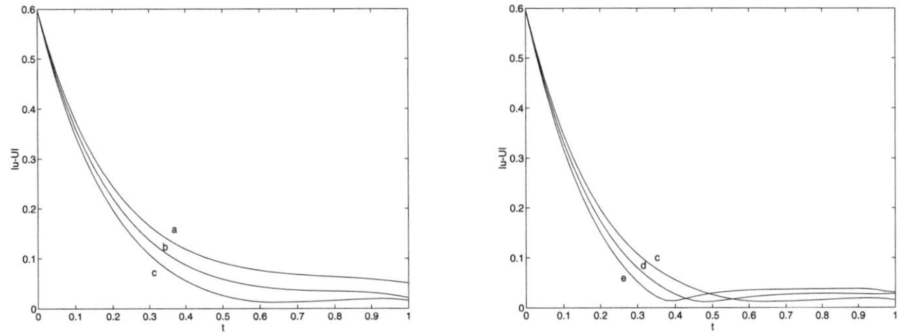

Figure 3.18. *The norm* $\|\mathbf{u} - \mathbf{U}\|$ *of the difference between the optimal flow and the target flow vs. time for different values of the bound* κ *on the control;* $\kappa = 3.0$ (a), 3.2 (b), 3.4 (c), 3.6 (d), *and* 3.8 (e).

so that it has an opposite rotational sense to and larger magnitude than that of the target flow. For this computation, we set $\beta = 0.5$ and $\kappa = 1.6$.

Figure 3.19 provides the norm $\|\mathbf{u} - \mathbf{U}\|$ of the difference between the optimally controlled flow $\mathbf{u}$ and the target flow $\mathbf{U}$ vs. time. At the beginning, the error rapidly decreases, but afterwards, the error increases due to changes in the target flow, which cannot be followed well using a control having the bound $\kappa = 1.6$ we have chosen. Evidently, this does not allow enough "control power" for the optimally controlled flow to match well the time evolution of the target flow. The evolutionary history is given in Figure 3.20. Visually, we see that a good match is achieved at $t = 0.3$. The match remains good at $t = 0.7$, but close inspection reveals that the match at the terminal time $t = T = 1$ has deteriorated from that at $t = 0.3$ and $t = 0.7$. This is, of course, also revealed by Figure 3.19.

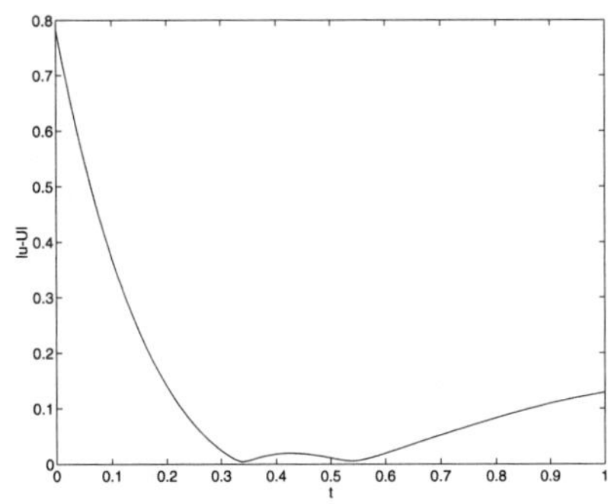

Figure 3.19. *The norm* $\|\mathbf{u} - \mathbf{U}\|$ *of the difference between the optimal flow and the target flow vs. time.*

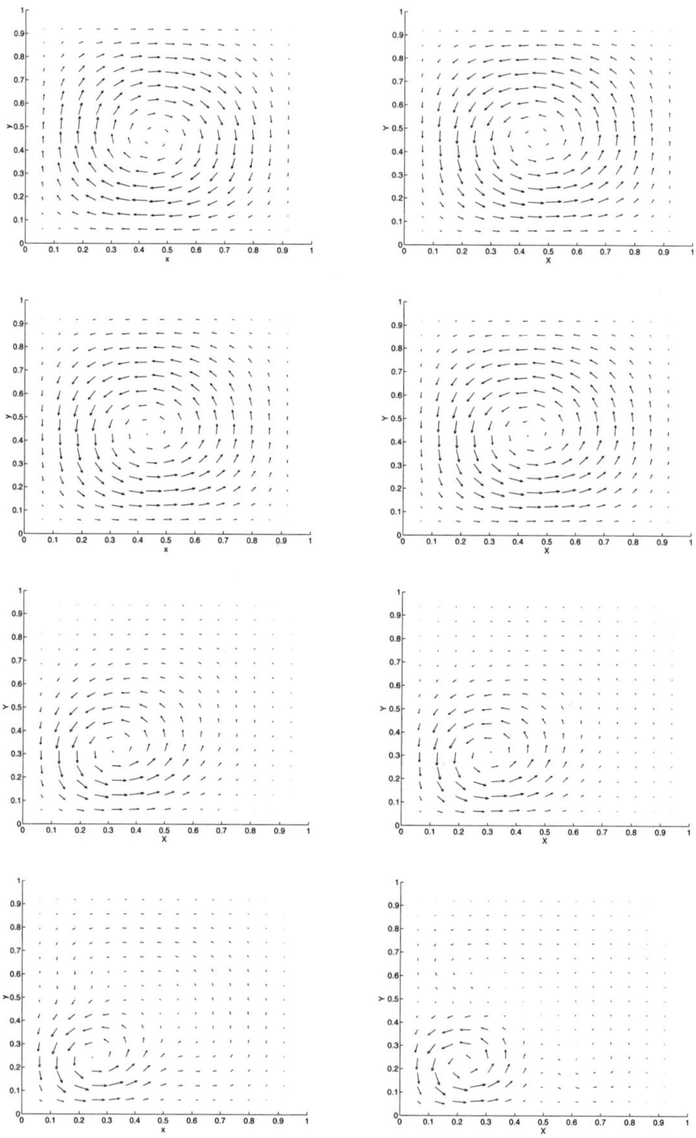

Figure 3.20. *Optimally controlled (left) and target (right) flows at* $t = 0, 0.3, 0.7$, *and* 1.

3.3.3 Boundary control

We now consider the case of controls applied on a portion of the boundary; the size of the control will be limited by penalization of the cost functional. The state equations are the Navier–Stokes system with inhomogeneous boundary conditions:

$$\left\{\begin{aligned} \partial_t \mathbf{u} + (\mathbf{u}\cdot\nabla)\mathbf{u} - \nu\Delta\mathbf{u} + \nabla p = \mathbf{0} \quad & \text{in } (0,T)\times\Omega\,, \\ \nabla\cdot\mathbf{u} = 0 \quad & \text{in } (0,T)\times\Omega\,, \\ \mathbf{u} = \mathbf{g} \quad & \text{on } (0,T)\times\Gamma_c\,, \\ \mathbf{u} = \mathbf{0} \quad & \text{on } (0,T)\times(\Gamma\setminus\Gamma_c)\,, \\ \mathbf{u}|_{t=0} = \mathbf{u}_0 \quad & \text{on } \Omega\,, \\ \int_\Omega p\, d\Omega = 0 \quad & \text{on } (0,T)\,, \\ \int_{\Gamma_c} \mathbf{g}\cdot\mathbf{n}\, d\Omega = 0 \quad & \text{on } (0,T)\,, \end{aligned}\right. \tag{3.19}$$

where the last relation is needed in view of the incompressibility constraint $\nabla\cdot\mathbf{u} = 0$. The objective functional is given by

$$\mathcal{J}_{\beta\sigma}(\mathbf{u},\mathbf{g}) = \mathcal{J}_\beta(\mathbf{u}) + \sigma\int_0^T\int_{\Gamma_c}(|\mathbf{g}|^2 + \sigma_1|\mathbf{g}_t|^2 + \sigma_2|\nabla_s\mathbf{g}|^2)\,d\Gamma dt\,, \tag{3.20}$$

where $\mathcal{J}_\beta(\mathbf{u})$ is defined in (3.11); $\sigma > 0$, $\sigma_1 \geq 0$, and $\sigma_2 \geq 0$ are penalty parameters; and ∇_s denotes the gradient in a tangent plane to the boundary. Note that the penalty term in (3.20) involves not only the control function $\mathbf{g}$, but also its time derivative and its space derivatives tangential to the boundary.

The optimization problem we consider now is to find an optimal state $(\mathbf{u}, p)$ and an optimal control $\mathbf{g}$ such that the functional $\mathcal{J}_{\beta\sigma}(\mathbf{u},\mathbf{g})$ defined in (3.20) is minimized subject to $\mathbf{u}$, p, and $\mathbf{g}$ satisfying the Navier–Stokes system (3.19).

The optimality system for this problem is given by the Navier–Stokes system (3.19), the adjoint system (3.14), and the optimality condition

$$\left\{\begin{aligned} -\sigma_1\partial_{tt}\mathbf{g} - \sigma_2\Delta_s\mathbf{g} + \mathbf{g} + q\mathbf{n} \qquad & \\ = \frac{1}{\sigma}\big(-r\mathbf{n} + \nabla\mathbf{w}\cdot\mathbf{n}\big)\big|_{\Gamma_c} \quad & \text{in } (0,T)\times\Gamma_c\,, \\ \mathbf{g}|_{t=0} = \mathbf{u}_0|_{\Gamma_c} \quad & \text{on } \Gamma_c\,, \\ \mathbf{g}|_{t=T} = \mathbf{0} \quad & \text{on } \Gamma_c\,, \\ \mathbf{g} = \mathbf{0} \quad & \text{on } (0,T)\times\partial\Gamma_c\,, \\ \int_{\Gamma_c}\mathbf{g}\cdot\mathbf{n}\,d\Gamma = 0\,, \quad & \end{aligned}\right. \tag{3.21}$$

where $q(t)$ is a Lagrange multiplier introduced to enforce the constraint $\int_{\Gamma_c}\mathbf{g}\cdot\mathbf{n}\,d\Gamma = 0$ and Δ_s denotes the surface Laplacian operator. If the control is a tangential control, i.e., if $\mathbf{g}\cdot\mathbf{n} \equiv 0$, then, in the partial differential equation of the optimality condition (3.21), the terms involving q and the adjoint pressure r can be neglected.

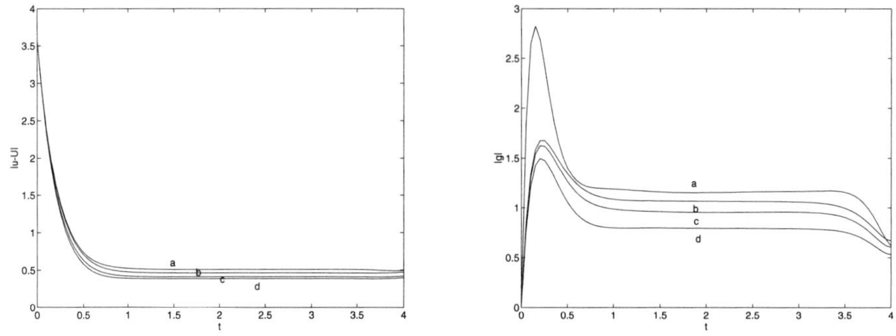

Figure 3.21. *The norm* $\|\mathbf{u} - \mathbf{U}\|$ *of the difference between the optimal flow and the target flow* (*left*) *and the norm* $\|\mathbf{g}\|$ *of the optimal control* (*right*) *vs. time for different number of sides of the boundary on which control is applied;* (a) *one-sided control,* (b) *two-sided control,* (c) *three-sided control, and* (d) *four-sided control.*

The fact that the penalization term in the functional (3.20) involves spatial and temporal derivatives of the control function **g** directly leads to the partial differential equation in the optimality condition (3.21). That partial differential equation is an elliptic partial differential equation in space-time posed along the control boundary Γ_c for $t \in (0, T)$.

If $\sigma_1 = \sigma_2 = 0$, i.e., there is no penalization of the derivatives of **g** in the objective functional (3.20), then the optimality condition reduces to the explicit formula

$$\mathbf{g} = \frac{1}{\sigma}\big(-r\mathbf{n} + \nabla\mathbf{w}\cdot\mathbf{n}\big)\big|_{\Gamma_c} - \frac{\mathbf{n}}{\sigma|\Gamma_c|}\int_{\Gamma_c}\big(-r\mathbf{n} + \nabla\mathbf{w}\cdot\mathbf{n}\big)\big|_{\Gamma_c}\cdot\mathbf{n}\, d\Gamma \quad \text{in } (0,T)\times\Gamma_c\,,$$

where $|\Gamma_c|$ denotes the area of Γ_c. However, both theoretically and practically, penalizing only the control function **g** and not its derivatives is not enough; one ends up with highly oscillatory controls.

With the obvious modification, Algorithm 3.1 can be applied to the optimization problem currently being considered.

Computational experiments

The computational set-up is very much as in Sections 3.3.1 and 3.3.2 except that now we apply boundary control. We choose the target velocity $\mathbf{U} = (U, V)$ given by

$$U(x,y) = 10\frac{d}{dy}\big(\phi(0.4,x)\phi(0.4,y)\big)\,, \qquad V(x,y) = -10\frac{d}{dx}\big(\phi(0.4,x)\phi(0.4,y)\big)$$

with $\phi(t,z) = (1-z)^2(1-\cos(2\pi t z))$. Control is applied tangential to the boundary. The initial velocity is chosen to be zero.

We examine the effects of applying control on a different number of sides of the boundary of the flow domain. In Figure 3.21, the norm $\|\mathbf{u} - \mathbf{U}\|$ of the difference between the optimal flow **u** and the target flow **U** and the norm $\|\mathbf{g}\|$ of the optimal control vs. time is

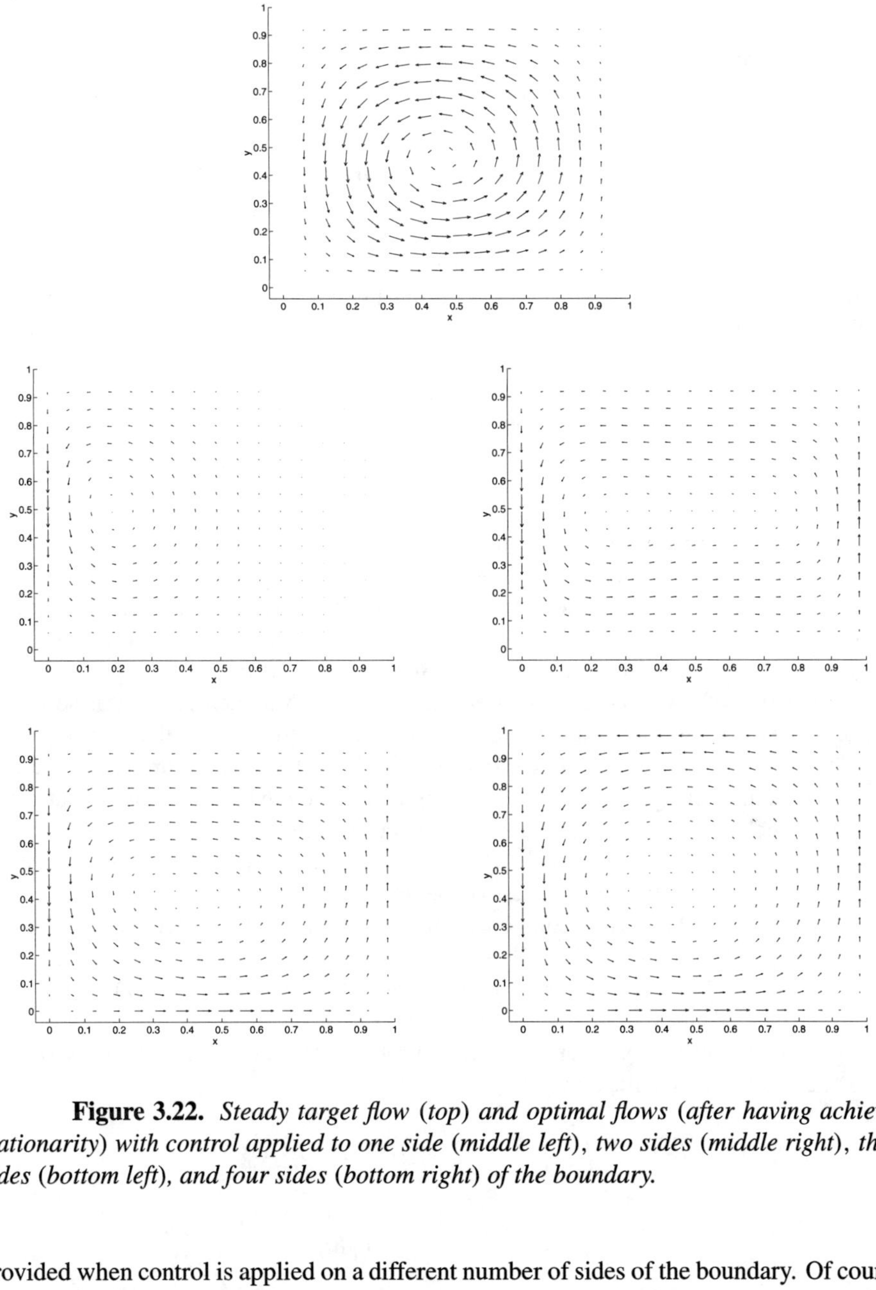

Figure 3.22. *Steady target flow (top) and optimal flows (after having achieved stationarity) with control applied to one side (middle left), two sides (middle right), three sides (bottom left), and four sides (bottom right) of the boundary.*

provided when control is applied on a different number of sides of the boundary. Of course, the match improves with an increasing number of controlled sides. As we can see from Figure 3.21, the control is better behaved and the maximum strength required decreases as the number of sides controlled increases. Visual evidence of the match between the optimal flow (after stationarity is achieved) and the target flow is provided in Figure 3.22. The improvement as the number of sides controlled increases is evident.

From the results of other computational experiments, it is found that changing σ_1 or σ_2 has little effect on the optimal solution. Changing the value σ does appreciably affect the optimal controls and states. Not surprisingly, it is also found that boundary velocity controls cannot control the interior of the domain if the desired flow changes rapidly.

3.4 Adjoint equation optimization approach to suppressing instabilities in boundary layers

We further illustrate the use of adjoint-based methods for the solution of optimal flow control and optimization problems by considering an optimization approach to suppressing instabilities in boundary layers. We discretize the continuous adjoint equations so that this example also falls into the differentiate-then-discretize approach to optimization and control. Details about this illustration can be found in [102]. Several other papers cited in the bibliography deal with similar problems.

The model problem

We consider the suppression of Tollmien–Schlichting (TS) instability waves [443] in boundary layer flows. Control is effected though the normal injection/suction of fluid through a single orifice along the wall. The constraint equations are the full, two-dimensional Navier–Stokes system. The objective functional we use measures, along a small section of the wall, the deviation in the normal stress from that for Blasius flow. The functional is penalized with a cost for control. We will not go into great detail for this example because it is somewhat similar to the examples of Section 3.3 for which we provided substantial details.

The geometric configuration for our problem is sketched in Figure 3.23. We have plane flow along a straight wall. There is a single orifice on the wall at which fluid can be injected or sucked; we assume that the fluid direction at the orifice is normal to the wall. Downstream of the "actuating" injection/suction orifice, we have a "sensing" strip along which we will monitor the normal stress of the flow.

The control is the normal component of the velocity at the orifice, i.e., we let $v = \mathbf{v} \cdot \mathbf{n} = g(t)Q(x)$ along the orifice. Note that we assume that the control function has the product form $g(t)Q(x)$, where $Q(x)$ is a specified quadratic function and $g(t)$ is the control to be determined. Thus, we have only the mass flow rate at any instant in time at our disposal in order to effect control, i.e., essentially, the control is only a scalar-valued function of time.

The functional is defined by

$$\begin{aligned} \mathcal{J}(\mathbf{v}, p, g) = &\int_{t_1}^{t_2} \int_{matching\ strip} |\tau_n(\mathbf{v}, p) - \widehat{\tau}_n|^2 \, dxdt \\ &+ \beta \int_{t_3}^{t_4} \int_{orifice} \left(\left| \frac{\partial g}{\partial t} \right|^2 + |g|^2 \right) dxdt \,, \end{aligned} \tag{3.22}$$

where β is a penalty parameter, $\tau_n(\mathbf{v}, p) = -p + \mathbf{n} \cdot (\nabla \mathbf{v} + \nabla \mathbf{v}^T) \cdot \mathbf{n}$ is the normal stress, $\widehat{\tau}_n$ is the normal stress of Blasius flow, (t_1, t_2) is the time interval over which we wish to match the normal stress, and (t_3, t_4) is the time interval over which we apply the control.

Due to the penalization of the time derivative of the control g in the functional (3.22), the optimality condition for this example is a two-point boundary value problem in time.

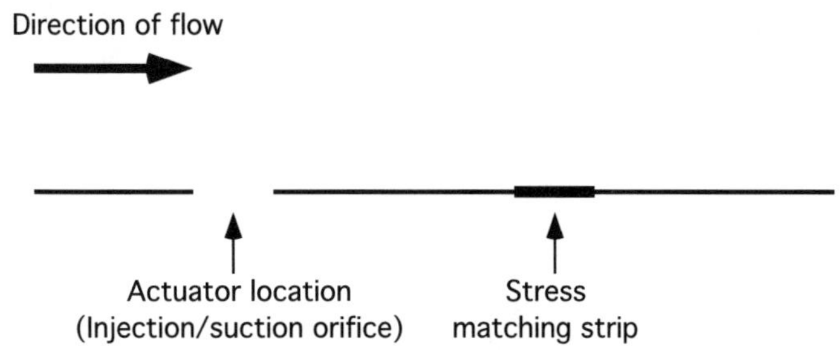

Figure 3.23. *Actuator injection/suction location relative to the small strip along the wall at which the normal stress is matched to that of Blasius flow.*

The adjoint system for this problem is, of course, posed backward in time, i.e., terminal conditions are given at the endpoint t_2 of the control interval and the adjoint system is well posed as a backward-in-time problem.

We use the *simplest possible uncoupling procedure* for the optimality system, i.e., we initialize the control function, then solve the state equations, i.e., the Navier–Stokes equations, forward in time to obtain the corresponding velocity and pressure fields. Then, we solve the adjoint equations backward in time to obtain the corresponding adjoint velocity and pressure fields which we substitute in the optimality condition to obtain a new guess for the control function. Surprisingly, this simple uncoupling algorithm converges well, i.e., in about four iterations, for this example.[47]

Computational results

The flow equations, i.e., the Navier–Stokes system, and the adjoint equations are discretized using a hybrid finite-difference spectral method. The flow is allowed to develop for one period T_p of the dominant TS instability wave. Control is then applied for two subsequent periods, i.e., $t_3 = T_p$ and $t_4 = 3T_p$. The stress matching is done for the same time interval as used for applying control, i.e., $t_1 = T_p$ and $t_2 = 3T_p$.

From Figure 3.24, we can see the convergence history of iterates of the optimal solution determined by the simple uncoupling algorithm applied to this example. The results for the fourth and fifth iterates, i.e., $i = 4$ and $i = 5$, are virtually indistinguishable. For the computations that led to Figure 3.24, the penalty parameter was set to $\beta = 10$.

In Figure 3.25, a comparison is provided between the horizontal and vertical velocity disturbance profiles at a downstream point and at the end of the control period $t = 3T_p$ for the uncontrolled flow and the optimal flow. We see that during the time that control is applied, a substantial reduction in the size of the disturbance is effected by applying the optimal control. If control is turned off at $t = 3T_p$, then the disturbances again grow; this can also be seen from Figure 3.25 which also provides the horizontal and vertical velocity disturbances at $t = 4T_p$, i.e., at a time after the control is turned off.

[47]This is not usually the case, i.e., the simple uncoupling strategy used for the optimality system of this example does not usually converge or, if it does, it converges slowly. In fact, as mentioned in Section 3.1, it can be shown that the simple uncoupling strategy is equivalent to a steepest descent method (with fixed step size) for the minimization of the objective functional. It is well known that such methods have very poor convergence properties.

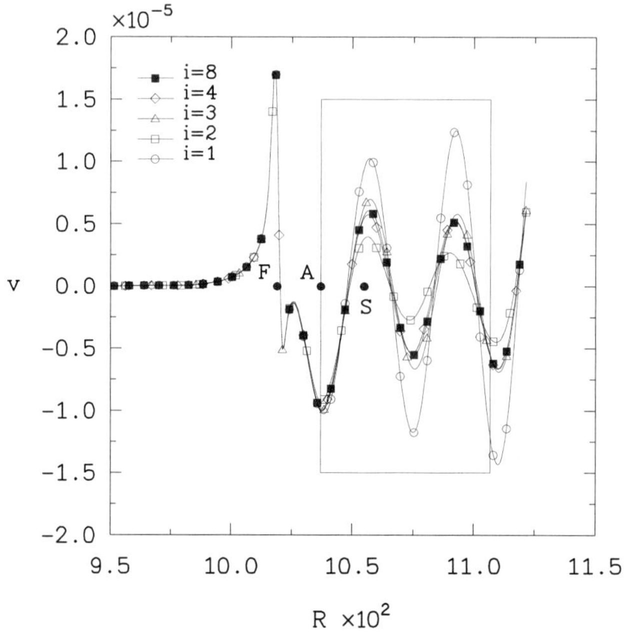

Figure 3.24. *Convergence of the vertical velocity at a downstream point near the wall; i denotes the iteration counter for the simple uncoupling algorithm.*

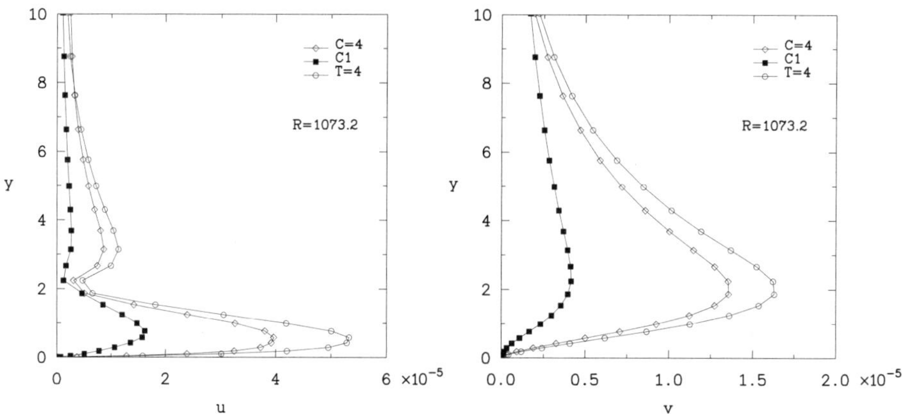

Figure 3.25. *Horizontal (left) and vertical (right) velocity disturbance profiles at a downstream point for the uncontrolled flow (open circles) and the optimal flow (filled squares) at the end of the control period $t = 3T_p$ and the horizontal and vertical velocity disturbances at $t = 4T_p$ after control is turned off (open diamonds). The optimal flow is determined using a penalty parameter $\beta = 10$.*

For TS waves, everything, e.g., the amplitude, phase, and frequency, is known analytically so that one can determine a "perfect" actuator that cancels out the instability wave almost completely; we refer to this as *wave cancellation.* It is interesting to compare the actuator (the control function $g(t)$) and sensing (the normal stress along a small section of the wall downstream of the actuator) histories for the wave-cancellation case with that obtained from the optimal control calculation of Figures 3.24 and 3.25. This comparison is made in Figure 3.26. We see that even though the functional (3.22) matches the normal stress over a very small portion of the wall, the optimal control has the correct frequency and phase; the amplitude, however, is not as large as that predicted by the analytical wave-cancellation solution. Note from Figure 3.26 that the sensing data that are used to effect the match in the functional (3.22) are also very close to the data obtained through the use of the analytically determined wave-cancellation actuation.

One can obtain better results by optimizing over the penalty parameter β; this is easy to do. In Figure 3.27, we repeat the plots of Figure 3.26 for the best value of the penalty parameter $\beta \approx 16.5$. Note that now the amplitude of the control function is much closer to that for the wave cancellation case while the frequency and phase remain in good agreement as do the sensing data downstream of the actuator. We also compare, in Figure 3.28, the vertical velocity disturbance for the uncontrolled flow, the wave cancellation flow, and the optimally controlled flow with the best value of the penalty parameter. We see that using the best value of the penalty parameter results in an optimal control that performs nearly as well as the control that is determined analytically using the known information about TS waves. It should be emphasized that the optimal control is determined without any foreknowledge of the nature of the instabilities present in the flow.

3.4.1 An observation about outflow boundary conditions

The problem we have just looked at required the truncation of an infinite domain to a finite computational domain and the imposition of some sort of artificial outflow boundary conditions for the state calculations. It is important to note that an outflow boundary for the flow equations is an inflow boundary for the adjoint system and, similarly, an inflow boundary for the flow is an outflow boundary for the adjoint flow. To see this, simply consider the terms

$$u_t + U u_x \qquad \text{for } 0 < x < 1 \text{ with } t \text{ increasing}$$

in one dimension. Suppose $U > 0$ is a constant for all x and t. Then, clearly,

$$\text{for } u, \qquad \begin{cases} x = 0 \text{ is a } \textit{state inflow} \text{ boundary}, \\ x = 1 \text{ is a } \textit{state outflow} \text{ boundary}. \end{cases}$$

The corresponding adjoint equations will then contain the terms

$$-\xi_t - U\xi_x \qquad \text{for } 0 \le x \le 1 \text{ with } t \textit{ decreasing}.$$

We still have $U > 0$ for all x and t and the characteristics for the state and adjoint terms are the same, i.e., $x - Ut = \text{constant}$. However, because of the "reversal" of the time direction,

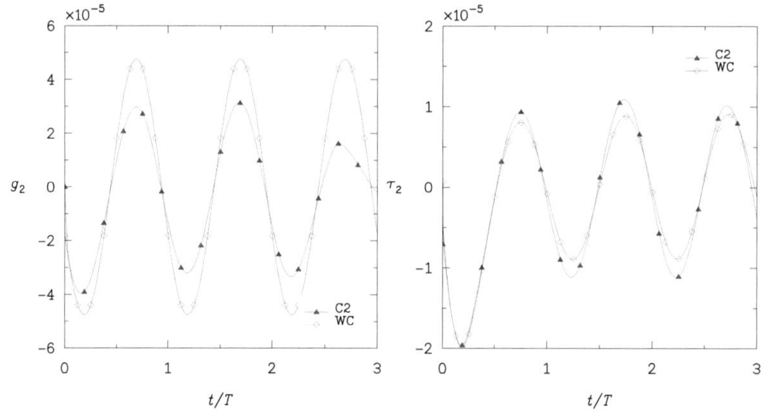

Figure 3.26. *Time history of the control (left) and normal stress (right) for the optimal control case (filled triangles) with $\beta = 10$ and for the wave cancellation case (open diamonds); there is excellent agreement in the frequency and phase but not in the amplitude.*

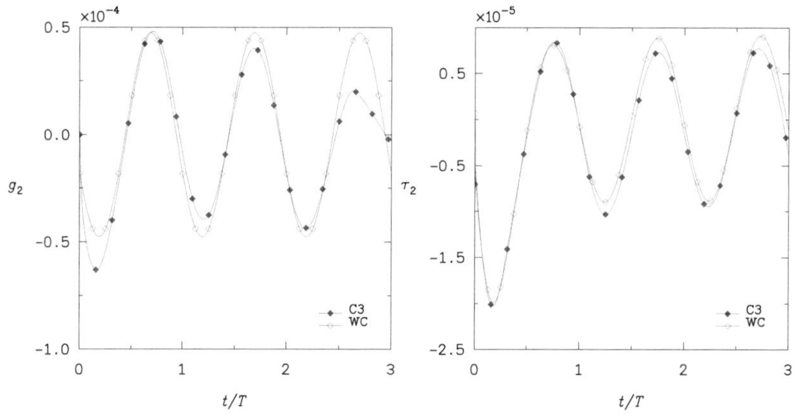

Figure 3.27. *Time history of the control (left) and normal stress (right) for the optimal control case (filled triangles) with $\beta = 16.5$ and for the wave cancellation case (open diamonds); there is good agreement in the frequency, phase, and amplitude.*

we now have that,

$$\text{for } \xi, \quad \begin{cases} x = 0 \text{ is an \textit{adjoint outflow} boundary,} \\ x = 1 \text{ is an \textit{adjoint inflow} boundary.} \end{cases}$$

A sketch depicting the state and adjoint state inflows and outflows and their relation to the characteristics is given in Figure 3.29.

We conclude that if the adjoint equations are to receive special outflow treatment, it should be applied at its outflow boundary, which is the fluid inflow boundary; no special treatment is needed for the adjoint equations at its inflow boundary, which is the fluid outflow boundary. For the computational results given in Figures 3.24–3.28, this is exactly what was done. Note that if automatic differentiation is used to derive the adjoint equations, then the outflow treatment for the state equations is differentiated as well, and no special treatment for the adjoint variable at its outflow boundary is applied.

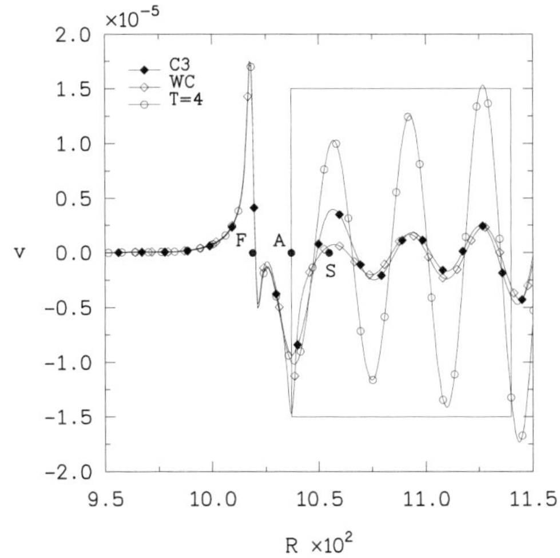

Figure 3.28. *Vertical velocity disturbance at a downstream point near the wall for uncontrolled flow (open circles), optimal flow with $\beta = 16.5$ and $t_2 = t_4 = 4T_p$ (filled triangles), and wave cancellation flow (open diamonds).*

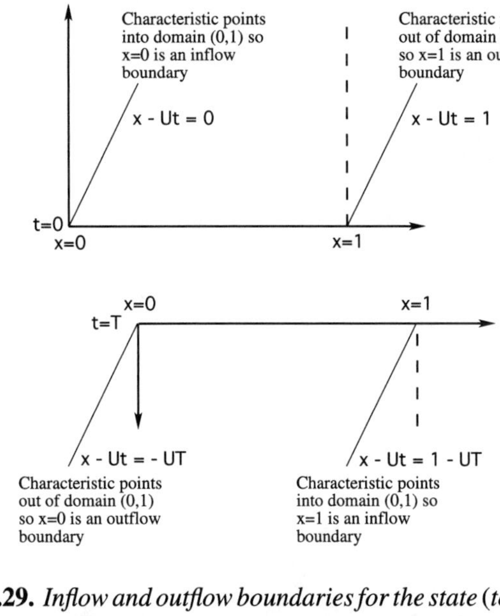

Figure 3.29. *Inflow and outflow boundaries for the state (top) and adjoint variables (bottom); the characteristics should be followed in the direction that time is followed, i.e., increasing time for the state and decreasing time for the adjoint.*

Chapter 4

Questions of Accuracy and Consistency

4.1 Insensitive functionals, inconsistent gradients, spurious minima, and regularized functionals in a viscous flow example

We use the simple context of Navier–Stokes flow in a channel with a bump to show how a number of pitfalls can arise in the implementation of flow control and optimization algorithms. The specific pitfalls we consider are

- the possible insensitivity of functionals with respect to design parameters,
- the possible inconsistency of functional gradient approximations, and
- the possible appearance of spurious minima in discretized functionals.

We also examine how regularization of the objective functional can help avoid or overcome the pitfalls and discuss the implications that our observations have to more complex, and therefore more useful, flow control and optimization problems. Along the way, we also compare the discretize-then-differentiate and differentiate-then-discretize approaches to optimization, especially in the context of the second pitfall. For details concerning this example, see [37, 38, 183, 274].

The model problem

We consider steady, incompressible, viscous flow in the channel $0 < x < 10$ and $0 < y < 3$ having a bump on the lower wall extending from $x = 1$ to $x = 3$. The flow is described by solutions of the stationary Navier–Stokes system

$$-\frac{1}{Re}\Delta\mathbf{u} + \mathbf{u}\cdot\nabla\mathbf{u} + \nabla p = \mathbf{0} \qquad \text{and} \qquad \nabla\cdot\mathbf{u} = 0 \qquad \text{in the channel} \tag{4.1}$$

along with the boundary conditions

$$\mathbf{u} = \mathbf{0} \qquad \text{on the lower and upper walls,} \tag{4.2}$$

$$\mathbf{u} = \begin{pmatrix} u \\ v \end{pmatrix} = \begin{pmatrix} \alpha_0 y(3-y) \\ 0 \end{pmatrix} \qquad \text{at the inflow } x = 0\,, \tag{4.3}$$

along with standard outflow conditions at $x = 10$, the downstream end of the channel. In (4.1)–(4.3), $\mathbf{u}$ and p denote the velocity and pressure fields, respectively, Re is the Reynolds number, and α_0 is a parameter that determines the mass flow rate at the inflow. We denote by u and v the horizontal and vertical components, respectively, of the velocity vector $\mathbf{u}$.

The bump is determined as a sum of Bezier polynomials [427]; some of the coefficients in the sum are determined by the requirement that the bump continuously meet the straight channel walls on either side of it; there remain three coefficients $\{\alpha_k\}_{k=1}^3$ at our disposal to effect changes in the flow field. The coefficient α_0 appearing in the inflow boundary condition (4.3) is another parameter at our disposal. The exact details of the description of the bump and of other aspects of the model problem are not crucial to the discussion we are about to undertake. Here, it suffices to assume that the bump is described by

$$y_B(x;\alpha_1,\alpha_2,\alpha_3) = \Psi(x) + \sum_{k=1}^{3}\alpha_k\Phi_k(x) \qquad \text{for } 1 \le x \le 3\,, \tag{4.4}$$

where $\Psi(x)$ and $\Phi_k(x)$, for $k = 1, 2, 3$, are given functions.

We compute a *target flow* $(\widehat{u}^h, \widehat{v}^h, \widehat{p}^h)$ by solving the Navier–Stokes system (4.1)–(4.3) (by a finite element method) with the parameters chosen to be

$$\widehat{\alpha}_0 = 0.5, \qquad \widehat{\alpha}_1 = 0.375, \qquad \widehat{\alpha}_2 = 0.5, \qquad \text{and} \qquad \widehat{\alpha}_4 = 0.375. \tag{4.5}$$

Although the bump is described as a sum of higher degree polynomials, for the target values of the bump parameters, the target bump reduces to a parabola.

The *objective functional* is given by, for some x_m in the interval $3 \le x \le 10$,

$$\mathcal{J}^h(\alpha_0,\alpha_1,\alpha_2,\alpha_3) = \frac{1}{2}\int_0^3 \Big(u^h(x_m, y;\alpha_0,\alpha_1,\alpha_2,\alpha_3) - \widehat{u}^h(x_m, y)\Big)^2 dy \tag{4.6}$$

so that it measures the discrepancy between the horizontal velocity component u^h of the discretized flow and the horizontal velocity component $\widehat{u}^h$ of the target flow along a vertical line across the channel located at the position $x = x_m$ downstream of the bump. Note that (4.6) is a *discretized* functional since the integrand involves the discretized target flow $\widehat{u}^h$ and u^h, a candidate flow determined by solving the discretized flow equations for a particular guess for the parameters. The functional could be (and usually is) further discretized by approximating the integral in (4.6) by a numerical integration rule. In the finite element setting, the integrand is a piecewise polynomial, and we can easily perform the integration in (4.6) exactly.

The *optimization problem* is then given as follows: For a given x_m, minimize the functional $\mathcal{J}^h$ with respect to the parameters $\{\alpha_k\}_{k=0}^3$ subject to (u^h, v^h, p^h) satisfying the *discretized* Navier–Stokes equations.

For all our calculations, we choose the value of the Reynolds number to be 10 and we apply a Taylor–Hood finite element discretization of the Navier–Stokes system [428, 430]. Unless otherwise noted, we use a logically Cartesian grid of size $h = 0.25$, with the obvious squeezing in the region above the bump.[48] Optimization is effected by a quasi-Newton BFGS/trust region method [382] with a very tight convergence tolerance $= 10^{-9}$. For all

[48] This level of grid refinement is sufficient to obtain accurate flow solutions for $Re = 10$.

cases, the initial values of the four parameters are chosen to be zero. Note that the target flow and parameters are feasible, i.e., an optimizer should be able to obtain exact values of the parameters and the value of the functional at the optimum is zero. Of course, gradient-based optimization algorithms are required to find a point in parameter space at which the gradient of the functional vanishes; in general, they will not make use of the fact that, in our case, the functional itself vanishes at such a point. However, the vanishing of the functional at the optimum certainly is information we can use to monitor the performance of optimization methods.

The number of optimization steps given in the tables below is relatively high since we are converging solutions to an extremely tight tolerance; in practice, it makes no sense to iterate beyond something a little smaller than the discretization error for the approximate solution of the state equations; in that case, the number of iteration steps needed will be drastically reduced.

It is difficult to come up with a more straightforward or elementary flow control or optimization problem than the one we consider here. However, as we shall see, even in this simple setting, all sorts of difficulties can arise.

Determining the gradient of the functional

The optimization method we use employs the gradient of the functional to determine new guesses of the parameters from old guesses. We use two approaches for determining the gradient: the first is a discretize-then-differentiate approach and the second is a differentiate-then-discretize approach.

The first approach is to approximate the gradient of the functional by a finite difference quotient approximation. For example, for the first parameter, we have

$$\frac{D\mathcal{J}^h}{D\alpha_0} \approx \frac{\mathcal{J}^h(\alpha_0+\delta\alpha_0,\alpha_1,\alpha_2,\alpha_3) - \mathcal{J}^h(\alpha_0,\alpha_1,\alpha_2,\alpha_3)}{\delta\alpha_0}, \tag{4.7}$$

where $\delta\alpha_0$ is a chosen (usually small) increment in the parameter α_0. Similar expressions hold for the other three parameters. Here, $D\mathcal{J}^h/D\alpha_k$, $k = 0, 1, 2, 3$, denote the components of the gradient of the functional with respect to the parameters $\{\alpha_k\}_{k=0}^3$. Note that, to determine the approximate first component of the gradient by (4.7), we need to solve the flow system (4.1)–(4.3) for the parameter set $\{\alpha_0+\delta\alpha_0, \alpha_1, \alpha_2, \alpha_3\}$. To determine approximations to all four components of the gradient of the functional requires four approximate flow solves in addition to the one needed to determine the approximate flow field corresponding to the base parameter values $\{\alpha_0, \alpha_1, \alpha_2, \alpha_3\}$.

An approximation to the gradient of the functional can be detemined more efficiently with the help of the *sensitivities*

$$\mathbf{u}_k = \begin{pmatrix} u_k \\ v_k \end{pmatrix} = \begin{pmatrix} \dfrac{\partial u}{\partial \alpha_k} \\ \dfrac{\partial v}{\partial \alpha_k} \end{pmatrix} \quad \text{and} \quad p_k = \frac{\partial p}{\partial \alpha_k} \quad \text{for} \quad k = 0, 1, 2, 3. \tag{4.8}$$

In general, we cannot determine the sensitivities exactly. In a differentiate-then-discretize approach that we refer to as the *sensitivity equation method*, we first differentiate the flow system (4.1)–(4.3) with respect to each of the design parameters $\{\alpha_i\}_{k=0}^{3}$ to obtain the four continuous sensitivity systems: for $k = 0, 1, 2, 3$,

$$\left\{\begin{array}{l} -\frac{1}{Re}\Delta \mathbf{u}_k + \mathbf{u}_k \cdot \nabla \mathbf{u} + \mathbf{u} \cdot \nabla \mathbf{u}_k + \nabla p_k = \mathbf{0} \\ \nabla \cdot \mathbf{u}_k = 0 \end{array}\right. \quad \text{in the channel,} \tag{4.9}$$

$$\mathbf{u}_k = \mathbf{0} \quad \text{on the lower and upper walls except along the bump,} \tag{4.10}$$

$$\mathbf{u}_k = \left\{\begin{array}{ll} \mathbf{0} & \text{if } k = 0 \\ -\left(\frac{\partial \mathbf{u}}{\partial y}\right)\Phi_k & \text{if } k = 1, 2, 3 \end{array}\right. \quad \text{along the bump,} \tag{4.11}$$

$$u_k = \left\{\begin{array}{ll} y(3-y) & \text{if } k = 0 \\ 0 & \text{if } k = 1, 2, 3 \end{array}\right. \quad \text{and} \quad v_k = 0 \quad \text{at the inflow } x = 0\,, \tag{4.12}$$

and the same outflow conditions as used for the velocity $\mathbf{u}$ and pressure p. We then determine approximate sensitivities by discretizing (4.9)–(4.12), e.g., by using the same finite element method as was used to discretize the flow system (4.1)–(4.3). We denote the approximate sensitivities determined by this process, e.g., by first differentiating the flow equations and then discretizing the resulting continuous sensitivity equations, by

$$\left(\frac{\partial u}{\partial \alpha_k}\right)^h, \quad \left(\frac{\partial v}{\partial \alpha_k}\right)^h, \quad \text{and} \quad \left(\frac{\partial p}{\partial \alpha_k}\right)^h \quad \text{for} \quad k = 0, 1, 2, 3\,. \tag{4.13}$$

One can instead use a discretize-then-differentiate approach for determining approximate sensitivities. In this approach, one first discretizes the flow equations, e.g., by a finite element method. Then one differentiates the discretized flow equations with respect to the design parameters to obtain four systems of discrete equations for the approximate flow sensitivities. We denote the approximate sensitivities determined by this process, e.g., by first discretizing the flow equations and then differentiating the resulting discretized flow equations, by

$$\frac{\partial u^h}{\partial \alpha_k}, \quad \frac{\partial v^h}{\partial \alpha_k}, \quad \text{and} \quad \frac{\partial p^h}{\partial \alpha_k} \quad \text{for} \quad k = 0, 1, 2, 3\,. \tag{4.14}$$

These are the approximate sensitivities obtained when one uses an automatic differentiation methodology to obtain sensitivities; see, e.g., [194, 197, 198, 199, 200, 375, 376, 377, 381, 384, 385, 386, 407].

Although both (4.13) and (4.14) are approximations to the exact sensitivities (4.8), they are in general not the same, e.g.,

$$\left(\frac{\partial u}{\partial \alpha_k}\right)^h \neq \left(\frac{\partial u^h}{\partial \alpha_k}\right);$$

the differentiation and discretization steps do not commute.

The gradient of the discretized functional (4.6) can be determined from either

$$\frac{D\mathcal{J}^h}{D\alpha_k} \approx \int_0^3 \left((u^h - \widehat{u}^h)\left(\frac{\partial u}{\partial \alpha_k}\right)^h \right)\Bigg|_{(x_m, y)} dy \tag{4.15}$$

using (4.13) or from

$$\frac{D\mathcal{J}^h}{D\alpha_k} = \int_0^3 \left((u^h - \widehat{u}^h) \frac{\partial u^h}{\partial \alpha_k} \right) \Bigg|_{(x_m, y)} dy \tag{4.16}$$

using (4.14). Note that (4.15) yields only an *approximation* to the gradient of the discretized functional because we use (4.13) instead of (4.14). On the other hand, (4.16) yields the exact gradient of the discretized functional (4.6). Both (4.15) and (4.16) may be viewed as approximations of the gradient of the same continuous functional, which involves exact solutions of the Navier–Stokes system (4.1)–(4.3); however, the fact that (4.16) is the exact gradient of the discretized functional while (4.15) is not the exact gradient of anything can and will have an important role to play in our discussions.

We will not consider both (4.7) and (4.16); both of these are discretize-then-differentiate approaches and have been found to have very much the same effects when used in gradient-based flow control and optimization methods, so long as the increments $\delta\alpha_k$ used in the finite difference quotient approach are small. Thus, we will use (4.7), i.e., finite difference quotient functional gradient approximations, as a representative discretize-then-differentiate approach and (4.15), i.e., the sensitivity equation method, as a representative differentiate-then-discretize approach.

We note that the gradient of the functional can also be determined or approximated through the use of solutions of *adjoint equations*. Both discretize-then-differentiate and differentiate-then-discretize strategies may again be employed. With regard to the issues, difficulties, and remedies discussed here, adjoint equation approaches do not offer any advantages or disadvantages over the finite difference quotient or sensitivity-based approaches. Thus, we will not consider adjoint equation approaches any further in this discussion.

4.1.1 Insensitive cost functionals

We first choose the matching line to be located at $x_m = 9$; see Figure 4.1.

We first use (4.15) to evaluate the gradient of the functional, i.e., we use solutions of the discretized continuous sensitivity equations. With a tolerance of 10^{-9}, the optimizer declared that satisfactory convergence was achieved after 20 iterations and returned the values given in Table 4.1.

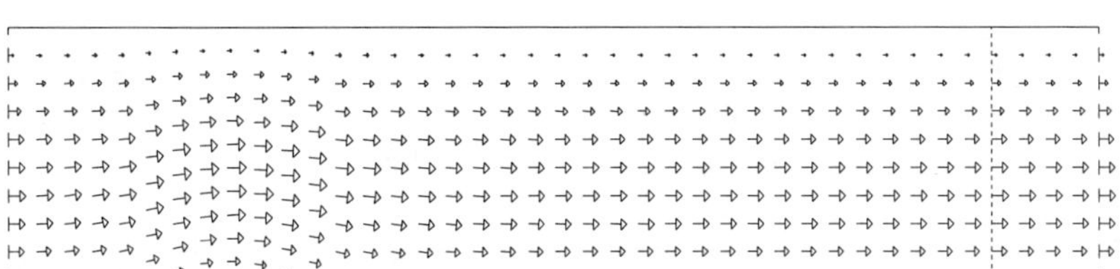

Figure 4.1. *The target flow $(\widehat{u}^h, \widehat{v}^h)$ in the channel and the matching line at $x = 9$.*

α_0	α_1	α_2	α_3	$\mathcal{J}^h$	$\max \lvert\nabla \mathcal{J}^h\rvert$
0.5000	0.0767	0.0656	0.2760	0.6×10^{-9}	0.3×10^{-5}

Table 4.1. *Parameter, functional, and gradient values after* 20 *optimizer iterations using solutions of discretized continuous sensitivity equations to evaluate the gradient of the functional.*

α_0	α_1	α_2	α_3	$\mathcal{J}^h$	$\max \lvert\nabla \mathcal{J}^h\rvert$
0.5000	0	0	0	0.3×10^{-8}	0.2×10^{-6}

Table 4.2. *Parameter, functional, and gradient values after* 10 *optimizer iterations using the finite difference quotient approach to evaluate the gradient of the functional.*

Note that the optimal value of α_0 is very close to its target value of 0.5 but that the three bump parameters are nowhere close to theirs, despite the fact that the functional and its gradient are very small. In fact, the value of the functional for the initial parameters is 0.4 so that the value of the functional has been reduced by 9 orders of magnitude!

If we instead determine the gradient of the functional using (4.7), i.e., a finite difference quotient approximation, we obtain an even more surprising set of parameter values after 10 iterations; see Table 4.2. Again, α_0 is very close to its target value and the functional and its gradient are very small, but the bump parameters are all zero, i.e., there is no bump! The bump shapes for the target parameters and for the parameters of Tables 4.1 and 4.2 are given in Figure 4.2.

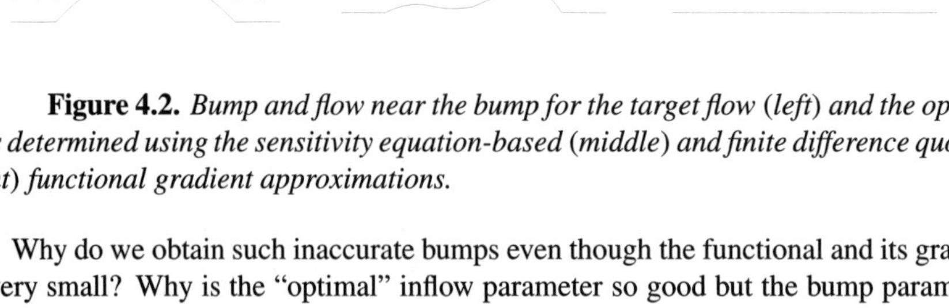

Figure 4.2. *Bump and flow near the bump for the target flow* (*left*) *and the optimal flows determined using the sensitivity equation-based* (*middle*) *and finite difference quotient* (*right*) *functional gradient approximations.*

Why do we obtain such inaccurate bumps even though the functional and its gradient are very small? Why is the "optimal" inflow parameter so good but the bump parameters so bad?

First, let's see why α_0 is so good. The role of optimization with respect to α_0 is to get the mass flow at the inflow to be the same as that of the target flow at the matching line. Thus, it is no wonder that the optimizer very quickly finds the right value for α_0; this is typical for parameters that determine the values of gross, e.g., integral, features of the flow.

For our choices of Reynolds number and matching plane location, once the mass flow at the inflow is equal to the mass flow across the matching plane (as is true once α_0 has converged), not only is the optimal profile at the matching plane that for Poiseuille flow, but it is in fact *essentially identical to that of the target flow*, irrespective of the shape of the bump. This must be so because, for low values of the Reynolds number, at the matching line both the target flow and the optimal flow are very nearly *Poiseuille flows with the same mass flow.* Thus, for any values of the bump parameters, the functional

$$\mathcal{J}^h(\alpha_0, \alpha_1, \alpha_2, \alpha_3) = \int_0^3 \left(u^h(9, y; \alpha_0, \alpha_1, \alpha_2, \alpha_3) - \widehat{u}^h(9, y)\right)^2 dy$$

is small because $u^h(9, y; \alpha_0, \alpha_1, \alpha_2, \alpha_3)$ is very close to $\widehat{u}^h(9, y)$.

Next, let's see why the computed approximation to the gradient of the discrete functional is small, even though the bump parameters are way off their correct values. Part of the answer lies with the sensitivities; the sensitivities with respect to the bump parameters are big near the bump but *are minute far downstream of the bump and, in particular, at the matching line* $x = 9$; see Figure 4.3.[49] This merely reflects the fact that at low values of the Reynolds number, no matter what shape the bump takes, the flow at a matching plane far downstream of the bump is going to be Poiseuille flow. The approximate gradient of the discrete functional with respect to the bump parameters given by (4.15) is small for any values of the bump parameters because, first, $u^h(9, y; \alpha_0, \alpha_1, \alpha_2, \alpha_3) \approx \widehat{u}^h(9, y)$, i.e., the flow and target flows are essentially the same, and second, the velocity sensitivity at the matching line is small. The second cause, i.e., the smallness of the sensitivities, is the important one since it remains in effect even if we choose a target flow that is not feasible.

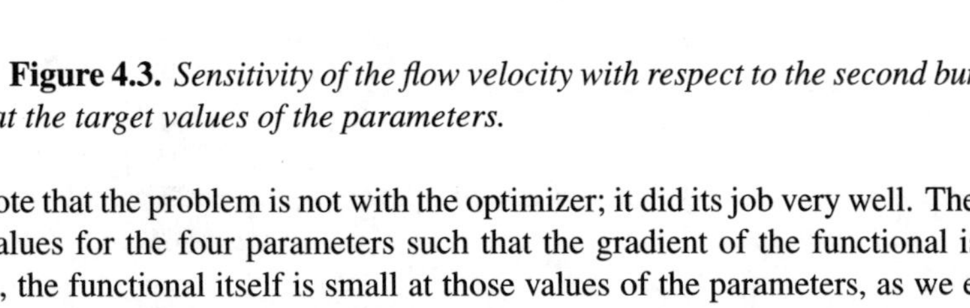

Figure 4.3. *Sensitivity of the flow velocity with respect to the second bump parameter* α_2 *at the target values of the parameters.*

Note that the problem is not with the optimizer; it did its job very well. The optimizer found values for the four parameters such that the gradient of the functional is small; in addition, the functional itself is small at those values of the parameters, as we expect it to be for the functional we are using. The problem is that *the functional is very insensitive to changes in the bump parameters.* Thus, care must be exercised in setting up optimization problems to make sure that the functional is not insensitive to some or all of the design parameters. Alternatively, if one has a functional that one wants to minimize and it is

[49] In contrast, the sensitivity of the flow field with respect to α_0 at the matching line is not small.

insensitive to changes in a design parameter, one might as well forget about that parameter and optimize with respect only to those parameters that can effect appreciable changes in the value of the functional.

4.1.2 Inconsistent functional gradients

In order to work with a more sensitive functional, from now on we set $x_m = 3$, i.e., we move the matching line to the back of the bump; see Figure 4.4. Otherwise, the problem specification remains the same. As we shall see, our troubles are just starting!

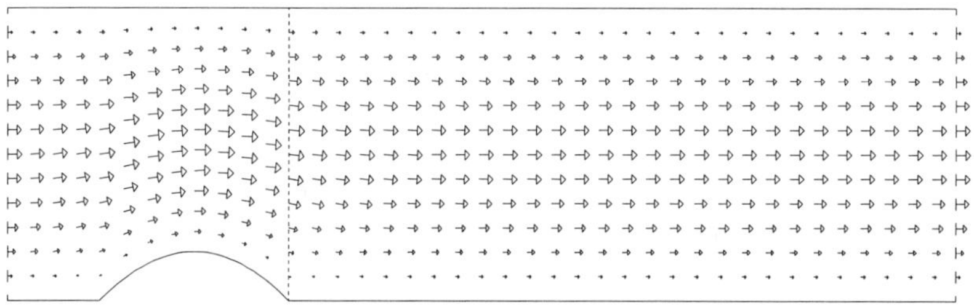

Figure 4.4. *The target flow $(\widehat{u}^h, \widehat{v}^h)$ in the channel and the matching line at $x = 3$.*

We determine the gradient of the functional using the approximate sensitivities determined by discretizing the continuous sensitivity equations. The optimizer gets confused after 33 iterations, at which point it quits at the parameter values given in Table 4.3. Now, not only are the bump parameters not good and the value of the functional not small, but more importantly, the computed gradient of the functional is not small. The shape of the bump and the corresponding velocity field for the parameters of Table 4.3 are given in Figure 4.5. Note that the oscillatory nature of the bump is very different from the target bump given in Figure 4.4.

α_0	α_1	α_2	α_3	$\mathcal{J}^h$	$\max \lvert\nabla\mathcal{J}^h\rvert$
0.5058	−0.0185	0.4352	−0.0448	0.4×10^{-3}	0.2×10^{-1}

Table 4.3. *Parameter, functional, and gradient values after* 33 *optimizer iterations using solutions of discretized continuous sensitivity equations to evaluate the gradient of the functional.*

The optimizer can handle functional values that are not small; after all, it doesn't know that, for the exact minimizer, the value of the functional is zero. However, the optimizer can't live with nonzero gradients since, at local or global minima, the gradient should be zero. Why did the optimizer quit even though the gradient isn't small?

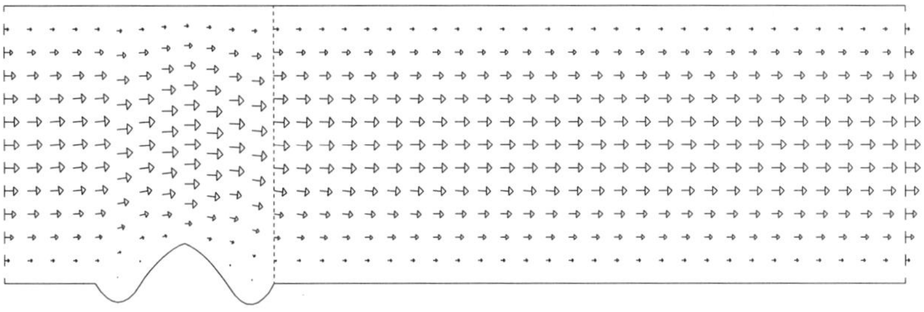

Figure 4.5. *Bump geometry and flow field resulting after* 33 *iterations with the functional gradient determined from discretizations of the continuous sensitivity equations.*

To get an indication of what has gone wrong, let's draw a line (in the four-dimensional parameter space) through the true minimizer (the parameter values for the target flow given by (4.5)) and the parameter values returned by the optimizer after 33 iterations as given in Table 4.3. Let's evaluate the functional along that line as well as the derivative of the functional in the direction of the line. The results are given in Figure 4.6.

It is clear from Figure 4.6 that the optimizer did not stop at a local minimum of the functional—so why did it stop? From the plots of the functional and its directional derivative along the line, we glean the following information about the behavior of the functional:

$$\begin{aligned} 0 \le s < 11: &\quad \text{the functional increases,}\\ 11 < s < 25: &\quad \text{the functional decreases,}\\ 25 < s \le 30: &\quad \text{the functional increases,} \end{aligned}$$

and its directional derivative:

$$\begin{aligned} les < 8: &\quad \text{the gradient of the functional is positive,}\\ 8 < s < 25: &\quad \text{the gradient of the functional is negative,}\\ 25 < s \le 30: &\quad \text{the gradient of the functional is positive,} \end{aligned}$$

as a function of arc length s along the line. There is a serious inconsistency in these figures: *in the interval* $8 < s < 11$, *the functional is increasing in s but the computed gradient of the functional determined by the differentiate-then-discretize sensitivity-based method is negative.*

Let's explore this a little more by examining a two-dimensional slice of parameter space that passes through the true minimum and the point in parameter space returned by the optimizer after 33 iterations. We compute the functional on that plane and the direction of the projection of the negative of the gradient onto that plane; these are displayed in Figure 4.7. Note that the negative of the gradient should be perpendicular to the level curves of the functional and should point in the direction of decreasing functional values.

From Figure 4.7 we see lots of places where the negative computed gradient is not perpendicular to the level curves. The optimizer doesn't care about this; all the optimizer cares about is that, when it is given a functional and its gradient, *the negative of the gradient*

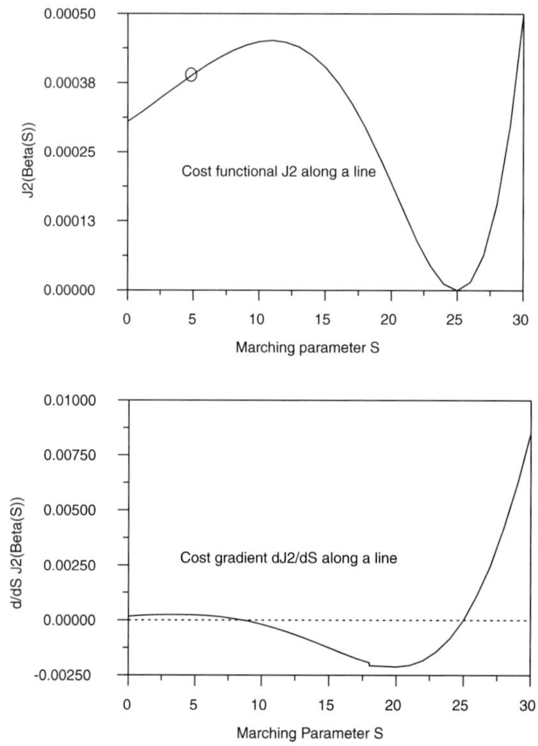

Figure 4.6. *Functional (top) and directional derivative of the functional (bottom) along the line in parameter space joining the true minimum and the values of the parameter returned by the optimizer after* 33 *iterations; the circle identifies the point at which the optimizer stopped.*

points downhill. Then it can guarantee that, for a sufficiently small step size, it can find new values of the parameters that reduce the value of the functional. Again, from Figure 4.7, we see that there certainly are locations in the two-dimensional slice at which *the negative of the computed gradient is pointing uphill.* Although examining a single two-dimensional slice out of a four-dimensional space is not definitive, what probably happened after 33 iterations is that the optimizer found a point in parameter space at which the functional it was given increases in the direction of the negative computed functional gradient it was also given. The optimizer cannot live with this inconsistency since *it cannot find a step along the direction of the negative of the computed gradient, no matter how small it chooses the step, which results in a decrease in the value of the functional.* At such a juncture, the optimizer just quits.

Let's now compute an approximation to the gradient of the functional using a finite difference quotient approximation, i.e., a discretize-then-differentiate approach. In Figure 4.8, using the same two-dimensional slice as was used for Figure 4.7, we again provide level curves of the functional and direction vectors for the projected negative of the approximate

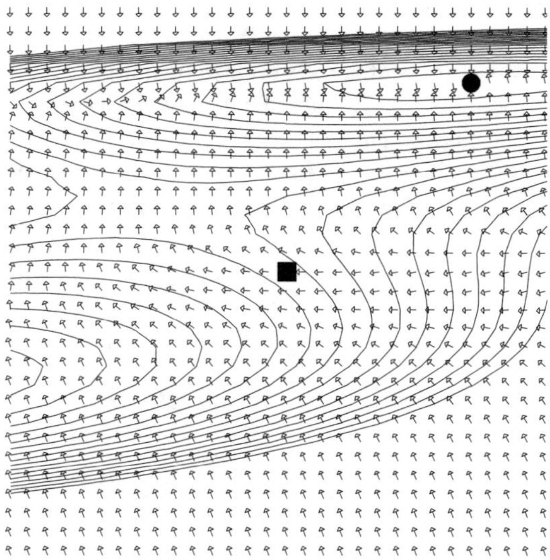

Figure 4.7. *Level curves of the functional and projected negative approximate gradient of the functional on a two-dimensional slice of parameter space containing the true minimum (circle) and the values of the parameter returned by the optimizer after* 33 *iterations (square); the approximate gradient of the functional is determined by the differentiate-then-discretize approach.*

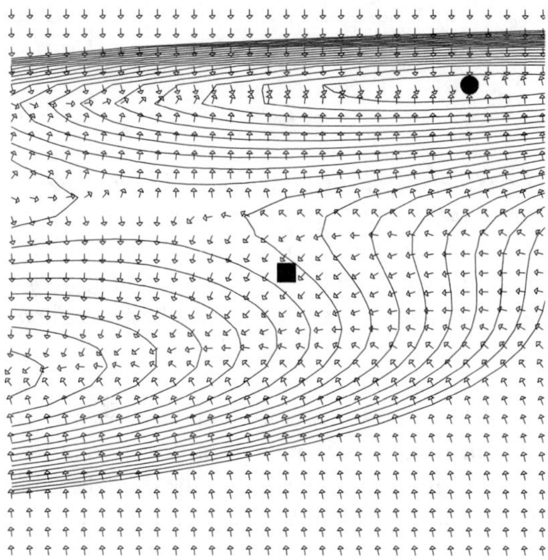

Figure 4.8. *Level curves of the functional and projected negative approximate gradient of the functional on the same two-dimensional slice of parameter space used for Figure* 4.7; *the gradient of the functional is determined by the finite difference quotient approach.*

gradient. Now there are no inconsistencies; the approximate gradient projected onto the two-dimensional slice is always nearly perpendicular to the level curves, and the negative of the projected approximate gradient always points downhill.

In Figure 4.9, we zoom in to the vicinity of the point in parameter space returned by the optimizer when it used the approximate gradients found by the differentiate-then-discretize approach. We see that there are lots of nearby locations at which the differentiate-then-discretize sensitivity-based negative gradient points in the wrong direction (uphill); the finite difference quotient based negative gradient points in the right direction (downhill). Note the points for which the angle between the two approximate gradients is larger than $\pi/2$.

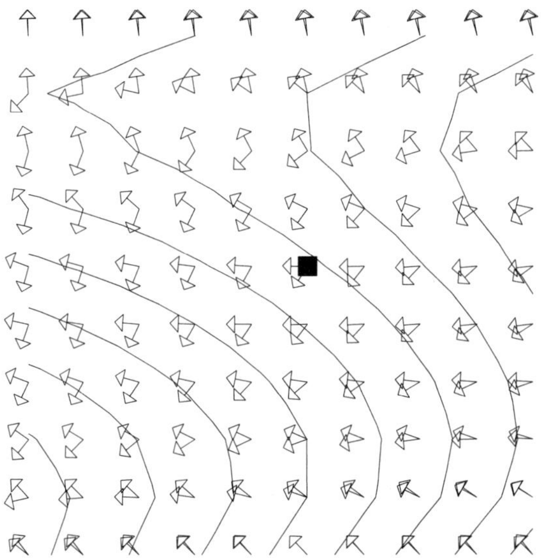

Figure 4.9. *Level curves of the functional and projected negative approximate gradients of the functional on the same two-dimensional slice of parameter space used for Figure* 4.7 *and in the vicinity of the point returned by the optimizer after* 33 *iterations* (*square*); *the direction of the approximate negative gradient of the functional determined by both a finite difference quotient approximation and by the differentiate-then-discretize sensitivity approach is displayed.*

The discretize-then-differentiate sensitivity-based approach, e.g., using automatic differentiation software, produces the exact gradient of the discrete functional that is being minimized, so, of course, that gradient also points in the right direction; using finite difference quotient approximations with small increments $\delta\alpha_k$ also produces consistent gradients. For this reason, some believe that the discretize-then-differentiate approach is superior to the differentiate-then-discretize approach. The differentiate-then-discretize approach does not produce an exact gradient of any functional (continuous or discrete) and is thus somewhat susceptible to producing inconsistent functional gradients. For the problem we are considering, if the grid size is sufficiently small, then all approaches must produce consistent gradients; however, the differentiate-then-discretize approach can produce an inconsistent gradient at a practical (from the point of view of flow simulation accuracy) grid size. Although this seems very bad, we shall see below that the solution of another difficulty also helps the differentiate-then-discretize approach produce consistent gradients. To this end, note from Figure 4.7 that even *the differentiate-then-discretize approach produces a consistent gradient near the true minimizer,* i.e., near the filled circle.

4.1.3 Spurious minima

Now that we know that using finite difference quotients to approximate the gradient of the functional yields consistent gradients, let's solve the optimization problem (with the matching line located at $x = 3$) using those gradient approximations instead of the differentiate-then-discretize approach used for Table 4.3. The optimizer declares convergence after 44 iterations and returns the data in Table 4.4 for the optimal solution. Note the very small gradient of the functional, which indicates that the solution provided by the optimizer is at least a local minimizer of the discretized functional. However, the bump parameters are again not close to those for the target flow. Note that, although the bump parameters are far from those of the target flow, the value of the functional is quite small, i.e., the target flow is well matched at the line $x = 3$ even with the wrong bump parameters, i.e., the wrong bump looks a lot like the target bump as far as the flow at $x = 3$ is concerned; see Figure 4.10.

α_0	α_1	α_2	α_3	$\mathcal{J}^h$	$\max \lvert\nabla \mathcal{J}^h\rvert$
0.5078	0.1407	0.5394	0.0599	0.3×10^{-6}	0.2×10^{-9}

Table 4.4. *Parameter, functional, and gradient values after* 44 *optimizer iterations using finite difference quotient approximations of the gradient of the functional.*

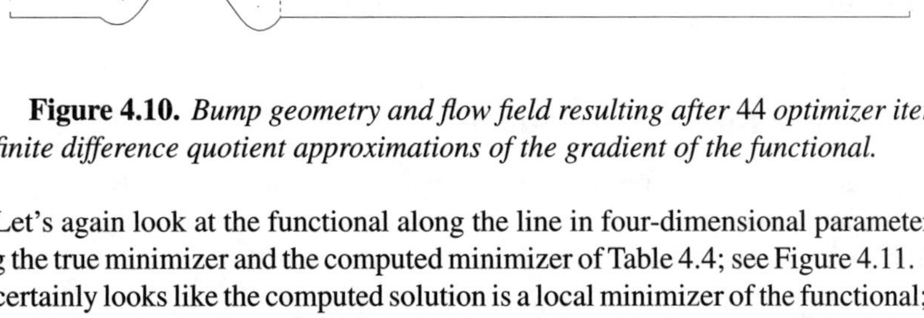

Figure 4.10. *Bump geometry and flow field resulting after* 44 *optimizer iterations using finite difference quotient approximations of the gradient of the functional.*

Let's again look at the functional along the line in four-dimensional parameter space joining the true minimizer and the computed minimizer of Table 4.4; see Figure 4.11. We see that it certainly looks like the computed solution is a local minimizer of the functional; in fact the magnitude of the gradient of the functional at this location is very small ($\approx 0.2 \times 10^{-9}$).

Results along lines in four dimensions are not very conclusive, so let's look again at the functional on a two-dimensional slice determined by a plane going through the true

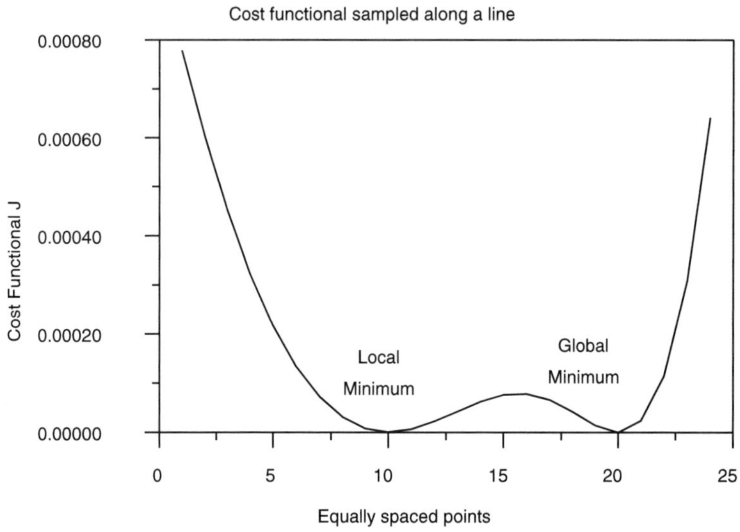

Figure 4.11. *Functional along the line in parameter space joining the true minimizer and the computed local minimizer of Table* 4.4.

minimizer and the optimal solution of Table 4.4. Figure 4.12 displays the level curves of the functional and the direction of the negative approximate gradient. Again, both the position of the computed solution along the level curves of the functional and the gradient of the functional indicate that the optimizer has indeed found a local minimizer of the functional.

Is this local minimizer for real, or is it a spurious one introduced by the discretization process? A good (but not foolproof) way to check if a suspect solution (especially one with oscillatory behavior) is really a solution or whether it is a numerical artifact is to refine the grid. If upon refinement the solution seems to converge, then there is a good chance that it is an actual solution; if upon refinement the suspect solution gets worse, e.g., oscillates with bigger amplitude or higher frequency, then chances are good that it is a numerical artifact. In Table 4.5 and visually in Figure 4.13, one can see the nonconvergence of the bump parameters as the grid size is reduced; the oscillations in the bump geometry have higher amplitude as the grid size is reduced.

Thus, we conclude that *the local minimizer found by the optimizer is a spurious, numerically induced one.*

4.1.4 Regularization of the functional

It is clear that, at least for some initial conditions, the optimizer leads us to the spurious local minimum. How can one get to the desired true minimum? To avoid the spurious minimum, in fact, to get rid of it altogether, one can use a penalized objective functional; since the solution corresponding to the spurious minimizer contains unwanted oscillations in the bump geometry, we add a penalty term that penalizes such oscillations. Thus, we choose *the penalized objective functional*

grid size	α_0	α_1	α_2	α_3	$\mathcal{J}^h$
0.250	0.5078	0.1408	0.5394	0.0600	0.3×10^{-6}
0.167	0.5087	−1.4610	0.3596	−0.0841	0.1×10^{-4}
0.125	0.5111	−1.9128	0.2984	−0.2906	0.1×10^{-4}

Table 4.5. *Converged parameter and functional values for different grid sizes; the optimizer uses finite difference quotient approximations of the gradient of the functional.*

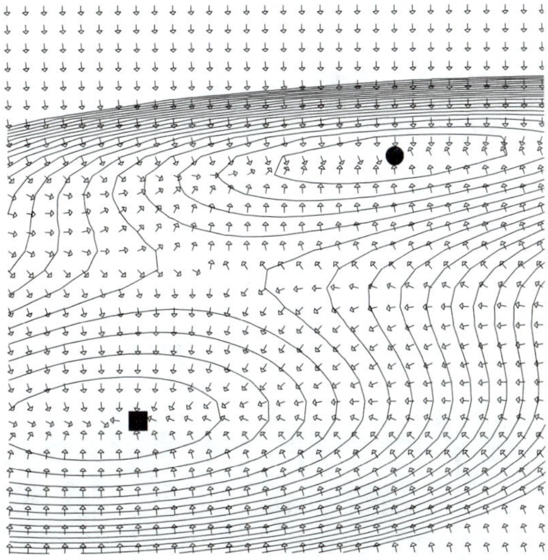

Figure 4.12. *Level curves of the functional and projected negative gradient of the functional on a two-dimensional slice of parameter space containing the true minimum (circle) and computed minimum returned by the optimizer after* 44 *iterations (square); the gradient of the functional is determined by a finite difference quotient approach.*

$$\begin{aligned}\mathcal{J}^h_\sigma(\alpha_0,\alpha_1,\alpha_2,\alpha_3) &= \mathcal{J}^h(\alpha_0,\alpha_1,\alpha_2,\alpha_3) + \frac{\sigma}{2}\int_1^3 \left(\frac{dy_B}{dx}\right)^2 dx \\ &= \frac{1}{2}\int_0^3 \left(u^h(x_m,y;\alpha_0,\alpha_1,\alpha_2,\alpha_3) - \widehat{u}^h(x_m,y)\right)^2 dy + \frac{\sigma}{2}\int_1^3 \left(\frac{dy_B}{dx}\right)^2 dx\,,\end{aligned} \tag{4.17}$$

where $y_B(x;\alpha_1,\alpha_2,\alpha_3)$ is the function that determines the bump (see (4.4)) and σ is a penalty parameter. Of course, the case $\sigma = 0$ corresponds to the unpenalized functional, i.e., $\mathcal{J}^h_0 = \mathcal{J}^h$.

First, in Table 4.6 we see what the optimizer does as we change σ. In that table, for various values of the penalty parameter, we give the number of steps taken by the optimizer to converge, the converged values of the parameters and functional found by the optimizer, and the value of

$$\mathcal{G} = \frac{1}{2}\int_1^3 \left(\frac{dy_B}{dx}\right)^2 dx,$$

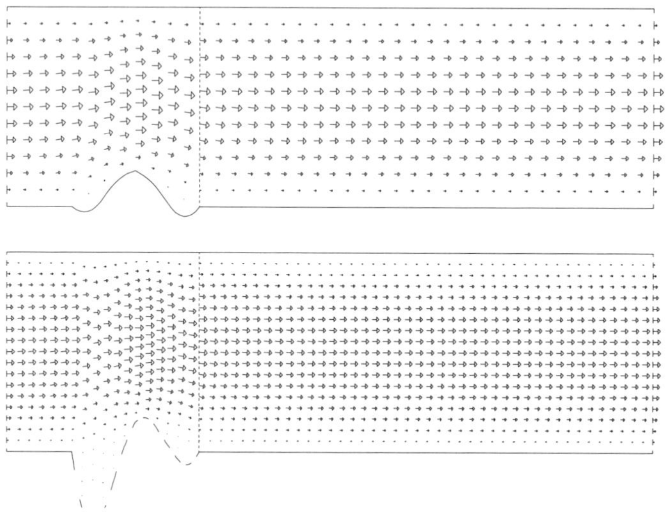

Figure 4.13. *Bump geometry and flow field for a grid size* $h = 0.25$ (*top*) *and* $h = 0.166$ (*bottom*).

σ	steps	α_0	α_1	α_2	α_3	$\mathcal{G}$	$\mathcal{J}^h$
1	22	0.5127	0.0026	0.0044	0.0032	0.4×10^{-4}	0.8×10^{-2}
10^{-1}	11	0.5122	0.0273	0.0467	0.0356	0.5×10^{-2}	0.8×10^{-2}
10^{-2}	19	0.5057	0.1740	0.3262	0.2706	0.2×10^{0}	0.4×10^{-2}
10^{-3}	31	0.5006	0.2515	0.4677	0.3665	0.5×10^{0}	0.5×10^{-3}
10^{-4}	40	0.5004	0.2849	0.5001	0.3636	0.5×10^{0}	0.6×10^{-4}
10^{-5}	47	0.5056	0.1305	0.5442	0.1527	0.1×10^{1}	0.1×10^{-4}
10^{-6}	47	0.5075	0.1387	0.5406	0.0712	0.6×10^{0}	0.1×10^{-5}
10^{-7}	43	0.5077	0.1405	0.5395	0.0611	0.6×10^{0}	0.4×10^{-6}
0	44	0.5077	0.1407	0.5394	0.0599	0.6×10^{0}	0.3×10^{-6}

Table 4.6. *Converged parameter and functional values and optimizer steps for different values of the penalty parameter* σ; $\mathcal{G}$ *is a measure of the amount of oscillation in the bump geometry.*

which is a measure of the amount of oscillation in the bump geometry.

Recall that the parameters for the matching function, i.e., the parameters we are trying to find, are $\alpha_0 = 0.5$, $\alpha_1 = 0.375$, $\alpha_2 = 0.5$, and $\alpha_3 = 0.375$. For all values of σ, the inflow parameter α_0 is pretty well matched. The best match for the bump parameters α_1, α_2, and α_3 occurs around $\sigma = 10^{-4}$. The match of the bump parameters gets worse if we raise or lower the value of σ from that value, but for different reasons. For large values of σ, the match deteriorates because the true minimizer of the functional $\mathcal{J}^h_\sigma$ moves further away from the true minimizer of the functional $\mathcal{J}^h$; the latter is what we are trying to find. In fact, as σ becomes large, the penalty term in the functional $\mathcal{J}^h_\sigma$ becomes dominant so that the minimization of that functional forces $dy_B/dx \to 0$, i.e., to a no bump situation; we already see evidence of this behavior even for $\sigma = 1$. For small values of σ the match deteriorates because the penalty term ceases to provide enough regularization, and a spurious local minimizer evidently captures the optimization iterates that start with the zero initial condition. Not surprisingly, the larger the value of σ, the better the penalty term does in reducing $\mathcal{G}$, the measure of the oscillations in the bump geometry.

Now, let's set $\sigma = 2 \times 10^{-4}$, which is in the range of values for the penalty parameter that appeared to give solutions that reasonably match the target values. Using this value, let's examine the level curves of the functional on a two-dimensional slice of parameter space; these are displayed in Figure 4.14. It seems that the spurious local minimizer is now not present. Of course, the true minimizer of $\mathcal{J}^h_{0.0002}$ is not the same as the true minimizer of $\mathcal{J}^h$; this is the price one has to pay for regularization.

A hybrid algorithm

We still have not accomplished our goal of finding the true minimizer of $\mathcal{J}^h$. Given that, we now suspect, for an appropriate value of σ, that the spurious minimum is not present and thus we can get reasonably close to the true minimizer of $\mathcal{J}^h$, we can develop the following

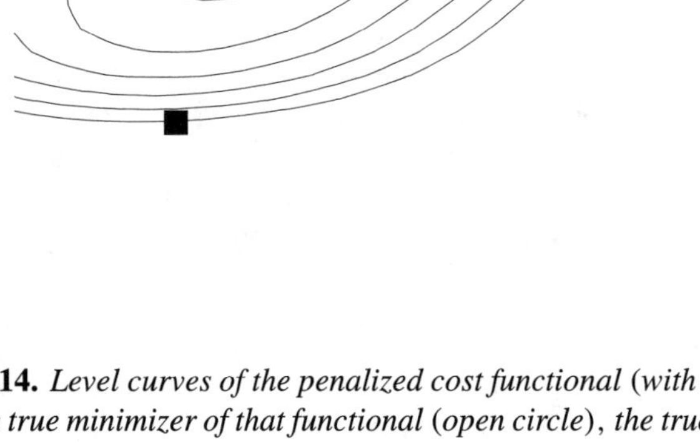

Figure 4.14. *Level curves of the penalized cost functional* (*with* $\sigma = 0.0002$) *and the positions of the true minimizer of that functional* (*open circle*), *the true minimizer of the unpenalized functional* (*filled circle*), *and the spurious local minimizer of the unpenalized functional* (*filled square*).

two-stage *hybrid algorithm*:

1. Step through the optimization method for the penalized functional $\mathcal{J}_\sigma^h$ until satisfactory convergence is achieved; we are now located at the true minimizer of the penalized functional;

2. using the results of stage 1 as an initial condition, step through the optimization for the unpenalized functional $\mathcal{J}^h$ until satisfactory convergence is achieved.

In this way, we trust the iterates will converge to the true minimizer of the unpenalized functional.

The optimal values of the parameters determined by the hybrid algorithm are

$$\alpha_0 = 0.5000, \qquad \alpha_1 = 0.3735, \qquad \alpha_2 = 0.5001, \qquad \text{and } \alpha_3 = 0.3747.$$

At last we have obtained a very good match to the parameters of the target flow!

The convergence tolerances used in the two stages of the hybrid algorithm need not be the same. In fact, it is usually more efficient to use a looser tolerance for stage 1 since one is not interested in obtaining accurate minimizers of the penalized functional. The role of the iteration using the penalized functional is merely to produce a good enough initial condition for the optimization of the unpenalized functional, where "good enough" means we are at a point in parameter space that is in the attraction region of a true, i.e., not spurious, minimizer of the unpenalized functional. In our example, optimization of the penalized functional produced the parameter values $\alpha_0 = 0.5004$, $\alpha_1 = 0.2764$, $\alpha_2 = 0.4952$, and $\alpha_3 = 0.3649$, which are very good initial guesses for the optimization of the unpenalized functional.

The hybrid algorithm requires the choice of two parameters in addition to whatever parameters are set for the optimization of the unpenalized functional. One must choose a penalty parameter, e.g., σ in (4.17), and a convergence tolerance for the optimization of the penalized functional.

4.1.5 Implications for practical flow control problems

In the context of a *very simple,* low Reynolds number, steady, incompressible viscous flow in a channel, we have shown how three pitfalls can be encountered in the numerical approximation of flow control and optimization problems. We have also shown how these pitfalls can be avoided or circumvented, most notably through the regularization of the objective functional. We now discuss some of the implications that our observations have on more *complex* and more *practical* flow control and optimization problems.

Insensitive functionals

The first pitfall we encountered was insensitive functionals, i.e., the functional is insensitive to changes in one or more of the design parameters. For example, the functional chosen in Section 4.1.1 was extremely insensitive to the parameters α_1, α_2, and α_3, which determine the shape of the bump.

There are actually two ways to view insensitive functionals, depending on whether the central goal of optimization is to extremize the objective functional as well as possible or to find, in some sense, "good" values for the design parameters. In many, if not most, practical flow control and optimization problems, the central goal is the first of these. The choice for the mathematical objective functional is dictated by the physical objective one wants to achieve. If the goal of optimization is, e.g., to make a given objective functional as small as possible, what does the insensitivity of the cost functional with respect to a design parameter tell us? The answer is simple: we can eliminate that design parameter from the problem since it is useless for meeting our objective. For example, if in the simple problem of Section 4.1.1 one really wanted to match the horizontal velocity component at the downstream position $x_m = 9$ to that of the target flow, one should not have used the bump parameters α_1, α_2, and α_3 as design parameters. Instead, one should have fixed these parameters to some convenient values, e.g., $\alpha_1 = \alpha_2 = \alpha_3 = 0$, and optimized with respect to the single inflow mass rate parameter α_0. Clearly, for any fixed values for α_1, α_2, and α_3, we can do a wonderful job of meeting the objective of making the functional (4.6) with $x_m = 9$ small using the single parameter α_0 for optimization. Of course, in practical situations, one may find that after eliminating any useless design parameters, the remaining parameters are not sufficient in number or effect to satisfactorily achieve the physical objective; in this case, one may be able to identify other effective design parameters that can be added to the optimization process; if not, then one must be content with what the remaining design parameters can do.

The second view of insensitive functionals for which the goal of optimization is to find a "good" set of design parameters does arise in practice. If the central goal of optimization is to determine "good" parameter values and one finds that the cost functional is insensitive to one or more of those parameters, then one must change the cost functional to one that is more sensitive. For example, in Section 4.1.1, we were not happy with the values of the bump parameters produced by the optimizer, even though the cost functional was clearly rendered as small as our tolerances allowed. We then went on, in Section 4.1.2, to change the objective functional so that it is more sensitive to the bump parameters.

For both views of insensitive cost functionals, one has to identify which design parameters have negligible or insufficient effect on the cost functional. Often, intuition can be used to eliminate such design parameters from the very start. For example, in Section 4.1.1, it should have been obvious to us that the bump parameters were useless in the context of the functional (4.6) with $x_m = 9$. Thus, we either change the cost functional (the second view) or don't use the bump parameters for optimization (the first view). Where intuition fails, sensitivity analyses can be very useful. For example, even if we had no intuition about the problem of Section 4.1.1, a cursory look at Figure 4.3 would convince us that the second bump parameter α_2 is useless for affecting the value of the functional (4.6) with $x_m = 9$.

Thus, insensitivities of the cost functional can be used to

- *reduce the number of design parameters* by eliminating from the design process those parameters that do not appreciably affect the cost functional,
- *induce changes in the choice of design parameters* by replacing the useless parameters with others that have a greater effect on the cost functional, or
- *induce changes in the cost functional* so that it becomes more sensitive to the design parameters.

All three of these will result in more efficient use of the optimizer, e.g., fewer iterations, and/or better results, e.g., lower functional values or better design parameter values.

Inconsistent gradients

The second pitfall we encountered was inconsistent approximate gradients of the objective functional.[50] This occurred when we used the differentiate-then-discretize approach. If one instead uses a discretize-then-differentiate approach, this is very unlikely to happen. In fact, if exact differentiations are used, as is the case when one employs automatic differentiation software, one determines the exact gradient of the discretized cost functional so that gradients cannot be inconsistent. If one uses a finite difference quotient approximation to the gradient of the functional, then one can always obtain consistent gradients by choosing small enough variations in the values of the parameters, e.g., by making $\delta\alpha_k$ small enough in formulas such as (4.7).[51]

The approximate gradients found by the differentiate-then-discretize approach are not the exact gradients of any functional. However, they are approximations to the gradients of both the continuous and discretized functionals. If the approximate sensitivities are sufficiently accurate, e.g., if a sufficiently fine grid is used to determine sensitivities, then functional gradients found by formulas such as (4.15) should be consistent. After all, as the grid size tends to zero, approximate gradients found this way should converge to the exact gradient of the continuous functional. However, it may be the case, as it was in Section 4.1.2, that a grid resulting in acceptable flow approximations may not result in sensitivities that are sufficiently accurate to avoid inconsistencies in the gradient of the functional. If, on the other hand, one uses a grid that is optimized for the sensitivities to calculate approximate sensitivities, one is much less likely to encounter inconsistent gradients.

At first glance, it would seem that the possibility of encountering inconsistent approximate gradients renders differentiate-then-discretize approaches much less desirable than discretize-then-differentiate approaches. There are two factors that should be taken into consideration before one jumps to this conclusion. The first is that the two most commonly used discretize-then-differentiate approaches are, for the same computational parameters, more costly than the sensitivity equation method. Finite difference quotient approximations to the gradient of the functional require additional expensive nonlinear flow solutions, in fact, at least one additional solution for each design parameter. Automatic differentiation software usually involves substantial overhead which makes it more costly to run than does hand-coded sensitivity equations software. However, there is a more important reason not to dismiss differentiate-then-discretize approaches, namely that regularization of the cost functional usually eliminates the occurrence of inconsistent approximate gradients. Furthermore, in the vicinity of true minimizers, computed gradients, even for differentiate-then-discretize approaches, are usually consistent. As a result, the hybrid algorithm on page 118 can be used not only to avoid spurious minima, but also to avoid inconsistent gradients.

Spurious minima

The third pitfall we encountered was artificial, numerically induced spurious minima. We also saw in Table 4.5 and Figure 4.13 an indication that as the grid size is reduced, the spurious minimizer does not disappear but instead represents a point in parameter space

[50] Of course, we assume that the functional is differentiable.

[51] Of course, in finite precision arithmetic, $\delta\alpha_k$ cannot be too small because of potential problems due to round-off errors.

that moves farther away from the true minimizer. The implication of this observation is that for a sufficiently small grid size, the spurious minimizer will be sufficiently far away from the true minimum so that a reasonable starting guess, e.g., the origin in parameter space, would belong to the attraction region of the true minimizer. Thus, it seems that one can circumvent the spurious minima pitfall by using a sufficiently small grid size. This conclusion is of little practical use. As we saw in our simple example, a good optimizer can yield iterates that converge to a spurious minimizer for grids that are adequate to obtain good flow approximations. On the other hand, for the sake of efficiency, one does not want to use a grid size smaller than that needed to obtain acceptable flow approximations. Thus, reducing the grid size is not usually a viable way to circumvent spurious minima. Fortunately, as in Section 4.1.4, we saw that a more practical means of avoiding the spurious minima pitfall is through the regularization of the cost functional.

4.2 Spurious minima and regularized functionals for flows with discontinuities

Spurious functional minima can occur for other types of flows and for other reasons. We illustrate this point in the context of a simple nozzle design problem for Euler flows containing shock waves. Additional results and detailed discussions may be found in [266].

4.2.1 Spurious minima

We consider a two-dimensional nozzle design problem for Euler flows. Flow solutions are obtained by shock fitting so that very sharp shock waves are obtained. The shape of the nozzle is determined by a given function containing an optimization parameter q_1. The inflow is assumed to be supersonic and the inflow conditions are given and fixed during the optimization. The outflow is assumed to be subsonic so that the flow field in the nozzle contains a shock wave; the outflow horizontal velocity is determined by a given function containing an optimization parameter q_2.

A target flow $\widehat{\mathbf{U}}$ is determined using a different radius for the outflow, a given nozzle shape, and a given outflow horizontal velocity; the target flow is not attainable by candidate optimal flows. The functional to be minimized is the matching functional

$$\mathcal{K}(q_1, q_2) = \frac{1}{2}\int_{\widetilde{\Omega}} |\mathbf{U} - \widehat{\mathbf{U}}|^2 d\Omega\,,$$

where the primitive variables $\mathbf{U}(q_1, q_2)$ are determined from the design parameters q_1 and q_2 by solving the Euler equations and where $\widetilde{\Omega}$ is a portion of the flow domain. Optimization is actually effected by minimizing a discrete functional $\mathcal{K}^h(q_1, q_2)$ which is determined from $\mathcal{K}(q_1, q_2)$ by using a standard numerical integration rule and by using solutions of the discretized Euler equations in the integrand. *The discrete functional contains many spurious local minima*, i.e., local minima that are not present in the exact functional; see the plots on the left of Figure 4.15.

What is the cause of the spurious minima in the discretized functional? As the design parameters change, the position of the shock wave changes as well; in particular, the shock wave may cross grid points as the design parameters change. To see how this causes spurious

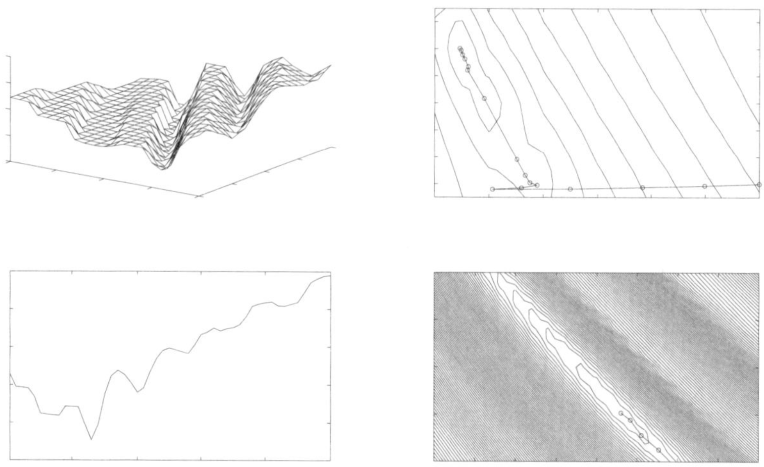

Figure 4.15. *Discretized functional before (left) and after (right) smoothing; note the local minima present without smoothing; the bottom-left plot is for a fixed value of the boundary design parameter; in the plots on the right, the circles indicate the values of the iterates of the design parameters; the bottom-right plot is for the vicinity of the minimum.*

minima, let's examine what happens to a typical term in the discrete functional; such a term is part of a sum and contains a factor such as (we have simplified to one design parameter α)

$$|u^h(x_j, y_j; \alpha) - \widehat{u}(x_j, y_j)|^2,$$

where α is the design parameter, (x_j, y_j) is a point in the grid used to discretize the flow, and $u^h(x_j, y_j; \alpha)$ is a discrete flow variable evaluated at the point (x_j, y_j) determined from the discrete Euler equations for the value α of the design parameter. Also, $\widehat{u}(x_j, y_j)$ is the target flow evaluated at the point (x_j, y_j); this value is independent of the value of the design parameter. Suppose that for the design parameter value α_1 the point (x_j, y_j) is on the opposite side of the shock for the discrete flow u^h determined from α_1 and the target flow $\widehat{u}$, e.g., $\widehat{u}$ is subsonic and u^h is supersonic or vice versa. Then the contribution to the functional from the single term containing $|u^h(x_j, y_j; \alpha_1) - \widehat{u}(x_j, y_j)|^2$ will be relatively large. Now suppose that, for the nearby value of the design parameter $\alpha_1 + \Delta\alpha$, the shock position changes so that now the point (x_j, y_j) is on the same side of the shock for the discrete flow u^h determined from $\alpha_1 + \Delta\alpha$ and the target flow $\widehat{u}$, e.g., both $\widehat{u}$ and u^h are subsonic or supersonic; see Figure 4.16. Then, the contribution to the functional from the single term containing $|u^h(x_j, y_j; \alpha_1 + \Delta\alpha) - \widehat{u}(x_j, y_j)|^2$ will be relatively small. Thus, if a change in the value of the design parameter causes the shock wave to cross the point (x_j, y_j), then the contribution to the functional from the term evaluated at that point changes significantly. This is the cause of the spurious minima in the discretized functional.

Note that the better the flow solver is at resolving shocks, the worse the problem with

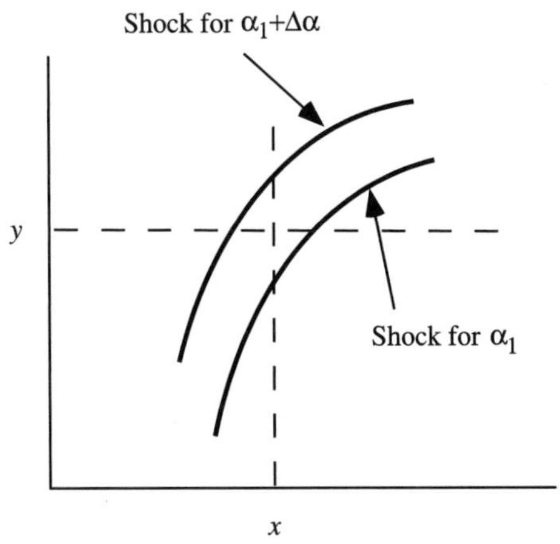

Figure 4.16. *As the value of a design parameter changes from* α_1 *to* $\alpha_1 + \Delta\alpha$*, the position of the shock wave crosses the point* (x, y).

numerically induced spurious minima, i.e., *the lousier the CFD solver is, the less serious the spurious minima problem becomes!* Note also that, in this case, *refining the grid doesn't make things better*, in fact, it makes it worse!

4.2.2 Regularization of discretized functionals

In order to avoid getting the optimizer stuck at these numerically induced local minima, one can again *regularize the discrete functional.* One way to effect regularization is to plug a "lousy" discrete flow, e.g., one having very smeared shocks, into the discretized functional.[52]

If one has a good discrete flow, i.e., with sharp shocks, one can make it "lousy" by replacing the solution at each node by the average of the solution at neighboring points; this has the effect of smoothing the discrete flow. The right-hand plots in Figure 4.15 show the effects of using such smoothed-out flow solutions for the evaluation of the functional.

One should only use the smoothed discrete flow in evaluating the functional; for other purposes, e.g., calculating sensitivities, one should use the "good" discrete flow without smoothing. Even though one uses "lousy" discrete flows to evaluate the functional, one can still obtain good approximations to optimal solutions; see Figure 4.17.

4.3 Sensitivities for flows with discontinuities

We next explore *sensitivities for flows with shock waves* and other singular surfaces by using the simple context of the *Riemann problem*, i.e., the one-dimensional shock tube problem. Since the exact solution of the Riemann problem can be written down, we will be able to compare approximate sensitivities to exact ones. We will use three solution methods for the flow and sensitivity calculations to illustrate the effects of artificial viscosity in the

[52]Of course, another way to regularize the functional is through the addition of penalty terms involving the design parameters, as was done in Section 4.1.4.

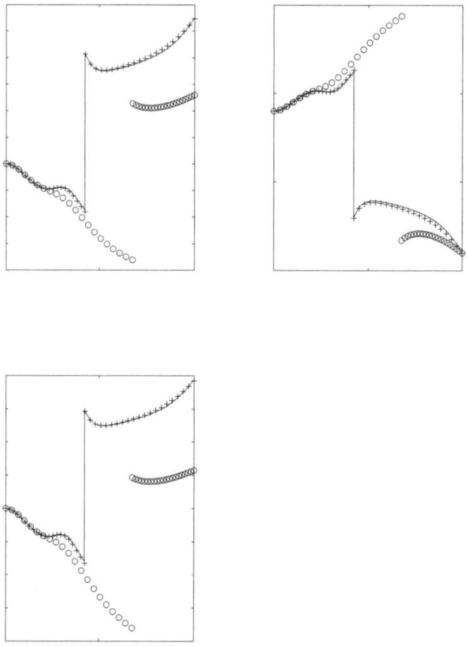

Figure 4.17. *Comparisons of the target flow (solid curve), the initial flow for the optimization (circles), and the optimal flow (plus signs) along the center line of the nozzle; top left—pressure; top right—horizontal component of the velocity; bottom—density.*

calculations of sensitivities for flows having discontinuities. Additional results and detailed discussions about this example may be found in [269, 270].

The model problem

The state equations we consider are the Euler equations for one-dimensional, compressible, inviscid flows

$$\mathbf{Q}_t + \mathbf{F}(\mathbf{Q})_x = \mathbf{0}, \tag{4.18}$$

where $\mathbf{Q}$ and $\mathbf{F}$ denote the conservative variables and flux function, respectively, and are defined by

$$\mathbf{Q} = \begin{pmatrix} \rho \\ m \\ e \end{pmatrix} \quad \text{and} \quad \mathbf{F}(\mathbf{Q}) = \begin{pmatrix} m \\ p + \dfrac{m^2}{\rho} \\ \dfrac{m}{\rho}(e+p) \end{pmatrix},$$

where ρ, m, and e denote the density, momentum, and internal energy per unit mass, respectively. The primitive variables

$$\mathbf{U} = \begin{pmatrix} p \\ \rho \\ u \end{pmatrix},$$

where ρ and u denote the density and velocity, respectively, are related to the conservative variables by

$$\rho = \rho\,, \qquad m = \rho u\,, \qquad \text{and} \qquad e = p/(\gamma - 1) + \rho u^2\,.$$

We assume that two fluids at rest having different densities and pressures are separated by a diaphragm located at $x = c$ and that, at the initial instant, the diaphragm is ruptured so that the initial conditions for the flow variables are given by

$$\begin{cases} \mathbf{Q}(x,0) = \mathbf{Q}_4 & \text{for } x < c, \\ \mathbf{Q}(x,0) = \mathbf{Q}_1 & \text{for } x > c, \end{cases} \qquad \text{or} \qquad \begin{cases} \mathbf{U}(x,0) = \mathbf{U}_4 & \text{for } x < c, \\ \mathbf{U}(x,0) = \mathbf{U}_1 & \text{for } x > c\,. \end{cases}$$

We assume initially $p_4 > p_1$ so that the subscript 4 denotes the initial high-pressure side of the diaphragm and the subscript 1 denotes the low-pressure side. We will focus on the sensitivity of the Riemann flow with respect to p_4, the initial pressure on the high-pressure side of the diaphragm.

The exact solution of the Riemann problem is given as follows [422, 436]:

- in the undisturbed high-pressure region,

$$\mathbf{U} = \mathbf{U}_4 \qquad \text{for} \quad x \le -a_4 t + c\,;$$

- in the rarefaction wave,

$$\mathbf{U} = \mathbf{U}_R \qquad \text{for} \quad -a_4 t + c \le x \le \left(\frac{\gamma+1}{2} u_3 - a_4\right) t + c\,;$$

- in the region between rarefaction wave and contact discontinuity,

$$\mathbf{U} = \mathbf{U}_3 \qquad \text{for} \quad \left(\frac{\gamma+1}{2} u_3 - a_4\right) t + c \le x < u_2 t + c\,;$$

- in the region between the shock wave and the contact discontinuity,

$$\mathbf{U} = \mathbf{U}_2 \qquad \text{for} \quad u_2 t + c < x < a_1 \left(\frac{\gamma-1}{2\gamma} + \frac{\gamma+1}{2\gamma}\frac{p_2}{p_1}\right)^{\frac{1}{2}} t + c\,;$$

- in the undisturbed low-pressure region,

$$\mathbf{U} = \mathbf{U}_1 \qquad \text{for} \quad a_1 \left(\frac{\gamma-1}{2\gamma} + \frac{\gamma+1}{2\gamma}\frac{p_2}{p_1}\right)^{\frac{1}{2}} t + c < x\,,$$

where $a_i^2 = \gamma p_i/\rho_i$ for $i = 1$ or 4. In the above expressions, p_2 is determined implicitly as the solution of

$$\frac{p_4}{p_1} = \frac{p_2}{p_1}\left(1 - \frac{(\gamma-1)(a_1/a_4)(p_2/p_1 - 1)}{\sqrt{2\gamma}\sqrt{2\gamma + (\gamma+1)(p_2/p_1 - 1)}}\right)^{\frac{-2\gamma}{\gamma-1}} \tag{4.19}$$

and the remaining variables are determined by

$$\begin{aligned}
\rho_2 &= \rho_1 \frac{p_2}{p_1}\left(1 + \frac{\gamma-1}{\gamma+1}\frac{p_1}{p_2}\right) \Big/ \left(1 + \frac{\gamma-1}{\gamma+1}\frac{p_2}{p_1}\right),\\
u_2 &= u_3 = a_1\left(\frac{p_2}{p_1} - 1\right)\sqrt{2/\gamma}\left((\gamma+1)\frac{p_2}{p_1} + (\gamma-1)\right)^{-\frac{1}{2}},\\
p_3 &= p_2,\\
\rho_3 &= \rho_4\left(\frac{p_3}{p_4}\right)^{\frac{1}{\gamma}},\\
u_R &= \left(\frac{u_3 - u_4}{\frac{\gamma+1}{2}u_3}\right)\left(\frac{x-c}{t}\right) + \frac{a_4 u_3 - \left(a_4 - \frac{\gamma+1}{2}u_3\right)u_4}{\frac{\gamma+1}{2}u_3},\\
p_R &= p_4\left(1 - \frac{\gamma-1}{2}\frac{u_R}{a_4}\right)^{\frac{2\gamma}{\gamma-1}},\\
\rho_R &= \rho_4\left(1 - \frac{\gamma-1}{2}\frac{u_R}{a_4}\right)^{\frac{2}{\gamma-1}}.
\end{aligned} \tag{4.20}$$

A plot of the solution of the Riemann problem at a specific instant of time is given in Figure 4.18. Note the five distinct regions characterizing the solution:

- left (4) and right (1) constant flow regions in which the solutions are given by the high- and low-pressure initial states, respectively;
- two other constant flow regions between the shock and the contact discontinuity (2) and the contact discontinuity and the rarefaction wave (3);
- a rarefaction (R) wave adjacent to the left high-pressure region.

Exact flow sensitivities

One cannot differentiate the flow (with respect to space, time, or design parameters) across the shock wave or the contact discontinuity across which the flow is discontinuous. If one differentiates across these surfaces, δ-functions result. Furthermore, although the flow is continuous at the edges of the rarefaction wave, it is not differentiable there. Differentiation across the edges of the rarefaction wave results in jump discontinuities in the sensitivities. However, the flow solution can be differentiated within each of the five regions in which the flow is smooth, i.e., in which the flow is differentiable.

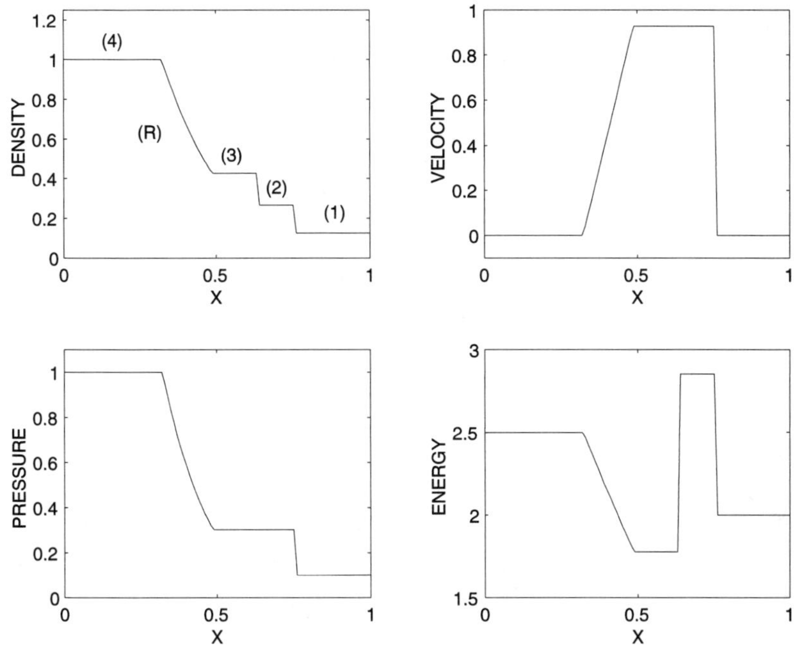

Figure 4.18. *Exact solution of the one-dimensional Riemann problem at* $t = 0.148$.

We denote the sensitivity of the flow $\mathbf{U}$ with respect to the high-pressure value p_4 by

$$\mathbf{U}' = \begin{pmatrix} \dfrac{\partial p}{\partial p_4} \\ \dfrac{\partial \rho}{\partial p_4} \\ \dfrac{\partial u}{\partial p_4} \end{pmatrix} = \begin{pmatrix} p' \\ \rho' \\ u' \end{pmatrix}.$$

Differentiation of the flow with respect to the design parameter p_4 yields the exact sensitivities within the smooth regions of the flow:

$$\begin{aligned}
\mathbf{U}' = \mathbf{U}'_4 &= \begin{pmatrix} 1 \\ 0 \\ 0 \end{pmatrix} \text{ in the high-pressure region,} \\
\mathbf{U}' = \mathbf{U}'_R &\quad \text{in the rarefaction wave,} \\
\mathbf{U}' = \mathbf{U}'_3 &\quad \text{between rarefaction wave and contact discontinuity,} \\
\mathbf{U}' = \mathbf{U}'_2 &\quad \text{between contact discontinuity and shock wave,} \\
\mathbf{U}' = \mathbf{U}'_1 &\quad \text{in the low-pressure region.}
\end{aligned}$$

A linear equation for p'_2 is found by differentiating the implicit equation (4.19) for p_2 with respect to p_4. The other variables are then given by differentiating the explicit formulas

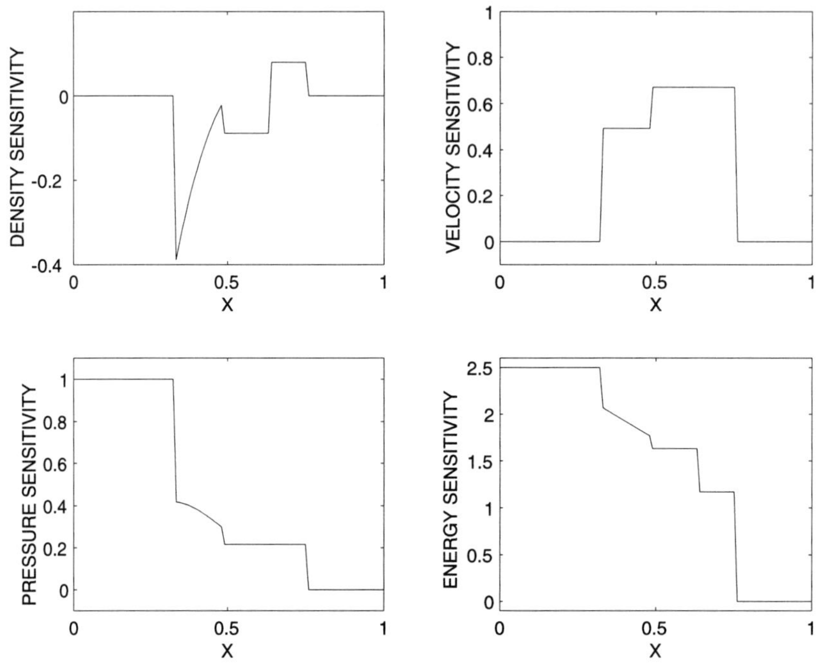

Figure 4.19. *Exact continuous sensitivities in the smooth regions of the flow, i.e., without δ-functions, with respect to p_4 at $t = 0.148$.*

(4.20). A plot of the exact sensitivities within each of the five smooth regions of the flow is given in Figure 4.19.

The exact sensitivities are the solutions of the *continuous sensitivity equations.* Switching now to conservative variables,[53] we have the sensitivities

$$\mathbf{S} = (\rho', m', e')^T, \qquad \text{where} \qquad \rho' = \frac{\partial \rho}{\partial p_4}, \quad m' = \frac{\partial m}{\partial p_4}, \quad e' = \frac{\partial e}{\partial p_4},$$

and then the sensitivity equations for the sensitivities are derived by differentiating the flow equation (4.18) with respect to p_4:

$$\mathbf{S}_t + \frac{\partial \mathbf{G}(\mathbf{Q}, \mathbf{S})}{\partial x} = \mathbf{0}, \tag{4.21}$$

[53]The sensitivities of the primitive and conservative variables are related by

$$\rho' = \rho', \qquad m' = \rho' u + \rho u', \qquad \text{and} \qquad e' = \frac{p'}{\gamma - 1} + \frac{1}{2}\rho' u^2 + \rho u u'.$$

where the sensitivity flux is given by

$$\mathbf{G}(\mathbf{Q},\mathbf{S}) = \begin{pmatrix} m' \\ \dfrac{2mm'}{\rho} - \dfrac{m^2}{\rho^2}\rho' + p' \\ \left(\dfrac{m'}{\rho} - \dfrac{m}{\rho^2}\rho'\right)(e+p) + \left(\dfrac{m}{\rho}\right)(e'+p') \end{pmatrix}.$$

The initial condition for the sensitivity **S**, which is found by differentiating the piecewise constant initial condition for the flow with respect to p_4, is given by

$$\mathbf{S}(x,0) = \begin{pmatrix} 0 \\ 0 \\ 1/(\gamma-1) \end{pmatrix} \quad \text{for } x < c \qquad \text{and} \qquad \mathbf{S}(x,0) = \mathbf{0} \quad \text{for } x > c.$$

The continuous sensitivity equation only holds inside each of the five regions within which the flow field is smooth. It does not hold across the shock wave, the contact discontinuity, or even the two edges of the rarefaction wave since the flux function $\mathbf{F}(\mathbf{Q})$ is not differentiable at those locations. It is important to note that *the sensitivity equations are linear, hyperbolic equations with discontinuous coefficients.*

Approximate sensitivities

We next want to examine and compare approximate sensitivities determined by different approaches, ignoring completely the fact that the sensitivity equations do not hold across the shock, the contact discontinuity, and the edges of the rarefaction wave.

We use three numerical schemes to approximate the flow: a two-step *Lax–Wendroff*-type method, a *Godunov*-type method, and a *Roe* scheme. See, e.g., [433], for a detailed discussion of these schemes.

We determine approximate sensitivities in three ways: finite difference quotient approximations based on incrementing the design parameter and recomputing the approximate flow solution; automatic differentiation, i.e., differentiating the discrete flow equations; and the sensitivity equation method, i.e., discretizing the continuous sensitivity equations. Specifically, the finite difference quotient approximation of the sensitivity with respect to p_4 is determined by the forward difference quotient

$$(\mathbf{Q}^h)_{p_4} \approx \frac{\mathbf{Q}^h(p_4+\Delta p_4) - \mathbf{Q}^h(p_4)}{\Delta p_4} \tag{4.22}$$

with $\Delta p_4 = 0.1$. The automatic differentiation approximate sensitivity is calculated using the software package ADIFOR [375, 376]. The sensitivity equation method approximate sensitivity is determined using the same three discretization methods, i.e., Lax–Wendroff, Godunov, and Roe, as are used for approximating the flow field.

Figures 4.20–4.22 give the results using the three different flow solvers and the three different approaches for determining sensitivities; for comparison, the exact sensitivities in the smooth regions of the flow are also provided. For Figure 4.20, the discrete flow equations and thus discrete flow fields are determined using the Lax–Wendroff scheme. The automatic differentiation sensitivity is determined by solving discrete sensitivity equations derived by differentiating the discrete flow equations. The finite difference quotient sensitivity is determined from (4.22) using multiple solutions of the discrete flow equations. The sensitivity equation sensitivities are determined by solving the discrete sensitivity equations derived by discretizing the exact sensitivity equations (4.21) by the same Lax–Wendroff scheme used for the flow equations. The same relation holds between Figures 4.21 and 4.22 and the Godunov and Roe schemes, respectively.

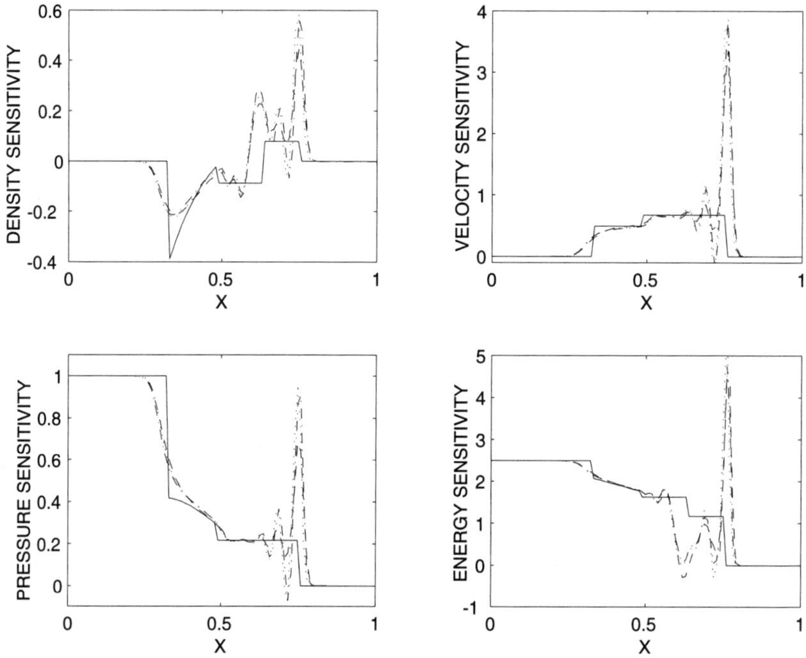

Figure 4.20. *Comparison of the three different sensitivity calculation methods using the Lax–Wendroff scheme for the flow versus the exact sensitivity (without the δ-functions); (· · ·) ADIFOR sensitivity; (-·-) finite difference sensitivity; (- -) sensitivity equation method sensitivity using the Lax–Wendroff scheme; (—) exact sensitivity.*

The most important observation that can be made from Figures 4.20–4.22 is that in all cases, i.e., for all three flow discretization schemes and for all three methods of determining sensitivities, an attempt is made implicitly within the algorithms to approximate the delta functions at the shock wave and the contact discontinuities.[54] We also see that the appearance of the spike-like approximations to the delta functions pollute the approximate sensitivities in the smooth regions of the flow. It seems that the Roe scheme does the best job and the Lax–Wendroff scheme does the worst job of approximating the sensitivities in those regions;

[54]Recall that in all cases the discontinuities in the flow and therefore the delta functions in the sensitivities were ignored in determining approximate sensitivities, i.e., no special treatment of those points at which the delta functions occur was invoked.

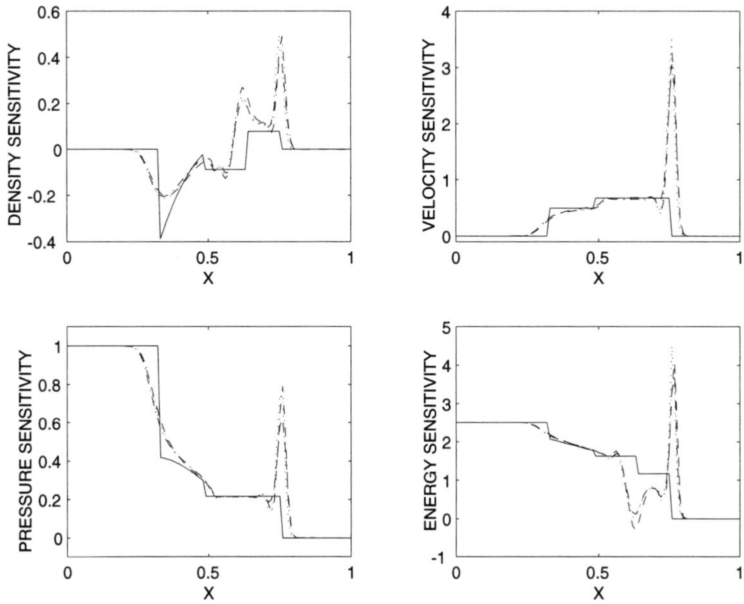

Figure 4.21. *Comparison of the three different sensitivity calculation methods using the Godunov scheme for the flow versus the exact sensitivity (without the δ-functions); (···) ADIFOR sensitivity; (-·-) finite difference sensitivity; (- -) sensitivity equation method sensitivity using the Godunov scheme; (—) exact sensitivity.*

this is entirely consistent with the ordering of the quality of flow solutions obtained by the three discretization schemes.

Figure 4.20 for the Lax–Wendroff scheme is the result of a fine-grid calculation. In Figures 4.23 and 4.24 the effects of using two coarsenings of the grid are presented. Comparing with Figure 4.20, we see that the finer the grid, the larger the height and the narrower the width of the spikes approximating the delta functions. Also, the finer the grid, the better the sensitivity approximations in the smooth regions of the flow. These observations hold as well for the other two approximation schemes.

Thus, the conclusions we draw from the numerical experiments for the Riemann problem are that sensitivities are not well approximated and that finer grids and better flow solvers seem to produce better sensitivity approximations.

Flow sensitivities and shock waves

In the calculations that led to Figures 4.20–4.24, no explicit effort was made to approximate the δ-function that occurs for the exact sensitivities. Nevertheless, we see from those figures that the approximate sensitivities do contain spikes that should be viewed as approximations to the δ-functions. We now discuss this phenomenon.

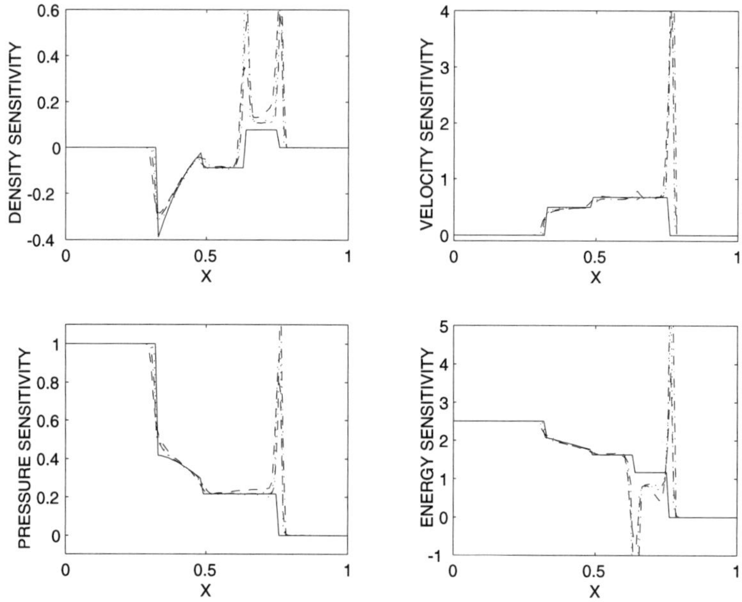

Figure 4.22. *Comparison of the three different sensitivity calculation methods using the Roe scheme for the flow versus the exact sensitivity (without the δ-functions); (···) ADIFOR sensitivity; (-·-) finite difference sensitivity; (- -) sensitivity equation method sensitivity using the Roe scheme; (—) exact sensitivity.*

Suppose that a one-dimensional flow is characterized by a single design parameter α, e.g., the inflow Mach number or a Bezier parameter that determines the shape of the boundary. Then, the flow field may be viewed as a function of x, t, and α, i.e., $\mathbf{Q} = \mathbf{Q}(x, t; \alpha)$. For a flow simulation, one would pick a particular value α_0 for the parameter and then proceed to determine $\mathbf{Q}$ as a function of x and t, i.e., $\mathbf{Q}(x, t; \alpha_0)$. We denote the location of the shock wave by x_s; in general, if the value of the parameter α is changed, the location of the shock wave changes so that the shock location is a function of the parameter, i.e., at least locally, we can write $x_s = x_s(t; \alpha)$.

For a fixed value of the parameter $\alpha = \alpha_0$, the flow variable $\mathbf{Q}(x, t; \alpha_0)$ is a discontinuous function of x and certainly cannot be differentiated with respect to x at the shock $x_s(t; \alpha_0)$. Likewise, if the value of x is fixed at, say, $x = x_0$, then, for at least some x_0, $\mathbf{Q}(x_0, t; \alpha)$, viewed as a function of α, is discontinuous and hence cannot be differentiated with respect to α at the shock $x_s(t_0; \alpha) = x_0$. *Flow sensitivity derivatives do not exist at shock waves* in much the same way that spatial and temporal flow derivatives do not exist at those locations. A schematic of this situation is given in Figure 4.25.

In the calculation of flow sensitivities by all three numerical approaches used for Figures 4.20–4.24, the presence of shock waves was ignored, e.g., the same algorithm was applied at points in $(x, t; \alpha)$-space where a shock wave is present as was applied at points

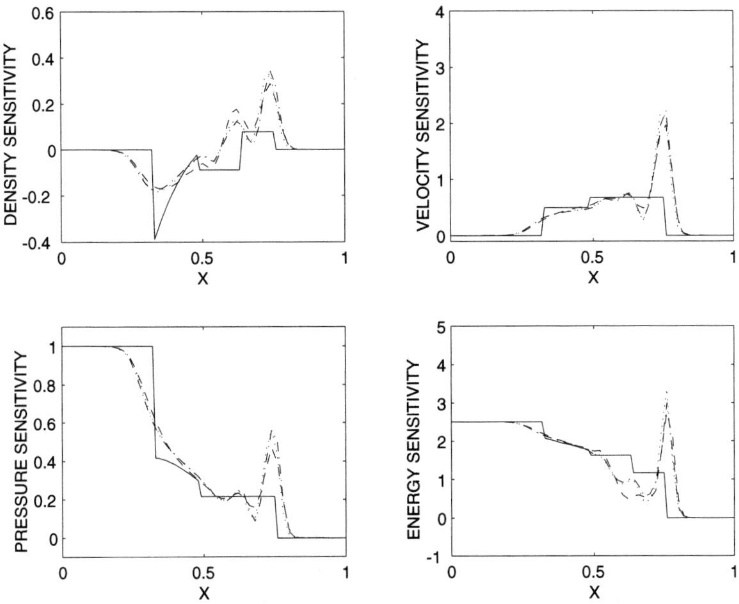

Figure 4.23. *Comparison of the three different sensitivity calculation methods using the Lax–Wendroff scheme on a medium grid versus the exact sensitivity (without the δ-functions*); ($\cdots$) *ADIFOR sensitivity;* (-·-) *finite difference sensitivity;* (- -) *sensitivity equation method sensitivity using the Lax–Wendroff scheme;* (—) *exact sensitivity.*

where the flow is smooth. As a result, approximate flow sensitivities with *large spikes* at the shock were obtained; these spikes approximate the δ-functions that the exact flow sensitivities contain at that location.

To see how the spikes arise, consider the *finite difference quotient* approach to approximating sensitivities. Suppose the flow is determined at (x_1, t_1) for two values α_1 and $(\alpha_1 + \Delta\alpha)$ of the parameter. We approximate the flow sensitivity at (x_1, t_1) by

$$\frac{\partial \mathbf{Q}(x_1, t_1; \alpha_1)}{\partial \alpha} \approx \frac{\mathbf{Q}(x_1, t_1; \alpha_1 + \Delta\alpha) - \mathbf{Q}(x_1, t_1; \alpha_1)}{\Delta\alpha}.$$

If the flow is smooth and $\Delta\alpha$ is small, then both the numerator and the denominator are small, and, in fact, as $\Delta\alpha \to 0$, both approach zero in such a way that their ratio converges to $\partial\mathbf{Q}/\partial\alpha(x_1, t_1; \alpha_1)$. But suppose instead that the points $(x_1, t_1; \alpha_1)$ and $(x_1, t_1; \alpha_1 + \Delta\alpha)$ in $(x, t; \alpha)$-space lie on opposite sides of the shock wave. Then, even if $\Delta\alpha$ is small, the numerator is relatively large. Hence, the solution of the finite difference sensitivity is large and a spike occurs when differencing across a shock. A schematic of this situation is given in Figure 4.26.

Using the *sensitivity equation* approach, the same kind of spikes develop. Suppose again that one design parameter α determines the flow so that $\mathbf{Q} = \mathbf{Q}(x, t; \alpha)$. The equation

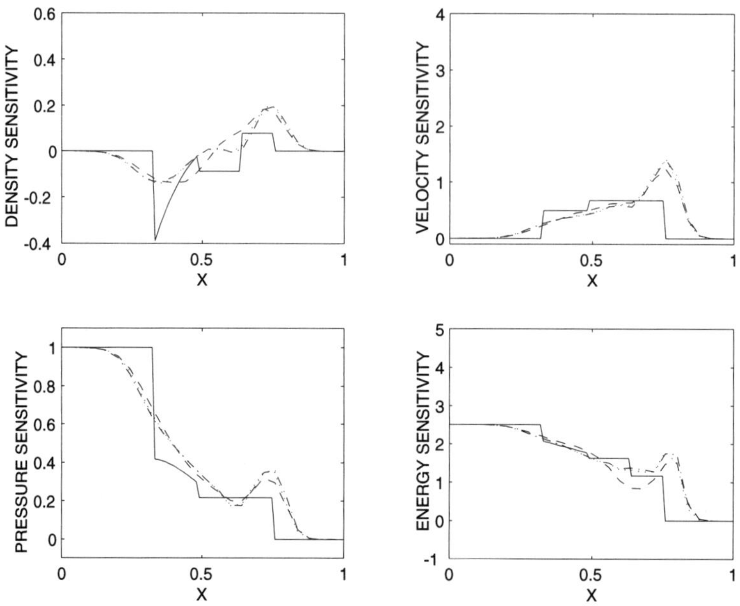

Figure 4.24. *Comparison of the three different sensitivity calculation methods using the Lax–Wendroff scheme on a coarse grid versus the exact sensitivity (without the δ-functions)*; (· · ·) *ADIFOR sensitivity;* (-·-) *finite difference sensitivity;* (- -) *sensitivity equation method sensitivity using the Lax–Wendroff scheme;* (—) *exact sensitivity.*

for the sensitivity $\mathbf{S} = \partial\mathbf{Q}/\partial\alpha$ is obtained by differentiating the flow equation with respect to α. In full detail, this process proceeds as follows:

$$\begin{aligned}\mathbf{0} &= \frac{\partial}{\partial\alpha}\left(\frac{\partial\mathbf{U}}{\partial t} + \frac{\partial\mathbf{F}(\mathbf{Q})}{\partial x}\right) = \frac{\partial^2\mathbf{Q}}{\partial\alpha\partial t} + \frac{\partial^2\mathbf{F}(\mathbf{Q})}{\partial\alpha\partial x} \\ &= \frac{\partial}{\partial t}\left(\frac{\partial\mathbf{Q}}{\partial\alpha}\right) + \frac{\partial}{\partial x}\left(\frac{\partial\mathbf{F}(\mathbf{Q})}{\partial\alpha}\right) = \frac{\partial\mathbf{S}}{\partial t} + \frac{\partial}{\partial x}\mathbf{G}(\mathbf{Q},\mathbf{S}),\end{aligned}$$

where the flux function $\mathbf{G}(\mathbf{Q},\mathbf{S})$ is defined by

$$\mathbf{G}(\mathbf{Q},\mathbf{S}) = \frac{\partial\mathbf{F}(\mathbf{Q})}{\partial\alpha} = \frac{d\mathbf{F}}{d\mathbf{Q}}(\mathbf{Q})\mathbf{S}.$$

In this derivation, it is important to note that *the order of differentiation has been interchanged.* In smooth regions of the flow, there is no difficulty in interchanging derivatives; however, at a shock wave, the derivation is not allowable due to the nonexistence of the appropriate derivatives at that location. One cannot use the flux function $\mathbf{G}$ as defined above unless one is willing to realize that the derivation has a certain amount of error built in and deal with this in the numerical scheme.

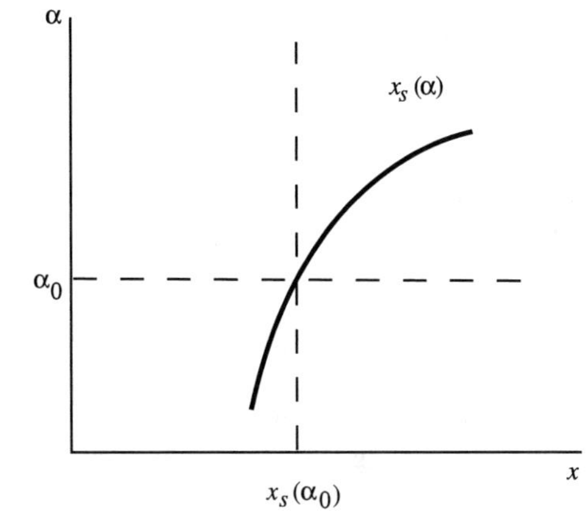

Figure 4.25. *A shock wave in (x, α)-space for a fixed value of t; the flow variables are discontinuous when one crosses the shock wave $x_s(\alpha)$ at the point $(x_s(\alpha_0), \alpha_0)$ along either the horizontal or vertical dashed lines.*

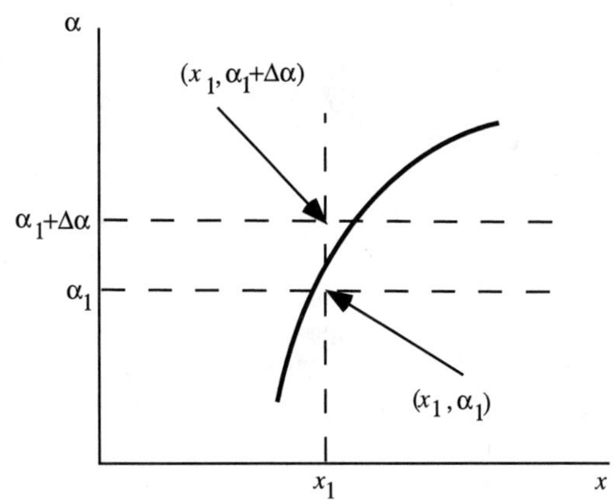

Figure 4.26. *The points (x_1, α_1) and $(x_1, \alpha_1 + \Delta\alpha)$ are on opposite sides of the shock wave in the (x, α)-plane; the flow variables at the two points differ significantly even if the two points are close together, i.e., even if $\Delta\alpha$ is small.*

It can also be seen that the same spikes arise in the *ADIFOR automatic differentiation* sensitivities. These spikes arise using the automatic differentiation method because, even though the exact solution is discontinuous, the discrete solution is "continuous", e.g., shock waves are smeared out. Hence, ADIFOR has no problem differentiating the function, but the result is spikes in the neighborhood of discontinuities.

Flow sensitivities and contact discontinuities

Contact discontinuities are another source of error in sensitivity calculations. Across a contact discontinuity, the velocity u and the pressure p are continuous while the density ρ and hence other conserved quantities experience a jump discontinuity. Thus, for the conserved quantities, the same problems that arise for shock waves and spikes are present in the sensitivities at a contact discontinuity.

Flow sensitivities and rarefaction waves

Unlike for shock waves and contact discontinuities, the flow field is continuous not only within the rarefaction wave, but also across its edges. However, the flow is not differentiable across those edges. As a result, the flow sensitivities experience a jump discontinuity at both edges of the rarefaction wave. Numerical schemes, even if they are designed to do a good job capturing shock-type discontinuities, have difficulty capturing the essentially different jumps in the sensitivities occurring at the edges of the rarefaction wave. This can be clearly seen in the computational results.

What must one do to compute accurate sensitivities?

Of course, if one is interested in the sensitivities themselves, e.g., to compute nearby flows, then sensitivities obtained by ignoring flow discontinuities are seemingly not usable. To improve the accuracy of the computed sensitivities, one must respect and pay attention to the discontinuities in the flow.

The sensitivity equations for Euler flows are *linear* hyperbolic partial differential equations with *discontinuous coefficients*. These equations have a solution for *any jump in the dependent variables* at points where the coefficients are discontinuous. What are the physically correct jump conditions to impose for the sensitivities? Not surprisingly, the correct jump conditions are determined by *differentiating the Rankine–Hugoniot conditions* with respect to the design parameter. The differentiated Rankine–Hugoniot conditions also yield the *shock position sensitivity*.

There does not seem to exist a good, general purpose method for solving the sensitivity equations with the correct jump conditions. It seems that the position of the discontinuities must be known, and there is no reason to believe that artificial dissipation will capture the correct jumps. However, there is some evidence that grid refinement near the discontinuities improves the sensitivity approximations; see, e.g., Figures 4.20, 4.23, and 4.24 and also [277].

What about adjoint variables across shock waves?

It can be shown that the adjoint variables are *continuous across shock waves*; see, e.g., [12, 226]. Derivatives of the adjoint variables are discontinuous, i.e., suffer a jump discontinuity, across shock waves. The continuity of adjoint variables across shock waves may be reason enough to use them instead of sensitivities to determine objective functional gradients.

4.3.1 Effects of inaccuracies in the sensitivities on optimization problems

We now consider the effects that inaccuracies in the sensitivities have on functional gradient approximations and thus on algorithms for flow control and optimization problems. We consider, in the Riemann flow setting, the problem of, at a specified time t_0, matching as well as possible the flow to a desired flow by altering the initial conditions of the problem. We allow ourselves a single design parameter, namely the initial left high pressure p_4. The initial left density, the initial right pressure, and the initial right density are kept fixed throughout the optimization procedure.

The cost functional is of matching type and is given by

$$\mathcal{J}(\mathbf{Q}) = \frac{1}{2}\int_{\Omega} \left|\mathbf{Q}(x, t_0; p_4) - \widehat{\mathbf{Q}}(x, t_0)\right|^2 dx\,, \tag{4.23}$$

where $\widehat{\mathbf{Q}}(x, t_0)$ is the target flow, $\mathbf{Q}(t_0; p_4)$ is the candidate flow, and Ω is the computational domain. In the cost functional, both the target and candidate flows have been evaluated at a specific time $t = t_0$.

The flow matching optimization problem is then defined as follows.

Problem 1. Find the initial left pressure p_4 such that the functional $\mathcal{J}(\mathbf{Q})$ of (4.23) is minimized, where $\mathbf{Q}(x, t; p_4)$ is a solution of the flow equations (4.18).

For this problem with the single design parameter p_4, the gradient of the cost functional is simply its derivative with respect to p_4:

$$\frac{D\mathcal{J}}{Dp_4} = \int_{\Omega} \Big(\mathbf{Q}(x, t_0; p_4) - \widehat{\mathbf{Q}}(x, t_0)\Big)\frac{\partial \mathbf{Q}(x, t_0; p_4)}{\partial p_4}\,dx\,. \tag{4.24}$$

The candidate flow for any value of the design parameter p_4 must satisfy the one-dimensional Euler equations (4.18).

In practice, one usually minimizes a *discrete approximation* to the cost functional. For example, the discrete functional

$$\mathcal{J}^h(\mathbf{Q}^h) = \frac{1}{2}\sum_{j=0}^{J} \Delta x \left|\mathbf{Q}_j^h(t_0; p_4) - \widehat{\mathbf{Q}}(x_j, t_0)\right|^2 \tag{4.25}$$

is obtained from $\mathcal{J}$ defined in (4.23) by a simple composite one-point quadrature rule. For the discrete functional, Ω is subdivided into J intervals of equal length Δx; $\mathbf{Q}_j^h(t_0; p_4)$ denotes an approximation to the exact flow $\mathbf{Q}(x_j, t_0; p_4)$, where $x_j = j\Delta x$. The target flow $\widehat{\mathbf{Q}}(x_j, t_0)$ is assumed known and may or may not have to be approximated at the grid points. We then pose an optimization problem with respect to the discrete functional.

Problem 2. Find the initial left pressure p_4 such that the functional $\mathcal{J}^h(\mathbf{Q}^h)$ of (4.25) is minimized, where $\mathbf{Q}^h(x, t; p_4)$ is a solution of an approximation to the flow equations.

The approximation $\mathbf{Q}^h(x, t; p_4)$ can be determined by one's favorite CFD algorithm. The gradient of the discrete functional (4.25) with respect to the design parameter p_4 is given by

$$\frac{D\mathcal{J}^h}{Dp_4} = \sum_{j=0}^{J} \Delta x \left(\mathbf{Q}_j^h(t_0; p_4) - \widehat{\mathbf{Q}}(x_j, t_0)\right) \frac{\partial}{\partial p_4} \mathbf{Q}_j^h(t_0; p_4) \,. \tag{4.26}$$

If the sensitivity $\partial \mathbf{Q}_j^h(t_0; p_4)/\partial p_4$ of the approximate flow solution $\mathbf{Q}_j^h(t_0; p_4)$ is determined exactly, e.g., by using automatic differentiation, then the gradient (4.26) of the discrete cost functional (4.25) is also determined exactly. In this sense, such an approach leads to a "consistent gradient," i.e., (4.26) is the true gradient of the discrete functional. In fact, if one takes the view, as is often done, that Problem 2 is the optimization problem one really wants to solve, then one can obtain exactly the gradient one really wants. Thus, if the exact sensitivity of the approximate flow is available, e.g., by automatic differentiation, then, up to round-off error, Problem 2 can be solved exactly; of course, Problem 1 is not solved exactly by such an approach. The consistency of the gradient in the sense discussed here is viewed as an important advantage of the discretize-then-differentiate approach; see Section 2.9.

If one instead takes the differentiate-then-discretize approach, then one does not obtain an exact gradient of either the continuous functional (4.23) or the discretized functional (4.25). To see this, first consider the exact gradient (4.24) of the continuous functional (4.23). It can be approximated by

$$\left(\frac{D\mathcal{J}}{Dp_4}\right)^h = \sum_{j=0}^{J} \Delta x \left(\mathbf{Q}_j^h(t_0; p_4) - \widehat{\mathbf{Q}}(x_j, t_0)\right) \left(\frac{\partial \mathbf{Q}}{\partial p_4}\right)_j^h (t_0; p_4) \,, \tag{4.27}$$

where $(\partial \mathbf{Q}/\partial p_4)_j^h$ denotes an approximation to the exact sensitivity $(\partial \mathbf{Q}/\partial p_4)$ at the grid point x_j, which may be determined from a discretization of the sensitivity equation (4.21). A functional gradient determined by the differentiate-then-discretize approach is not the true gradient of anything. It is not the true gradient of the continuous functional (4.23) nor the discretized functional (4.25). It is possible that this can cause difficulties due to inconsistent gradients in the solution of Problem 2; however, should they arise, they are usually not difficult to overcome; see, e.g., Section 4.1.5.

Note that both (4.26) and (4.27) are approximations to the true gradient (4.24) and, if adequate discretization schemes are used, one expects that, as the grid size tends to zero, both the former converge to the latter.

Observations about optimization experiments

Optimization experiments were performed using the finite difference quotient, the automatic differentiation, and the sensitivity-equation methods for obtaining sensitivities, and also using the exact sensitivities, without paying any attention to the presence of the δ-functions or the fact that for discontinuous flows, the functional is not truly differentiable. Optimization was effected using a BFGS/trust region algorithm [382]. The performance of all four methods for finding sensitivities is about the same, with the exact sensitivities seeming to do a little better.

For nonattainable target flows, the optimal solution for p_4 is not equal to the value used to generate the target flow and is slightly different for each method of finding sensitivities. The results of these experiments and numerous other optimization studies of more practical problems indicate that *the inaccurate sensitivities obtained by ignoring the discontinuities in the flow do not seriously affect the performance of optimization methods.* It seems that the *inaccurate sensitivities are usually good enough to compute a gradient of the functional that points in the right direction*, i.e., the negative of the computed gradient is a descent direction.

Actually, sharp spikes and even smeared-out spikes have a beneficial effect on the optimization process, at least in the initial stages. The spikes indicate large sensitivities near a shock wave or other discontinuities in the flow. As a result, the optimizer quickly places the shock in the correct position, typically in just one or two iterations. Subsequently, the optimizer works hard to get the continuous parts of the flow right. In this second stage, the presence of spikes, even sharp ones, may slow down the convergence of the optimizer.

A possible hybrid method is to use sensitivities with large spikes for a few (no more that two or three are usually needed) iterations of the optimizer to get the shock waves in the correct positions and then lop off the spikes for subsequent iterations. One way to effect the removal of the spikes is to set some approximate flow variable $\phi_k(\mathbf{x}_i)$ to some limited value by a predetermined algorithm whenever both $|\phi_k(\mathbf{x}_i)|$ and $|\phi_k(\mathbf{x}_i) - \phi_k(\mathbf{x}_j)|$ are "large," where $\phi_k(\mathbf{x}_i)$ denotes a sensitivity at the grid point $\mathbf{x}_i$ and $\phi_k(\mathbf{x}_j)$ denotes the sensitivity at a neighboring grid point $\mathbf{x}_j$.

4.4 Inaccuracies at the boundaries

Recall from (3.8) the no-penetration condition for the sensitivity of Euler flows with respect to a shape parameter:

$$\begin{gathered} -(m_1)_k \frac{\partial \Psi}{\partial x} + (m_2)_k = \Big(\frac{\partial m_1}{\partial y} \frac{\partial \Psi}{\partial x} - \frac{\partial m_2}{\partial y} \Big) \Psi_k + m_1 \frac{\partial \Psi_k}{\partial x} \,, \\ k = 1, \dots, K \,, \quad \text{on } y = \Psi(x; \alpha_1, \dots, \alpha_K), \end{gathered} \tag{4.28}$$

where $y = \Psi(x; \alpha_1, \dots, \alpha_K)$ denotes the unknown portion of the boundary. Also, recall from (3.21) the optimality condition for a boundary control problem for time dependent viscous flows:

$$-\gamma_1 \partial_{tt} \mathbf{g} - \gamma_2 \Delta_s \mathbf{g} + \mathbf{g} + q\mathbf{n} = \frac{1}{\gamma} \big(-r\mathbf{n} + \nabla \mathbf{w} \cdot \mathbf{n} \big) \big|_{\Gamma_c} \quad \text{in } (0, T) \times \Gamma_c \,, \tag{4.29}$$

where $\mathbf{w}$ and r denote the adjoint velocity and pressure, respectively. These are but two examples of *the appearance of derivatives of the state or adjoint variables evaluated at the boundary*, something that is typical of many flow control and optimization problems. In (4.28), we have $\partial m_1/\partial y$ and $\partial m_2/\partial x$ evaluated on the obstacle $y = \Psi$, while in (4.29), we have $\nabla \mathbf{w} \cdot \mathbf{n}$ evaluated on the control part of the boundary Γ_c. In a discrete setting, both the steps of taking derivatives and then restricting them to the boundary can *introduce inaccuracies.* We explore and illustrate this issue in more detail in the context of a very simple example involving a linear ordinary differential equation.

The state system is given by

$$\begin{cases} -\dfrac{d^2\phi}{dx^2} + c\phi = f \quad \text{on } (0, L), \\ \phi(0) = a \quad \text{and} \quad \phi(L) = 0\,, \end{cases} \tag{4.30}$$

where $c > 0$ and f are given functions; a and L are to be determined through the optimization process, i.e., *a and L are the controls or design parameters.* We see that a is a *value control* since it appears as right-hand-side data in the state system and L is a *shape control* since it determines the "shape," in this case, the length, of the domain. It is useful to view the state ϕ, which is a solution of the state system, explicitly as a function of three variables, i.e., $\phi = \phi(x; a, L)$.

The optimization problem is to find a, L, and $\phi(x; a, L)$ such that

$$\mathcal{J}(\phi) = \frac{1}{2}\int_0^L (\phi - \Phi)^2\, dx$$

is minimized and such that the state system (4.30) is satisfied, where $\Phi(x)$ is a given function.

The corresponding adjoint system is given by

$$\begin{cases} -\dfrac{d^2\lambda}{dx^2} + c\lambda = (\phi - \Phi) \quad \text{on } (0, L), \\ \lambda(0) = 0 \quad \text{and} \quad \lambda(L) = 0\,. \end{cases} \tag{4.31}$$

Given values for the design parameters a and L, the solution λ of the adjoint system (4.31) as well as the solution ϕ of the state system (4.30) can be used to evaluate the components of the gradient of the functional with respect to the design parameters

$$\frac{D\mathcal{J}}{Da} = \frac{d\lambda}{dx}(0) \quad \text{and} \quad \frac{D\mathcal{J}}{DL} = \frac{1}{2}\Big(\phi(L) - \Phi(L)\Big)^2 + \frac{d\lambda}{dx}(L)\frac{d\phi}{dx}(L)\,. \tag{4.32}$$

The components of the gradient of the functional may be used in a gradient-based optimization algorithm to update the values of the design parameters.

One can also determine the components of the gradient of the functional in terms of the sensitivities

$$\phi_a = \frac{\partial\phi}{\partial a} \quad \text{and} \quad \phi_L = \frac{\partial\phi}{\partial L}\,.$$

These may be determined by solving the sensitivity equations

$$\begin{cases} -\dfrac{d^2\phi_a}{dx^2} + c\phi_a = 0 \quad \text{on } (0, L), \\ \phi_a(0) = 1, \quad \phi_a(L) = 0 \end{cases} \tag{4.33}$$

and[55]

$$\begin{cases} -\dfrac{d^2\phi_L}{dx^2} + c\phi_L = 0 \quad \text{on} \quad (0, L), \\ \phi_L(0) = 0, \qquad \phi_L(L) = -\dfrac{d\phi}{dx}\Big|_{x=L} . \end{cases} \tag{4.34}$$

Then, the components of the gradient of the functional in terms of the sensitivities are given by

$$\begin{aligned} \frac{D\mathcal{J}}{Da} &= \int_0^L (\phi - \Phi)\phi_a \, dx, \\ \frac{D\mathcal{J}}{DL} &= \frac{1}{2}\Big(\phi(L) - \Phi(L)\Big)^2 + \int_0^L (\phi - \Phi)\phi_L \, dx . \end{aligned} \tag{4.35}$$

Note that

- the adjoint equations (4.31) for the *adjoint variable* λ do not involve derivatives of the state ϕ evaluated at the boundary;
- the sensitivity equations (4.33) for the *sensitivity with respect to a value control* ϕ_a do not involve derivatives of the state ϕ evaluated at the boundary;
- the sensitivity equations (4.33) for the *sensitivity with respect to a shape control* ϕ_L explicitly involve $d\phi/dx$ evaluated at $x = L$, i.e., the derivative of the state evaluated at the boundary.

On the other hand, note that

- from (4.35), neither component of the gradient of the functional written in terms of sensitivities involves the derivatives of the state or the sensitivities evaluated at the boundary;
- from (4.32), the derivative of the functional with respect to the shape parameter L written in terms of the adjoint variable explicitly involves $d\phi/dx$ evaluated at $x = L$, i.e., the derivative of the state evaluated at the boundary;
- from (4.32), both components of the gradient of the functional written in terms of the adjoint variable explicitly involve $d\lambda/dx$ evaluated at the boundary, i.e., the derivative of the adjoint variable evaluated at the boundary.

For more general problems in higher dimensions, the relevant derivatives that appear as data in the boundary conditions for the sensitivities with respect to shape parameters or that appear in adjoint-based formulas for the components of the gradient of the functional are the *natural boundary conditions* for the constraint equations, e.g., for the Laplace operator, they would be normal derivatives on the boundary and, for the Navier–Stokes operator, they would be the components of the stress force on the boundary.

At the discretized level, errors can be introduced in two ways when approximating derivatives along boundaries:

---

[55]The last boundary condition in (4.34) follows since there are two dependencies on L that should be accounted for in the boundary condition $\phi(L) = 0$ for the state. To make this clear, we write the state as $\phi(x; L)$ to indicate that if the design parameter L changes, then so does the state ϕ. Then, the boundary condition for the state takes the form $\phi(x = L; L) = 0$, which clearly shows the two dependencies on L. If we differentiate this boundary condition, we obtain

$$\frac{d\phi}{dx}\Big|_{x=L} + \frac{\partial\phi}{\partial L} = 0,$$

from which the boundary condition at $x = L$ in (4.34) follows.

- Errors may be introduced when derivatives are not as accurately approximated as are the functions, e.g., one usually loses one order of approximation when differencing a discrete function;
- errors may be introduced when evaluating derivatives along the boundary, e.g., if one uses one-sided differences.

As a result, *functional gradients obtained using adjoint variables or sensitivities may be compromised* for shape controls in both cases and also for value controls in the adjoint case.

There are some possible solutions to the inaccuracy problem for boundary conditions in sensitivity and adjoint systems. One could, for example, *refine the grid* or use *higher-order discretizations* in the vicinity of the offending portion of the boundary, e.g., $x = L$ in our example. Adaptive grid strategies seem to be especially useful in this context; see, e.g., [219, 220, 218, 273, 285, 370]. Grid refinement, grid adaptation, or higher-order discretization for the adjoint or sensitivity equations means that one does not solve for them in the same way one solves for the flow. As a result, writing codes for the adjoint or sensitivity equations requires different grids. One cannot, for example, use automatic differentiation to get an adjoint or sensitivity code that uses different grids or discretization strategies specially suited for those variables; in that case, one implicitly accepts the grid and discretization methods used for the flow when calculating sensitivities or adjoints.

In many practical optimization calculations, the difficulty with inaccurate boundary data is completely ignored and yet the calculations turn out to be successful. The key to the *convergence* of a gradient-based optimization method is to make sure that one is really going downhill when one travels in the direction (in parameter space) opposite to that of the computed gradient of the functional. One does not need fabulously accurate gradients in an optimization setting; one just needs the negative of the computed gradient to point in a direction for which the functional is truly decreasing. It seems that in many practical optimization calculations this is indeed the case. However, the inaccuracies in the gradient of the functional could slow down the convergence of the optimization iteration. Of course, if one is interested in sensitivities in their own right, then the issue of inaccurate boundary conditions for sensitivities becomes more important.

Chapter 5

Reducing the Costs of Optimization and Control Calculations

Due to the high costs of the computations, especially for three-dimensional problems, one cannot routinely solve flow control and optimization problems. There have been numerous ways suggested to make flow control and optimization algorithms more efficient with respect to CPU and storage costs. Many cost reduction schemes have been proposed, tested, and sometimes actually employed to solve specific yet useful flow control and optimization problems; however, very few of them are rooted in firm mathematical ground or have otherwise been shown to be of general use.

Storage costs are obviously higher for flow control and optimization problems than they are for simulation problems. More fields have to be stored, i.e., in addition to the state variables, one has sensitivity or adjoint variables to keep. If the number of design parameters is high, the number of sensitivities that have to be stored is also high. However, by far the situation that requires the greatest storage is encountered in adjoint-based optimization algorithms for time-dependent optimal control problems; indeed, in this case, straightforward implementations result in storage costs that are prohibitively high. In Section 5.1, we examine some storage savings schemes for this situation.

CPU costs, even for stationary problems, are much too high for both sensitivity- and adjoint-based approaches to optimization. Each iteration of an optimization algorithm requires at least one (nonlinear) flow solve; *it is the flow solves that usually dominate the CPU costs in flow optimization and control calculations.* One can try to cut down on CPU costs by reducing the number of iterations in the optimization process, reducing the cost of the individual flow solves, or a combination of the two. In principle, improving the optimization algorithms used, e.g., using a trust region/quasi-Newton method instead of a simple descent method, would likely help cut down on the number of iterations needed to obtain optimal solutions. Optimization algorithms are well studied and their properties well known so that we do not discuss them further here other than to say that there are a large variety of methods available. See, e.g., [367, 368, 371, 372, 373, 379, 382, 383, 390, 393, 394, 395, 396, 397, 402, 403, 411, 412].

Many ideas have been advanced for reducing the cost of the flow solves necessary at each iteration of an optimization algorithm; these include using reduced-order models for the flow, regularizing the objective functional, following nonfeasible paths to the optimum, parallel computing, and fixed grid, e.g., fictitious domain, methods for shape optimization. In Section 5.2, we examine the use of reduced-order methods for reducing CPU costs of flow solves. Below we also briefly discuss the CPU cost savings that can be realized by regularizing the objective functional or by following nonfeasible paths to the optimum. Of course, parallel computing algorithms, if they do well for simulation problems, have the potential to help reduce the costs of solving optimization problems as well; also, fixed grid methods for shape control problems have the potential to remove the need to regrid after each iteration of the optimizer, which obviously would reduce costs; see, e.g., [187, 391].

We note that usually a price has to be paid for most schemes used to reduce the CPU costs associated with flow control and optimization problems; one may have to give up some accuracy or settle for only approximate optimality.

Regularization of the cost functional

In Sections 4.1.4 and 4.2.2, we twice encountered situations in which we regularized the cost functional. What we saw was that functionals to be minimized are not "well behaved"; this happens in practice because, e.g., functionals may not be convex, may depend very weakly on some of the design variables, or may depend on discontinuous flow variables. As we saw, in some cases discretized functionals may exhibit spurious numerical local minima; these can slow down the convergence to the desired minima or, more likely, prevent such convergence. In such cases, regularization of the objective functional, e.g., by adding penalty terms, can improve the performance of the optimizer, at least at the early stages of the optimization.

Of course, regularization changes the position of the minima. The penalty parameter can be reduced in size or it can be eliminated once one is "near" the minimum so that one can converge to the true minimum.

This type of cost reduction technique may effect a reduction in the number of optimization iterations needed since one works, at least at the beginning, with an "easier" or "nicer" functional; however, it may result in inaccuracies and requires the choice of a number of parameters, e.g., the penalty parameters and the number of iterations after which the values of the penalty parameters are lowered.

Feasibility and optimality

A *feasible iterate* in an optimization process is one that satisfies the constraints, i.e., the nonlinear flow equations. Most implementations of optimization algorithms expect that a feasible iterate is obtained at each step of the optimization process. This means that at each step of the optimization method, one obtains a fully converged flow solution as well as sensitivity or adjoint solutions corresponding to the current values of the controls or design parameters.

One can go to the other extreme as well. For example, one can use some linearization for the flow equations, e.g., by Newton's method, and define and solve an optimal control or optimization problem for the linearized flow equations acting as constraints. Then, one can update the linearization method to obtain a new set of linearized equations for which one repeats the optimization process. In this way, one follows an optimal path to the optimal solution of the nonlinear problem. At each step, one has an optimal solution for the wrong problem, i.e., for the linearized state equations, so that the iterates do not satisfy the real, nonlinear constraint equations.

Instead, when one is far from the optimum, one can opt to obtain a "poor" flow solution, which we hope is "good enough" for the optimizer to use to perform a single step toward the minimum. For example, if one is using Newton's method in the flow solver, one can perform one or two Newton iterations instead of the many Newton steps that may be required to get a fully converged flow solution; one can possibly also use a coarser grid to obtain approximate flow solutions when one is far from the optimum. As one approaches the optimum, one can improve the quality of the flow solution. In this way, one approaches the optimal solution on a path on which the iterates are neither feasible, i.e., they do not satisfy the flow equations, nor optimal, i.e., they are not optimal solutions of any problem. This type of cost reduction technique effects a reduction in the cost of each optimization iteration; however, it may result in the need for more (or fewer) such iterations or it may even prevent an optimizer from converging. However, when it works, it can effect a substantial cost reduction. See [395] for an example of this approach.

5.1 Storage saving schemes for time-dependent adjoint calculations

In the time-dependent setting, optimization methods that use adjoint equation solutions to determine the gradient of the objective functional require forward-in-time solutions of the state equations and backward-in-time solutions of the adjoint equations. The state variables appear in the coefficients and right-hand sides of the adjoint equations and have to be available to the adjoint equation solver as it marches backward in time. This means that one must store the state variables for every time step of the discretization. This is a huge amount of information to store; for example, in three dimensions, if one has four scalar fields (the velocity vector and pressure), 10^m grid points, and 10^n time steps, one has to store a total of $4 \times 10^{m+n}$ numbers. This motivates one to look for ways to reduce the storage requirements for time-dependent adjoint-based optimization solvers.

Frozen coefficient and right-hand-side scheme

Let $[0, T]$ denote the time interval over which optimization is to take place. Choose time values $0 = T_0 < T_1 < T_2 < \cdots < T_K = T$; this partitioning is much coarser than that used in the time-stepping algorithm for solving the flow or adjoint equations. During the forward-in-time state solve, one stores the state variables needed for the adjoint computation only at the times T_k, $k = 0, 1, \ldots, K$. See Figure 5.1 for a sketch of the relation between the time step for the flow and adjoint solutions and for flow solution storage.

During the adjoint computation over the time interval $[T_{k-1}, T_k]$, one uses the state variables corresponding to T_k or the average of the state variables at T_{k-1} and T_k as constant (in time) coefficients and in the right-hand sides of the adjoint equations; one can even use the linear interpolant (in time) of the state variables at T_{k-1} and T_k. Thus, instead of having to store the state variables at each time step of the forward-in-time state solve, one stores the state at only a "few" selected times. This approach has "worked" in some specific computations, e.g., the computational example of Section 3.4, but its generality is open to question.

Figure 5.1. *Time steps for the flow and adjoint solutions and the large time steps for the storage of the flow solution.*

Recomputing-the-state scheme

We begin as in the first storage saving algorithm. Let $[0, T]$ denote the time interval over which optimization is to take place. Then, choose time values $0 = T_0 < T_1 < T_2 < \cdots < T_K = T$; this partitioning is much coarser than that used in the time-stepping algorithm. Then, during the forward-in-time state solve, store the state variables needed for the adjoint computation only at the times T_k, $k = 0, 1, \ldots, K$.

At the start of the adjoint computation over the time interval $[T_{k-1}, T_k]$, one recomputes and stores the state over that time interval, using the previously stored state at T_{k-1} as an initial condition. Thus, one needs to store the state at only a "few" selected times and store the complete state over the interval $[T_{k-1}, T_k]$ over which the state is being recomputed.

This approach obviously yields the same answer as that for the basic method for which one stores the complete state solution at all time steps of the flow and adjoint solvers; however, there is the additional cost of a second state solve, i.e., essentially a doubling of the CPU requirements.

A locally optimal scheme

Let $[0, T]$ denote the time interval over which optimization is to take place. Choose time values $0 = T_0 < T_1 < T_2 < \cdots < T_K = T$; this partitioning may be chosen so that each subinterval contains one, few, or many time steps of the time-stepping scheme used for the state solution. We then pose and solve the optimal control problem sequentially over the subintervals $[T_k, T_{k+1}]$, $k = 0, \ldots, K-1$, using the solution from the previous subinterval $[T_{k-1}, T_k]$ evaluated at T_k as an initial condition.

For example, if the optimal control problem is

minimize the functional

$$\mathcal{J}(\mathbf{u}) = \frac{1}{2}\int_0^T \int_\Omega |\mathbf{u} - \mathbf{U}|^2 \, d\Omega dt$$

subject to the Navier–Stokes system holding over $(0, T) \times \Omega$ *with initial condition* $\mathbf{u}(0, \cdot) = \mathbf{u}_0(\cdot)$,

one would instead sequentially solve the K problems

for $k = 0, 1, \ldots, K-1$, minimize the functional

$$\mathcal{J}_k(\mathbf{u}_{k+1}) = \frac{1}{2}\int_{T_k}^{T_{k+1}}\int_\Omega |\mathbf{u}_{k+1} - \mathbf{U}|^2\, d\Omega dt$$

subject to the Navier–Stokes system holding over $(T_k, T_{k+1}) \times \Omega$ with initial condition

$$\mathbf{u}_{k+1}(T_k, \cdot) = \mathbf{u}_k(T_k, \cdot)\,,$$

where $\mathbf{u}_{k+1}$ denotes the solution over the interval $[T_k, T_{k+1}]$.

Clearly, $\mathbf{u}_{k+1} \neq \mathbf{u}$ on $(T_k, T_{k+1}) \times \Omega$, where $\mathbf{u}$ denotes the solution of the original optimization problem and thus, of course, $\mathbf{u}_k$ is *suboptimal* with respect to the original optimization problem. Note that, in general, the adjoint variables will be discontinuous at the break times T_k. On the other hand, *each of the subinterval problems is small* compared to the original problem, i.e., they involve relatively few time steps, so that storage costs are much lower.

If one chooses the time partitioning T_k, $k = 0, \ldots, K$, to correspond to that of the time-stepping algorithm for the flow and adjoint solvers, i.e., each subinterval consists of a single time step for that algorithm, one then essentially solves a sequence of stationary problems.

This method has been shown (computationally and analytically) to work well for low Reynolds number problems and tracking functionals such as the one just considered; see, e.g., [129]. On the other hand, it cannot possibly work well for all problems; for example, consider the case of the "controllability" functional

$$\mathcal{K}(\mathbf{u}) = \frac{1}{2}\int_\Omega |\mathbf{u}(T, \cdot) - \mathbf{U}(\cdot)|^2\, d\Omega\,,$$

i.e., a functional involving only the solution at the terminal time T. Clearly, such a functional will have no influence on the first $K-1$ subinterval optimization solutions so that, in a local-in-time optimization scheme, all control is effected in the last subinterval (T_{K-1}, T_K).

Iterated sequential-in-time methods

The sequential-in-time suboptimal control scheme can be thought of as the first step of an iterative method. Following the application of the forward sequential-in-time method, one can solve the subinterval optimal control problems backward in time through the time intervals, using the results from previous subintervals to define terminal conditions for the adjoint variables. The forward/backward subinterval solves can then be repeated. A number of similar approaches have been examined and found to be effective; see, e.g., group [C] in the bibliography.

5.2 Reduced-order methods

Recently, much attention has been paid to reducing the costs of the nonlinear flow solutions by using reduced-order models for the flow. These are low-dimensional approximations to the flow, where the low dimensionality is relative to, e.g., a high-fidelity finite element flow approximation.

For a flow simulation, a reduced-order method would proceed as follows. One chooses a *reduced basis* ϕ_i, $i = 1, \dots, \widetilde{M}$, where $\widetilde{M}$ we hope is very small compared to the usual number of functions used in a finite element approximation or the number of grid points used in a finite difference approximation. Next, one defines an approximation to the flow of the form

$$\widetilde{\phi} = \sum_{i=1}^{\widetilde{M}} c_i \phi_i$$

so that $\widetilde{\phi} \in V = \text{span}\{\phi_1, \dots, \phi_{\widetilde{M}}\}$. Then, one determines the coefficients c_i, $i = 1, \dots, \widetilde{M}$, by solving the flow equations in the set V, e.g., one could find a Galerkin solution of the flow equations in a standard way, using V for the space of approximations. The cost of such a computation is very small if $\widetilde{M}$ is small. It has been shown through computational experimentation that some reduced basis techniques can be effective in obtaining good flow solutions cheaply; see, e.g., [442].

In a flow control or optimization setting, one is faced with multiple flow solves. If one uses standard discretization techniques, e.g., finite element or finite volume methods, by far this represents the dominant part of the computational costs. If one approximates the flow in the reduced, low-dimensional set V, then each flow solution will be relatively very cheap. In an adjoint-equation–based optimization method, one would also determine the adjoint equations for the low-dimensional discrete flow equations. Thus, if $\widetilde{M}$ is small, the cost of each iteration of the optimizer would be very small relative to that using full, high-fidelity flow solutions.

Two big questions need to be answered:

- How does one determine the reduced basis or spanning set $\{\phi_1, \dots, \phi_{\widetilde{M}}\}$?
- Do reduced-order flow approximations work in an optimization setting?

Let us address the second question. Some impressive optimal (and feedback) flow calculations have been performed using reduced-order flow solution methods; see, e.g., group [H] in the bibliography. It is clear that reduced-order methods should work in an *interpolatory setting*. If the optimal solution and the path to the optimal solution can be well approximated in the reduced set V, then one should expect that things will work well. Thus, the reduced basis spanning V should contain all the features, e.g., the dynamics, of the optimal flow solution as well as of the intermediate flow solutions obtained during the optimization iteration.

What happens in an *extrapolatory setting*, i.e., when the reduced basis does not approximate the optimal solution well, is not so clear. Most of the reduced-order optimization computations that have been done so far are in the interpolatory regime. It does seem clear that if the reduced set V does not contain a good approximation to the optimal solution, then one cannot hope to successfully determine that solution. However, there is some hope that reduced-order techniques can be used in the extrapolatory setting; for example, in the structural mechanics setting, it has been shown how to use some reduced-basis techniques to pass through bifurcation points in a solution path; see, e.g., [438, 439, 440]. Certainly, in the extrapolatory setting, some updating of the basis is necessary.

There is no adequate theoretical foundation for reduced-order methods, even in the flow simulation setting. Certainly, these methods deserve more study from computational and theoretical points of view; without an inexpensive method for reducing the cost of flow computations, it is unlikely that the solution of three-dimensional optimization and control problems involving the Navier–Stokes system will become routine any time soon.

With regard to the first question, we point out that a number of reduced basis sets have been proposed.

- *Lagrange bases* consist of flow solutions corresponding to several different values of the parameters (Reynolds number, design parameters, etc.). These solutions are obtained by standard (and expensive) techniques such as finite element or finite volume methods. Thus, for example, if one has the design parameters $\{\alpha_k\}_{k=1}^K$, one obtains $\widetilde{M}$ approximate flow solutions for $\widetilde{M}$ sets of parameter values to form the $\widetilde{M}$-dimensional Lagrange reduced basis.
- *Hermite bases* consist of the flow variables and the first derivatives of the flow variables with respect to parameters (the sensitivities) determined for different values of the parameters. The flow and sensitivity approximations are obtained through standard (and expensive) techniques such as finite element or finite volume methods. Thus, again, if one has the design parameters $\{\alpha_k\}_{k=1}^K$, one chooses m sets of parameter values and then obtains the corresponding m approximate flow solutions and the corresponding mK sensitivity derivative approximations. The $\widetilde{M} = m(K+1)$ flow and sensitivity approximations form the $\widetilde{M}$-dimensional Hermite reduced basis.
- *Taylor bases* consist of the flow variables and derivatives of the flow variables with respect to parameters (sensitivities and higher-order sensitivities) determined for a fixed set of design parameters. The flow and derivative approximations are obtained through standard (and expensive) techniques such as finite element or finite volume methods. The Taylor basis may be somewhat complicated to program due to the complexity of the partial differential equations that determine the higher-order sensitivities. In addition, the number of higher-order derivatives grows very rapidly with the number of design parameters. For example, if one has 10 design parameters, there are 55 different second derivative sensitivities. Thus, the dimension of the Taylor reduced basis grows quickly with the number of parameters and the number of derivatives used.
- *Snapshot bases* consist of a flow solution for a fixed set of design parameter values (or several flow solutions corresponding to different sets of parameter values) evaluated at several time instants during the evolution process. Snapshot bases are usually determined by solving the full discretized system obtained by, e.g., a finite volume or finite element discretization. In principle, one can even use experimental data to determine a snapshot basis.

Three observations are important to note. First, we again emphasize that, for any of the reduced bases, one should make sure that they can be used to approximate well the portion of parameter space in which one expects to find the optimal parameters. This, of course, requires some intuition about where in parameter space the optimal set of parameters are located.

Second, we note that all the reduced bases require the solution of high-fidelity and therefore very expensive discrete flow and/or sensitivity equations. The idea is that these expensive calculations can be done off line before the optimization of the design parameters or controls is attempted. Moreover, one hopes that a single reduced basis can be used in several design settings.

Finally, we note that all the basis sets are global in nature, i.e., the support of the basis functions is global, so that solving the flow or sensitivity or adjoint equations with respect to any of the reduced bases requires the solution of dense linear and nonlinear systems. Thus, unless the dimension of the reduced basis is "small," it cannot be used without some further processing. Unfortunately, it is often the case that, in order to obtain meaningful approximations, the use of the above reduced basis sets requires the use of a relatively large number of basis functions. However, the reduced basis sets often contain "redundant" information in the sense that the dynamics of the flow should be approximated well by a set of functions of much lower dimension. The question then arises of how one can extract a smaller reduced basis set that contains all the essential information of the larger reduced basis set. One approach to this end is discussed in the next section.

5.2.1 Proper orthogonal decomposition

The proper orthogonal decomposition (POD), or the Karhunen–Loève, basis is usually used with snapshot bases but can be used with any reduced basis. It is in some sense an optimal basis and, it seems, is becoming the reduced-order method of choice for control and optimization problems.

Generating the POD basis

Given a snapshot or other basis, the POD basis is determined in the following manner.[56]

1. Starting with any reduced basis $\{\phi_1, \ldots, \phi_{\widetilde{M}}\}$, e.g., a snapshot basis, where $\widetilde{M}$ is not necessarily very small (usually of order 100 or more), compute the mean

$$\overline{\phi} = \frac{1}{\widetilde{M}} \sum_{j=1}^{\widetilde{M}} \phi_j \,.$$

2. Build the $\widetilde{M} \times \widetilde{M}$ correlation matrix C, where

$$c_{ij} = \frac{1}{\widetilde{M}} (\phi_i - \overline{\phi})^T (\phi_j - \overline{\phi}) \,.$$

3. Compute the eigenvalues $\{\lambda_1, \ldots, \lambda_{\widetilde{M}}\}$ and the corresponding eigenvectors $\{v_1, \ldots, v_{\widetilde{M}}\}$ of C, ordering the eigenvalues by decreasing size.

4. Choose M so that

$$M = \text{minimum value of } m \text{ such that } \quad \frac{\sum_{j=1}^{m} \lambda_j}{\sum_{j=1}^{\widetilde{M}} \lambda_j} \geq \gamma \,,$$

where $0 \leq \gamma \leq 1$ is a chosen tolerance.

[56]We assume that the snapshot basis has been determined from, e.g., a finite element simulation (or perhaps a few simulations) of the flow. Thus, the snapshot basis is made up of vectors in $\mathbb{R}^N$, where $N \gg 1$ is the dimension of, e.g., the finite element space used in the simulations. In general, N is related to the number of grid points used in the high-fidelity flow simulation code.

5. Set

$$\Phi_i = \sum_{j=1}^{\widetilde{M}} (v_i)_j(\phi_j - \overline{\phi}) \qquad \text{for } i = 1, \ldots, M\,,$$

where $(v_i)_j$ denotes the jth component of the ith eigenvector.

6. Normalize according to

$$\Phi_i = \frac{\Phi_i}{|\Phi_i|}, \qquad \text{where} \qquad |\Phi_i|^2 = \Phi_i^T \Phi_i\,.$$

7. The set $\{\Phi_i, \ldots, \Phi_M\}$ is the POD basis.

The following observations about the above process are relevant. The matrix C is symmetric and positive semidefinite so that the eigenvalues $\{\lambda_1, \ldots, \lambda_{\widetilde{M}}\}$ are all real and nonnegative. The POD basis $\{\Phi_i, \ldots, \Phi_M\}$ is an orthonormal one, i.e.,

$$\Phi_i^T \Phi_j = \begin{cases} 0 & \text{if } i \neq j, \\ 1 & \text{if } i = j. \end{cases}$$

The POD basis is optimal in the following sense [425, 434]. Let $\{\Psi_i\}_{i=1}^{\widetilde{M}}$ denote an arbitrary orthonormal basis for the span of the modified snapshot set $\{\phi_j - \overline{\phi}\}_{j=1}^{\widetilde{M}}$. Let $P_{\Psi,M}(\phi_j - \overline{\phi})$ denote the projection of $(\phi_j - \overline{\phi})$ into the subspace spanned by $\{\Psi_i\}_{i=1}^{M}$ and let the *error* be defined by

$$\mathcal{E} = \sum_{j=1}^{\widetilde{M}} |(\phi_j - \overline{\phi}) - P_{\psi,M}(\phi_j - \overline{\phi})|^2\,. \tag{5.1}$$

Then, the minimum error is obtained when $\Psi_i = \Phi_i$ for $i = 1, \ldots, M$, i.e., when the Ψ_i's are the POD basis vectors.

The tolerance γ is chosen to be close to unity, e.g., $\gamma > 0.99$, so that the POD basis captures most of the "energy" in the snapshot basis. It is found (computationally) that, in many cases, the eigenvalues decay very rapidly so that even if γ is close to unity, $M \ll \widetilde{M}$, i.e., M is small; typically $M \approx 10$. If the eigenvalues do not decay rapidly, chances are the snapshot basis used to generate the POD basis is not useful anyway.

The key to the effectiveness of the POD basis is the generation of a "good" snapshot basis, which means the snapshot basis does a good job in capturing the dynamics of the flow. Generating good snapshot bases is still an art—and not a very well developed one at that.

Using the POD basis in control problems

The POD basis is determined *before* one starts to solve the optimal control problem by taking snapshots of a high-fidelity, e.g., large-size finite element, approximate state calculation for a given set of control parameters. One may want to include in the snapshot set snapshots taken from multiple state calculations using different sets of control parameters. One then discretizes, e.g., by a Galerkin method, the state equations, e.g., the Navier–Stokes equations, using the POD basis; this results in dense systems since the POD basis elements have global support, but the size of the problem is very small because M is presumably very small. If sensitivities or adjoint variables are needed, they are defined in terms of the POD-based discretization of the state equations so that the sensitivity or adjoint equations themselves form small discrete systems. *Throughout the optimization, only the small POD-based discrete state and adjoint or sensitivity systems are used*; this results in tremendous savings in CPU (and storage) costs when compared to the use of, say, a finite element discrete system within the optimization process.

Of course, the only way POD-based optimization is going to work is if the snapshot basis used to generate the POD basis contains enough information to model all the dynamical behavior of the flows that are encountered throughout the optimization process. For example, if the control parameters used to generate the snapshot basis are "close" to the optimal parameters, then there is hope that POD-based optimization is going to work. There have been numerous computational demonstrations (most in the incompressible, viscous flow setting and most for relatively simple problems) of the effectiveness of POD-based optimization; see, e.g., the papers listed in group [H] in the bibliography.

If, during the optimization process, the snapshot basis can no longer accurately approximate the dynamics of the flows encountered, then one must determine a new snapshot basis and a new resulting POD basis. For example, if one feels that the POD basis is no longer accurately approximating the state, one can again solve the large state system using the current values of the control parameters determined by the optimizer, take snapshots of that solution, add them to the snapshots previously determined, and recompute the POD basis before proceeding with the optimization process; see, e.g., [293, 310]. One may have to update the snapshot basis more than once. Of course, each update of the snapshot basis requires a large state solve and is expensive. Moreover, recomputing the POD basis requires the solution of an eigenvalue problem of a size equal to the number of snapshots, so this may become an expensive undertaking as well. However, the hope is that the great majority of optimization iterations can be done using the small POD-based systems, i.e., that snapshot basis updates have to be done infrequently relative to the total number of optimization iterations, so that substantial cost savings are still effected. Furthermore, there have been proposed new reduced-order methods based on centroidal Voronoi tessellations (CVTs). CVTs can also be combined with the POD process. These approaches have the promise of providing effective reduced-order bases that are cheaper to determine than the standard POD basis and may therefore be attractive in flow control and optimization settings; see [425]. Also note that since CVT or CVT–POD reduced bases are cheaply determined, one can also use much larger snapshot sets, which may also be beneficial in producing effective reduced bases.

Chapter 6

A Peek at the Analysis and Numerical Analysis of Optimal Flow Control Problems

In this chapter, we discuss some of the analytical results that are obtainable for optimal flow control problems and for their approximation by finite element methods. One reason for taking a brief look at analytical results is to reveal their value in the design of practical computational algorithms. The context for the three examples we consider is the Navier–Stokes system for incompressible, viscous flows. We do not provide any proofs; these may be found in the cited references. We also note that [7] provides an extensive and thorough study of the analysis of several flow control problems.

Each of Sections 6.1, 6.2, and 6.3 is notationally self-contained, i.e., all notation used in a section is defined within that section. Due to the different natures of the problems treated, there may be small conflicts in the notation used in the three sections.

6.1 Analysis and approximation of stationary optimal control problems

We consider an abstract framework for the analysis and approximation of a class of nonlinear optimal control and optimization problems. Proofs and other details may be found in [387]. Nonlinearities occur in both the objective functional and in the constraints. The framework includes an abstract nonlinear optimization problem posed on infinite-dimensional spaces and an approximate problem posed on finite-dimensional spaces, together with a number of hypotheses concerning the two problems. The abstract framework and the results derived from that framework can be applied to a large variety of optimal control and optimization problems; here we apply the framework, in Section 6.1.3, to a boundary control problem for the stationary Navier–Stokes equations for incompressible, viscous flows.

The major steps in the development and analysis of the framework are as follows:

- Define an abstract class of nonlinear control or optimization problems;
- show that, under certain assumptions, optimal solutions exist;
- show that, under certain additional assumptions, Lagrange multipliers exist that may be used to enforce the constraints;

- use the Lagrange multiplier technique to derive an optimality system from which optimal states and controls may be deduced;
- define algorithms for the approximation, in finite-dimensional spaces, of optimal states and controls;
- derive estimates for the error in the approximations to the optimal states and controls.

Two of the key ingredients used to carry out the above plan are a theory given in [410] for showing the existence of Lagrange multipliers and a theory first developed in [419] for the approximation of a class of nonlinear problems. In both of these theories, certain properties of compact operators on Banach spaces play a central role. We point out that the nonuniqueness of solutions of the nonlinear constraint equations deems it appropriate to employ Lagrange multiplier principles.

The abstract problem

We begin with the definition of the abstract class of nonlinear control or optimization problems that we study.

We introduce the spaces and control set as follows. Let G, X, and Y be reflexive Banach spaces whose norms are denoted by $\|\cdot\|_G$, $\|\cdot\|_X$, and $\|\cdot\|_Y$, respectively. Dual spaces will be denoted by $(\cdot)^*$. The duality pairing between X and X^* is denoted by $\langle\cdot,\cdot\rangle_X$; one similarly defines $\langle\cdot,\cdot\rangle_Y$ and $\langle\cdot,\cdot\rangle_G$. The subscripts are often omitted whenever there is no chance for confusion. Let Θ, the control set, be a closed convex subset of G. Let Z be a subspace of Y with a compact embedding. Note that the compactness of the embedding $Z \subset Y$ will play an important role.

We assume that the functional to be minimized takes the form

$$\mathcal{J}(v,z) = \lambda\,\mathcal{F}(v) + \lambda\,\mathcal{E}(z) \quad \forall\,(v,z) \in X \times \Theta\,, \tag{6.1}$$

where $\mathcal{F}$ is a functional on X, $\mathcal{E}$ is a functional on Θ, and λ is a given parameter that is assumed to belong to a compact interval $\Lambda \subset \mathbb{R}_+$.

The constraint equation $M(v,z) = 0$ relating the state variable v and the control variable z is defined as follows. Let N be a differentiable mapping from X to Y, K a continuous linear operator from Θ to Y, and T a continuous linear operator from Y to X. For any $\lambda \in \Lambda$, we define the mapping M from $X \times \Theta$ to X by

$$M(v,z) = v + \lambda\,TN(v) + \lambda\,TK(z) \quad \forall\,(v,z) \in X \times \Theta\,. \tag{6.2}$$

With these definitions we now consider the constrained minimization problem

$$\min_{(v,z)\in X\times\Theta} \mathcal{J}(v,z) \quad \text{subject to} \quad M(v,z) = 0\,. \tag{6.3}$$

In (6.3), we seek a global minimizer with respect to the set $\{(v,z) \in X \times \Theta : M(v,z) = 0\}$. Although, under suitable hypotheses, we will show that the problem (6.3) has a solution, in practice, one can only characterize local minima, i.e., points $(u,g) \in X \times \Theta$ such that, for some $\epsilon > 0$,

$$\begin{aligned} \mathcal{J}(u,g) \le \mathcal{J}(v,z) \quad \forall\,(v,z) \in X \times \Theta \text{ such that} \\ M(v,z) = 0 \text{ and } \|u - v\|_X \le \epsilon\,. \end{aligned} \tag{6.4}$$

Thus, when we consider algorithms for locating constrained minima of $\mathcal{J}$, we must be content to find local minima in the sense of (6.4).

After showing that optimal solutions exist and that one is justified in using the Lagrange multiplier rule, we will introduce some simplifications in order to render the abstract problem (6.3), or (6.4), more amenable to approximation. The first is to consider only the control set $\Theta = G$. The second is to consider only Fréchet differentiable functionals $\mathcal{E}(\cdot)$ such that their Fréchet derivative is of the form $\mathcal{E}'(g) = E^{-1}g$, where E is an invertible linear operator from G^* to G.

6.1.1 Analysis of an abstract nonlinear stationary control problem

Hypotheses concerning the abstract problem

The first set of hypotheses will be invoked to prove the existence of optimal solutions; they are given as follows.

H1. $\inf_{v\in X} \mathcal{F}(v) > -\infty$.

H2. *There exist constants* $\alpha, \beta > 0$ *such that* $\mathcal{E}(z) \geq \alpha \|z\|^\beta \quad \forall\, z \in \Theta$.

H3. *There exists a* $(v, z) \in X \times \Theta$ *satisfying* $M(v, z) = 0$.

H4. *If* $u^{(n)} \rightharpoonup u$ *in* X *and* $g^{(n)} \rightharpoonup g$ *in* G, *where* $\{(u^{(n)}, g^{(n)})\} \subset X \times \Theta$, *then* $N(u^{(n)}) \rightharpoonup N(u)$ *in* Y *and* $K(g^{(n)}) \rightharpoonup K(g)$ *in* Y.

H5. $\mathcal{J}(\cdot, \cdot)$ *is weakly lower semicontinuous on* $X \times \Theta$.

H6. *If* $\{(u^{(n)}, g^{(n)})\} \subset X \times \Theta$ *is such that* $\{\mathcal{F}(u^{(n)})\}$ *is a bounded set in* $\mathbb{R}$ *and* $M(u^{(n)}, g^{(n)}) = 0$, *then* $\{u^{(n)}\}$ *is a bounded set in* X.

The second set of assumptions will be used to justify the use of the Lagrange multiplier rule and to derive an optimality system from which optimal states and controls may be determined. The second set is given as follows.

H7. *For each* $z \in \Theta$, $v \mapsto \mathcal{J}(v, z)$ *and* $v \mapsto M(v, z)$ *are Fréchet differentiable.*

H8. $z \mapsto \mathcal{E}(z)$ *is convex, i.e.,*

$$\mathcal{E}\big(\gamma z_1 + (1-\gamma) z_2\big) \leq \gamma\, \mathcal{E}(z_1) + (1-\gamma)\, \mathcal{E}(z_2) \quad \forall\, z_1, z_2 \in \Theta\,, \ \forall\, \gamma \in [0, 1]\,.$$

H9. *For* $v \in X$, $N'(v)$ *maps* X *into* Z.

In (H9), N' denotes the Fréchet derivative of N.

A simplified optimality system may be obtained if one invokes the following additional assumption.

H10. $\Theta = G$ *and the mapping* $z \mapsto \mathcal{E}(z)$ *is Fréchet differentiable on* G.

Hypotheses (H7)–(H10) allow us to obtain a simplified optimality system for almost all values of the parameter $\lambda \in \Lambda$. In many cases, it is possible to show that the same optimality system holds for all values of λ. The following two additional assumptions, which will be invoked only in case $(1/\lambda)$ is an eigenvalue of $-TN'(u)$, each provides a setting in which this last result is valid.

H11. *If* $v^* \in X^*$ *satisfies* $\big(I + \lambda\,[N'(u)]^* T^*\big)v^* = 0$ *and* $K^*T^*v^* = 0$, *then* $v^* = 0$. *Or:*

H12. *The mapping* $(v, z) \mapsto v + \lambda\, TN'(u)v + \lambda\, TKz$ *is onto from* $X \times G$ *to* Y.

In order to make the optimality system more amenable to approximation and computation, we will invoke the following additional assumption.

H13. $\mathcal{E}'(g) = E^{-1}g$, *where* E *is an invertible linear operator from* G^* *to* G, *and* g *is an optimal control for the constrained minimization problem* (6.4).

Existence of an optimal solution

Assumptions (H1)–(H6) can be used to establish the existence of optimal solutions.

Theorem 6.1. *Assume that the functional* $\mathcal{J}$ *and mapping* M *defined by* (6.1) *and* (6.2), *respectively, satisfy the hypotheses* (H1)–(H6). *Then, there exists a solution to the minimization problem* (6.3).

Remark. The hypotheses (H1)–(H6) are not sufficient to guarantee that optimal solutions are unique. Indeed, in many applications to nonlinear problems, including optimal control problems for the Navier–Stokes system, optimal solutions in the sense of (6.4) are in general not uniquely determined.

Existence of Lagrange multipliers

The additional assumptions (H7)–(H9) can be used to show that the Lagrange multiplier rule may be used to turn the constrained minimization problem (6.4) into an unconstrained one. The result is based on the following useful abstract Lagrange multiplier rule whose proof can be found in [410].

Theorem 6.2. *Let* X_1 *and* X_2 *be two Banach spaces and* Θ *an arbitrary set. Suppose* $\mathcal{J}$ *is a functional on* $X_1 \times \Theta$ *and* M *a mapping from* $X_1 \times \Theta$ *to* X_2. *Assume that* $(u, g) \in X_1 \times \Theta$ *is a solution to the following constrained minimization problem:*

$$M(u, g) = 0 \text{ and there exists an } \epsilon > 0 \text{ such that } \mathcal{J}(u, g) \le \mathcal{J}(v, z) \text{ for all } (v, z) \in X_1 \times \Theta \text{ such that } \|u - v\|_{X_1} \le \epsilon \text{ and } M(v, z) = 0. \tag{6.5}$$

Let U *be an open neighborhood of* u *in* X_1. *Assume further that the following conditions are satisfied:*

For each $z \in \Theta$, $v \mapsto \mathcal{J}(v, z)$ *and* $v \mapsto M(v, z)$ *are Fréchet differentiable at* $v = u$; (6.6)

$$\begin{cases} \textit{for any } v \in U,\ z_1, z_2 \in \Theta, \textit{ and } \gamma \in [0,1], \textit{ there exists a } z_\gamma = z_\gamma(v, z_1, z_2) \\ \textit{such that} \\ \qquad M(v, z_\gamma) = \gamma M(v, z_1) + (1-\gamma) M(v, z_2) \\ \textit{and} \\ \qquad \mathcal{J}(v, z_\gamma) \le \gamma \mathcal{J}(v, z_1) + (1-\gamma)\mathcal{J}(v, z_2)\,; \end{cases} \tag{6.7}$$

and

$$\text{Range}\big(M_u(u,g)\big) \textit{ is closed with a finite codimension,} \tag{6.8}$$

where $M_u(u, g)$ denotes the Fréchet derivative of M with respect to u. Then, there exists a $k \in \mathbb{R}$ and a $\mu \in X_2^*$ that are not both equal to zero such that

$$k\,\langle \mathcal{J}_u(u,g), v\rangle - \langle \mu, M_u(u,g)v\rangle = 0 \quad \forall\, v \in X_1$$

and

$$\min_{z\in\Theta} \mathcal{L}(u, z, \mu, k) = \mathcal{L}(u, g, \mu, k)\,,$$

where $\mathcal{L}(u, g, \mu, k) = k\,\mathcal{J}(u, g) - \langle \mu, M(u, g)\rangle$ is the Lagrangian for the constrained minimization problem (6.5) *and where $\mathcal{J}_u(u, g)$ denotes the Fréchet derivative of $\mathcal{J}$ with respect to u. Moreover, if*

$$\textit{the algebraic sum } M_u(u,g)X_1 + M(u,\Theta) \textit{ contains } 0 \in X_2 \textit{ as an interior point,} \tag{6.9}$$

then we may choose $k = 1$, i.e., there exists a $\mu \in X_2^*$ such that

$$\langle \mathcal{J}_u(u,g), v\rangle - \langle \mu, M_u(u,g)v\rangle = 0 \quad \forall\, v \in X_1$$

and

$$\min_{z\in\Theta} \mathcal{L}(u, z, \mu, 1) = \mathcal{L}(u, g, \mu, 1)\,.$$

Theorem 6.2 can be applied to the optimization problem (6.4). Note that, in the following result, the existence of at least one pair (u, g) satisfying (6.4) is guaranteed by Theorem 6.1.

Theorem 6.3. *Let $\lambda \in \Lambda$ be given. Assume that* (H1)–(H9) *hold. Let $(u, g) \in X \times \Theta$ be an optimal solution satisfying* (6.4). *Then, there exists a $k \in \mathbb{R}$ and a $\mu \in X^*$ that are not both equal to zero such that*

$$k\,\langle \mathcal{J}_u(u,g), w\rangle - \langle \mu, M_u(u,g)\cdot w\rangle = 0 \quad \forall\, w \in X \tag{6.10}$$

and

$$\min_{z\in\Theta} \mathcal{L}(u, z, \mu, k) = \mathcal{L}(u, g, \mu, k)\,. \tag{6.11}$$

Furthermore, if $(1/\lambda) \notin \sigma\big(-TN'(u)\big)$, we may choose $k = 1$, i.e., for almost all λ, there exists a $\mu \in X^*$ such that

$$\langle \mathcal{J}_u(u,g), w\rangle - \langle \mu, M_u(\lambda, u, g)\cdot w\rangle = 0 \quad \forall\, w \in X \tag{6.12}$$

and

$$\min_{z\in\Theta} \mathcal{L}(u, z, \mu, 1) = \mathcal{L}(u, g, \mu, 1) \,. \tag{6.13}$$

So far, Θ has only been assumed to be a closed and convex subset of G. No smoothness condition on the control variable g has been assumed in the functional or in the constraint. Thus, the necessary condition of optimality with respect to variations in the control variable is expressed in the cumbersome relation (6.11). We now turn to the case where Θ contains a neighborhood of g, where (u, g) is an optimal solution. In particular, we assume that $\Theta = G$. In this case, (6.11) can be given a more concrete structure.

Theorem 6.4. *Let $\lambda \in \Lambda$ be given. Assume that* (H1)–(H10) *hold. Let $(u, g) \in X \times G$ be a solution of the problem* (6.4). *Then, there exists a $k \in \mathbb{R}$ and a $\mu \in X^*$ that are not both equal to zero such that*

$$k \, \langle \mathcal{J}_u(u, g), w\rangle - \langle \mu, \big(I + \lambda\, T N'(u)\big) w\rangle = 0 \quad \forall\, w \in X \tag{6.14}$$

and

$$k\langle \mathcal{E}'(g), z\rangle - \langle \mu, TKz\rangle = 0 \quad \forall\, z \in G \,. \tag{6.15}$$

Furthermore, if $(1/\lambda) \notin \sigma\big(-TN'(u)\big)$, we may choose $k = 1$, i.e., there exists a $\mu \in X^*$ such that

$$\langle \mathcal{J}_u(u, g), w\rangle - \langle \mu, \big(I + \lambda\, T N'(u)\big) w\rangle = 0 \quad \forall\, w \in X \tag{6.16}$$

and

$$\langle \mathcal{E}'(g), z\rangle - \langle \mu, TKz\rangle = 0 \quad \forall\, z \in G \tag{6.17}$$

hold.

Remark. If $k = 0$, then there exists a $\mu \neq 0$ such that

$$-\langle \mu, M_u(u, g) w\rangle = 0 \quad \forall\, w \in X$$

so that the optimality system necessarily has infinitely many solutions. In fact, for any $C \in \mathbb{R}$, $(C\mu)$ is a solution whenever μ is a solution. This creates both theoretical and numerical difficulties. Thus, it is of great interest to try to eliminate this situation. Fortunately, Theorems 6.3 and 6.4 tell us that we may set $k = 1 \neq 0$ for almost all values of $(1/\lambda)$, i.e., except for the at most countable set of values in $\sigma\big(-TN'(u)\big)$.

If the control g enters the constraint in a favorable manner, then we may take $k = 1$ even when $(1/\lambda) \in \sigma\big(-TN'(u)\big)$. Specifically, we invoke one of assumptions (H11) and (H12) to obtain the following result.

Theorem 6.5. *Assume that the hypotheses of Theorem* 6.4 *hold. Assume that if $(1/\lambda) \in \sigma\big(-TN'(u)\big)$, then either* (H11) *or* (H12) *holds. Then, for all $\lambda \in \Lambda$, there exists a $\mu \in X^*$ such that* (6.16) *and* (6.17) *hold.*

The optimality system

Under the assumptions of Theorem 6.5, an optimal state $u \in X$, an optimal control $g \in G$, and the corresponding Lagrange multiplier $\mu \in X^*$ satisfy the optimality system of equations formed by (6.2), (6.16), and (6.17). From (6.1), we have that $\mathcal{J}_u = \lambda \mathcal{F}'$ and $\mathcal{J}_g = \lambda \mathcal{E}'$, where $\mathcal{F}'$ denotes the obvious Fréchet derivative. Then, (6.16)–(6.17) may be rewritten in the form

$$\mu + \lambda\, [N'(u)]^* T^* \mu - \lambda\, \mathcal{F}'(u) = 0 \quad \text{in } X^* \tag{6.18}$$

and

$$\mathcal{E}'(g) - K^* T^* \mu = 0 \quad \text{in } G^* . \tag{6.19}$$

For purposes of numerical approximations, it turns out to be convenient to make the change of variable $\xi = T^*\mu$. Then, the optimality system (6.2), (6.18), and (6.19) for $u \in X$, $g \in G$, and $\xi \in Y^*$ takes the form

$$u + \lambda\, T N(u) + \lambda\, T K g = 0 \quad \text{in } X , \tag{6.20}$$

$$\xi + \lambda\, T^* [N'(u)]^* \xi - \lambda\, T^* \mathcal{F}'(u) = 0 \quad \text{in } Y^* , \tag{6.21}$$

and

$$\mathcal{E}'(g) - K^* \xi = 0 \quad \text{in } G^* . \tag{6.22}$$

It will also be convenient to invoke an additional simplifying assumption concerning the dependence of the objective functional on the control. Specifically, we assume that (H13) holds. Then, (6.20)–(6.22) can be rewritten as

$$u + \lambda\, T N(u) + \lambda\, T K g = 0 \quad \text{in } X , \tag{6.23}$$

$$\xi + \lambda\, T^* [N'(u)]^* \xi - \lambda\, T^* \mathcal{F}'(u) = 0 \quad \text{in } Y^* , \tag{6.24}$$

and

$$g - E K^* \xi = 0 \quad \text{in } G . \tag{6.25}$$

Remark. Note that the optimality systems, e.g., (6.23)–(6.25), are linear in the adjoint variable ξ. Also, note that the control g may be eliminated from the optimality system (6.23)–(6.25). Indeed, the substitution of (6.25) into (6.23) yields

$$u + \lambda\, T N(u) + \lambda\, T K E K^* \xi = 0 \quad \text{in } X . \tag{6.26}$$

Thus, (6.24) and (6.26) determine the optimal state u and adjoint state ξ; subsequently, (6.25) may be used to determine the optimal control g from ξ. This observation serves to emphasize the important, direct role that the adjoint state plays in the determination of the optimal control.

Remark. Given a $\xi \in Y^*$, it is not always possible to evaluate g exactly from (6.25). For example, the application of the operator E may involve the solution of a partial differential equation. Thus, although it is often convenient to devise algorithms for the approximation of optimal control and states based on the simplified optimality system (6.24) and (6.26), in some other cases it is best to deal with the full form (6.23)–(6.25). Thus, when we consider approximations of optimal controls and states, we will deal with the latter.

Remark. In many applications we have that $X^* = Y$. Since these spaces are assumed to be reflexive, we also have that $Y^* = X$. In this case, we have that both u and ξ belong to X.

6.1.2 Finite-dimensional approximations of an abstract nonlinear stationary control and optimization problem

We now define and analyze algorithms for the finite-dimensional approximation of solutions of the optimality system (6.23)–(6.25).

Formulation of finite-dimensional approximate problems

A finite-dimensional discretization of the optimality system (6.23)–(6.25) is defined as follows. First, one chooses families of finite-dimensional subspaces $X^h \subset X$, $(Y^*)^h \subset Y^*$, and $G^h \subset G$. These families are parameterized by a parameter h that tends to zero.[57] Next, one defines approximate operators $T^h : Y \to X^h$, $E^h : G^* \to G^h$, and $(T^*)^h : X^* \to (Y^*)^h$. Of course, one views T^h, E^h, and $(T^*)^h$ as approximations to the operators T, E, and T^*, respectively. Note that $(T^*)^h$ is not necessarily the same as $(T^h)^*$. The former is a discretization of an adjoint operator while the latter is the adjoint of a discrete operator.

Once the approximating subspaces and operators have been chosen, an approximate problem is defined as follows. We seek $u^h \in X^h$, $g^h \in G^h$, and $\xi^h \in (Y^*)^h$ such that

$$u^h + \lambda T^h N(u^h) + \lambda T^h K g^h = 0 \quad \text{in } X^h \,, \tag{6.27}$$

$$\xi^h + \lambda (T^*)^h [N'(u^h)]^* \xi^h - \lambda (T^*)^h \mathcal{F}'(u^h) = 0 \quad \text{in } (Y^*)^h \,, \tag{6.28}$$

and

$$g^h - E^h K^* \xi^h = 0 \quad \text{in } G^h \,. \tag{6.29}$$

Hypotheses concerning the approximate problem

We make the following hypotheses concerning the approximate operators T^h, $(T^*)^h$, and E^h.

H14. $\|(T - T^h)y\|_X = 0 \quad \forall\, y \in Y$.

H15. $\lim_{h \to 0} \|(T^* - (T^*)^h)v\|_{Y^*} = 0 \quad \forall\, v \in X^*$.

H16. $\lim_{h \to 0} \|(E - E^h)s\|_G = 0 \quad \forall\, s \in G^*$.

We also need the following additional hypotheses on the operators appearing in the definition of the abstract problem (6.4).

H17. *$N \in C^3(X; Y)$ and $\mathcal{F} \in C^3(X; \mathbb{R})$.*

H18. *N'', N''', $\mathcal{F}''$, and $\mathcal{F}'''$ are locally bounded, i.e., they map bounded sets to bounded sets.*

---

[57] For example, this parameter can be chosen to be some measure of the grid size in a subdivision of Ω into finite elements.

H19. *For $v \in X$, in addition to* (H9), *i.e., $N'(v) \in \mathcal{L}(X; Z)$, where $Z \hookrightarrow\hookrightarrow Y$, we have that $[N'(v)]^* \in \mathcal{L}(Y^*; \widehat{Z})$, where $\widehat{Z} \hookrightarrow\hookrightarrow X^*$; that for $\eta \in Y^*$, $[N''(v)]^* \cdot \eta \in \mathcal{L}(Y^*; \widehat{Z})$; and that for $w \in X$, $\mathcal{F}''(v) \cdot w \in \mathcal{L}(X; \widehat{Z})$.*

H20. *K maps G into Z.*

Here, $(\cdot)''$ and $(\cdot)'''$ denote second and third Fréchet derivatives, respectively.

Quotation of results concerning the approximation of a class of nonlinear problems

The error estimate to be derived makes use of results of [419, 423] (see also [428]) concerning the approximation of a class of nonlinear problems. These results imply that, under certain hypotheses, the error of approximation of solutions of certain nonlinear problems is basically the same as the error of approximation of solutions of related linear problems. Here, for the sake of completeness, we state the relevant results, specialized to our needs.

The nonlinear problems considered in [419, 423, 428] are of the following type. Given $\lambda \in \Lambda$, we seek $\psi \in \mathcal{X}$ such that

$$\mathcal{H}(\lambda, \psi) \equiv \psi + \mathcal{T}\mathcal{G}(\lambda, \psi) = 0\,, \tag{6.30}$$

where $\mathcal{T} \in \mathcal{L}(\mathcal{Y}; \mathcal{X})$, $\mathcal{G}$ is a C^2 mapping from $\Lambda \times \mathcal{X}$ into $\mathcal{Y}$, $\mathcal{X}$ and $\mathcal{Y}$ are Banach spaces, and Λ is a compact interval of $\mathbb{R}$. We say that $\{(\lambda, \psi(\lambda)) : \lambda \in \Lambda\}$ is a *branch of solutions* of (6.30) if $\lambda \to \psi(\lambda)$ is a continuous function from Λ into $\mathcal{X}$ such that $\mathcal{H}(\lambda, \psi(\lambda)) = 0$. The branch is called a *regular branch* if we also have that $\mathcal{H}_\psi(\lambda, \psi(\lambda))$ is an isomorphism from $\mathcal{X}$ into $\mathcal{X}$ for all $\lambda \in \Lambda$. Here, $\mathcal{H}_\psi(\cdot, \cdot)$ denotes the Fréchet derivative of $\mathcal{H}(\cdot, \cdot)$ with respect to the second argument. We assume that there exists another Banach space $\mathcal{Z}$, contained in $\mathcal{Y}$, with continuous embedding, such that

$$\mathcal{G}_\psi(\lambda, \psi) \in \mathcal{L}(\mathcal{X}; \mathcal{Z}) \quad \forall\, \lambda \in \Lambda \text{ and } \psi \in \mathcal{X}\,, \tag{6.31}$$

where $\mathcal{G}_\psi(\cdot, \cdot)$ denotes the Fréchet derivative of $\mathcal{G}(\cdot, \cdot)$ with respect to the second argument.

Approximations are defined by introducing a subspace $\mathcal{X}^h \subset \mathcal{X}$ and an approximating operator $\mathcal{T}^h \in \mathcal{L}(\mathcal{Y}; \mathcal{X}^h)$. Then, given $\lambda \in \Lambda$, we seek $\psi^h \in \mathcal{X}^h$ such that

$$\mathcal{H}^h(\lambda, \psi^h) \equiv \psi^h + \mathcal{T}^h\mathcal{G}(\lambda, \psi^h) = 0\,. \tag{6.32}$$

Concerning the operator $\mathcal{T}^h$, we assume the approximation properties

$$\lim_{h\to 0} \|(\mathcal{T}^h - \mathcal{T})\omega\|_{\mathcal{X}} = 0 \quad \forall\, \omega \in \mathcal{Y} \tag{6.33}$$

and

$$\lim_{h\to 0} \|(\mathcal{T}^h - \mathcal{T})\|_{\mathcal{L}(\mathcal{Z};\mathcal{X})} = 0\,. \tag{6.34}$$

Note that whenever the embedding $\mathcal{Z} \subset \mathcal{Y}$ is compact, (6.34) follows from (6.33) and, moreover, (6.31) implies that the operator $\mathcal{T}\mathcal{G}_\psi(\lambda, \psi) \in \mathcal{L}(\mathcal{X}; \mathcal{X})$ is compact.

We can now state the result of [419, 423] that is used in the study of approximations of the optimality system. In the statement of the theorem, $D^2\mathcal{G}$ represents any and all second Fréchet derivatives of $\mathcal{G}$.

THEOREM 6.6. *Let $\mathcal{X}$ and $\mathcal{Y}$ be Banach spaces and Λ a compact subset of $\mathbb{R}$. Assume that $\mathcal{G}$ is a C^2 mapping from $\Lambda \times \mathcal{X}$ into $\mathcal{Y}$ and that $D^2\mathcal{G}$ is bounded on all bounded sets of $\Lambda \times \mathcal{X}$. Assume that* (6.31), (6.33), *and* (6.34) *hold and that $\{(\lambda, \psi(\lambda)); \lambda \in \Lambda\}$ is a branch of regular solutions of* (6.30). *Then, there exists a neighborhood $\mathcal{O}$ of the origin in $\mathcal{X}$ and, for $h \leq h_0$ small enough, a unique C^2 function $\lambda \to \psi^h(\lambda) \in \mathcal{X}^h$ such that $\{(\lambda, \psi^h(\lambda)); \lambda \in \Lambda\}$ is a branch of regular solutions of* (6.32) *and $\psi^h(\lambda) - \psi(\lambda) \in \mathcal{O}$ for all $\lambda \in \Lambda$. Moreover, there exists a constant $C > 0$, independent of h and λ, such that*

$$\|\psi^h(\lambda) - \psi(\lambda)\|_{\mathcal{X}} \leq C\|(\mathcal{T}^h - \mathcal{T})\mathcal{G}(\lambda, \psi(\lambda))\|_{\mathcal{X}} \quad \forall\, \lambda \in \Lambda\,. \tag{6.35}$$

Error estimates for the approximation of solutions of the optimality system

Theorem 6.6 can be applied to study the approximation of solutions of the optimality system. Set $\mathcal{X} = X \times G \times Y^*$, $\mathcal{Y} = Y \times X^*$, $\mathcal{Z} = Z \times \widehat{Z}$, and $\mathcal{X}^h = X^h \times G^h \times (Y^*)^h$. (Recall that $\widehat{Z}$ was introduced in (H19).) By the hypotheses on Z and $\widehat{Z}$, we have that $\mathcal{Z}$ is compactly embedded into $\mathcal{Y}$. Let $\mathcal{T} \in \mathcal{L}(\mathcal{Y}; \mathcal{X})$ be defined in the following manner: $\mathcal{T}(\tilde{r}, \tilde{\tau}) = (\tilde{u}, \tilde{g}, \tilde{\xi})$ for $(\tilde{r}, \tilde{\tau}) \in \mathcal{Y}$ and $(\tilde{u}, \tilde{g}, \tilde{\xi}) \in \mathcal{X}$ if and only if

$$\tilde{u} + T\tilde{r} = 0\,, \tag{6.36}$$

$$\tilde{\xi} + T^*\tilde{\tau} = 0\,, \tag{6.37}$$

and

$$\tilde{g} - EK^*\tilde{\xi} = 0\,. \tag{6.38}$$

Similarly, the operator $\mathcal{T}^h \in \mathcal{L}(\mathcal{Y}; \mathcal{X}^h)$ is defined as follows: $\mathcal{T}^h(\tilde{r}, \tilde{\tau}) = (\tilde{u}^h, \tilde{g}^h, \tilde{\xi}^h)$ for $(\tilde{r}, \tilde{\tau}) \in \mathcal{Y}$ and $(\tilde{u}^h, \tilde{g}^h, \tilde{\xi}^h) \in \mathcal{X}^h$ if and only if

$$\tilde{u}^h + T^h\tilde{r} = 0\,, \tag{6.39}$$

$$\tilde{\xi}^h + (T^*)^h\tilde{\tau} = 0\,, \tag{6.40}$$

and

$$\tilde{g}^h - E^h K^*\tilde{\xi}^h = 0\,. \tag{6.41}$$

The nonlinear mapping $\mathcal{G} : \Lambda \times \mathcal{X} \to \mathcal{Y}$ is defined as follows: $\mathcal{G}\big(\lambda, (\tilde{u}, \tilde{g}, \tilde{\xi})\big) = (\tilde{r}, \tilde{\tau})$ for $\lambda \in \Lambda$, $(\tilde{u}, \tilde{g}, \tilde{\xi}) \in \mathcal{X}$, and $(\tilde{r}, \tilde{\tau}) \in \mathcal{Y}$ if and only if

$$\tilde{r} = \lambda\, N(\tilde{u}) + \lambda\, K\tilde{g} \tag{6.42}$$

and

$$\tilde{\tau} = \lambda\, [N'(\tilde{u})]^*\tilde{\xi} - \lambda\, \mathcal{F}'(\tilde{u})\,. \tag{6.43}$$

It is evident that the optimality system (6.23)–(6.25) and its finite-dimensional counterpart (6.27)–(6.29) can be written as

$$(u, g, \xi) + \mathcal{T}\mathcal{G}\big(\lambda, (u, g, \xi)\big) = 0$$

and

$$(u^h, g^h, \xi^h) + \mathcal{T}^h\mathcal{G}\big(\lambda, (u^h, g^h, \xi^h)\big) = 0\,,$$

respectively, i.e., with $\psi = (u, g, \xi)$ and $\psi^h = (u^h, g^h, \psi^h)$, in the form of (6.30) and (6.32), respectively.

Now we examine the approximation properties of $\mathcal{T}^h$.

Lemma 6.7. *Let the operators $\mathcal{T}$ and $\mathcal{T}^h$ be defined by* (6.36)–(6.38) *and* (6.39)–(6.41), *respectively. Assume that the hypotheses* (H14)–(H16) *hold. Then,*

$$\lim_{h\to 0} \|(\mathcal{T} - \mathcal{T}^h)(r, \tau)\|_{\mathcal{X}} = 0 \quad \forall\, (r, \tau) \in \mathcal{Y}. \tag{6.44}$$

Next, we examine the derivative of the mapping $\mathcal{G}$.

Lemma 6.8. *Let the mapping $\mathcal{G} : \lambda \times \mathcal{X} \to \mathcal{Y}$ be defined by* (6.42)–(6.43). *Assume that the hypotheses* (H9), (H17), *and* (H19)–(H20) *hold. Then, for every $\lambda \in \Lambda$ and every $(u, g, \xi) \in \mathcal{X}$, the operator $\mathcal{G}_{(u,g,\xi)}\big(\lambda, (u, g, \xi)\big) \in \mathcal{L}(\mathcal{X}; \mathcal{Z})$.*

A solution $\big(u(\lambda), g(\lambda), \xi(\lambda)\big)$ of the optimality system (6.23)–(6.25) is called *regular* if the system (for the unknowns $(\tilde{u}, \tilde{g}, \tilde{\xi})$)

$$\tilde{u} + \lambda\, T N'(u)\tilde{u} + \lambda\, T K \tilde{g} = \tilde{x}\,, \tag{6.45}$$

$$\tilde{\xi} + \lambda\, T^*[N''(u)]^*\tilde{u} \cdot \xi + \lambda\, T^*[N'(u)]^*\tilde{\xi} - \lambda\, T^*\mathcal{F}''(u)\tilde{u} = \tilde{y}\,, \tag{6.46}$$

and

$$\tilde{g} - E K^*\tilde{\xi} = \tilde{z} \tag{6.47}$$

is uniquely solvable for any $(\tilde{x}, \tilde{z}, \tilde{y}) \in \mathcal{X} = X \times G \times Y^*$.[58]

In the following theorem, we will assume that the solution $\big(u(\lambda), g(\lambda), \xi(\lambda)\big)$ of the optimality system (6.23)–(6.25) that we are trying to approximate is a regular solution. The assumptions we have made, in particular (H9), (H19), and (H20), are sufficient to guarantee that for almost all values of λ, this is indeed the case.

Lemma 6.9. *Assume the hypotheses of Lemma* 6.8. *Then, for almost all λ, solutions $\big(u(\lambda), g(\lambda), \xi(\lambda)\big)$ of the optimality system* (6.23)–(6.25) *are regular.*

Using Theorem 6.6, we can now provide an error estimate for approximations of solutions of the abstract optimal control problem.

Theorem 6.10. *Let $\big(u(\lambda), g(\lambda), \xi(\lambda)\big) \in \mathcal{X}$ for $\lambda \in \Lambda$ be a branch of regular solutions of the optimality system* (6.23)–(6.25). *Assume that the hypotheses* (H14)–(H20) *hold. Then, there exists a $\delta > 0$ and an $h_0 > 0$ such that for $h < h_0$, the discrete optimality system* (6.27)–(6.29) *has a unique solution $\big(u^h(\lambda), g^h(\lambda), \xi^h(\lambda)\big)$ satisfying*

$$\|\big(u(\lambda), g(\lambda), \xi(\lambda)\big) - \big(u^h(\lambda), g^h(\lambda), \xi^h(\lambda)\big)\|_{\mathcal{X}} < \delta\,.$$

[58] Note that the linear operator appearing on the left-hand side of (6.45)–(6.47) is obtained by linearizing the optimality system (6.23)–(6.25) about (u, g, ξ).

Moreover,

$$\lim_{h\to 0} \|\big(u(\lambda), g(\lambda), \xi(\lambda)\big) - \big(u^h(\lambda), g^h(\lambda), \xi^h(\lambda)\big)\|_{\mathcal{X}} = 0 \tag{6.48}$$

uniformly in $\lambda \in \Lambda$ *and there exists a constant* C*, independent of* h *and* λ*, such that*

$$\begin{aligned} \lim_{h\to 0} \|\big(u(\lambda), g(\lambda), \xi(\lambda)\big) - \big(u^h(\lambda), g^h(\lambda), \xi^h(\lambda)\big)\|_{\mathcal{X}} \\ \le C\lambda\Big\{ \|(T^h - T)\big(N(u(\lambda)) + Kg(\lambda)\big)\|_X + \|(E^h - E)K^*\xi(\lambda)\|_G \\ + \|\big((T^*)^h - T^*\big)\Big([N'(u(\lambda))]^*\xi - \mathcal{F}'(u(\lambda))\Big)\|_{Y^*}\Big\}. \end{aligned} \tag{6.49}$$

It is easily seen that (6.48) and (6.49) are equivalent to

$$\lim_{h\to 0}\Big\{ \|u(\lambda) - u^h(\lambda)\|_X + \|g(\lambda) - g^h(\lambda)\|_G + \|\xi(\lambda) - \xi^h(\lambda)\|_{Y^*}\Big\} = 0$$

uniformly in $\lambda \in \Lambda$ and that there exists a constant C, independent of h and λ, such that

$$\begin{aligned} \|u(\lambda) - u^h(\lambda)\|_X + \|g(\lambda) - g^h(\lambda)\|_G + \|\xi(\lambda) - \xi^h(\lambda)\|_{Y^*} \\ \le C\lambda\Big\{ \|(T^h - T)\big(N(u(\lambda)) + Kg(\lambda)\big)\|_X + \|(E^h - E)K^*\xi(\lambda)\|_G \\ + \|\big((T^*)^h - T^*\big)\big([N'(u(\lambda))]^*\xi(\lambda) - \mathcal{F}'(u(\lambda))\big)\|_{Y^*}\Big\}. \end{aligned}$$

If, in (6.35), the operator $\mathcal{T}$ is invertible, we have, using (6.30), that

$$\|\psi^h(\lambda) - \psi(\lambda)\|_{\mathcal{X}} \le C\|(\mathcal{T}^h\mathcal{T}^{-1} - I)\psi(\lambda)\|_{\mathcal{X}} \quad \forall\, \lambda \in \Lambda\,.$$

Thus, if the operator T from Y to X is invertible, we have that (6.49) is equivalent to

$$\begin{aligned} \|u(\lambda) - u^h(\lambda)\|_X + \|g(\lambda) - g^h(\lambda)\|_G + \|\xi(\lambda) - \xi^h(\lambda)\|_{Y^*} \\ \le C\Big\{ \|(T^hT^{-1} - I)u(\lambda)\|_X + \|(E^hE^{-1} - I)g(\lambda)\|_G \\ + \|\big((T^*)^h(T^*)^{-1} - I\big)\xi(\lambda)\|_{Y^*}\Big\}. \end{aligned} \tag{6.50}$$

6.1.3 Application to a boundary velocity control problem for the stationary Navier–Stokes equations

Notation

We first establish some notation. Throughout, C will denote a positive constant whose meaning and value change with the context. Also, $H^s(\mathcal{D})$ for $s \in \mathbb{R}$ denotes the standard real Sobolev space of order s with respect to the set $\mathcal{D}$, where $\mathcal{D}$ could be either a bounded domain $\Omega \in \mathbb{R}^d$, $d = 2, 3$, or part of the boundary Γ of such a domain. Of particular interest are the spaces $H^0(\mathcal{D}) = L^2(\mathcal{D})$ and

$$H^1(\mathcal{D}) = \left\{ \phi \in L^2(\mathcal{D}) \;\middle|\; \frac{\partial\phi}{\partial x_j} \in L^2(\mathcal{D}) \quad \text{for } j = 1, \dots, d \right\}.$$

Also of interest is the subspace

$$H_0^1(\mathcal{D}) = \left\{ \phi \in H^1(\mathcal{D}) \;\middle|\; \phi = 0 \quad \text{on } \partial\mathcal{D} \right\},$$

where $\partial\mathcal{D}$ denotes the boundary of $\mathcal{D}$.

Dual spaces will be denoted by $(\cdot)^*$. Duality pairings between spaces and their dual spaces will be denoted by $\langle\cdot,\cdot\rangle$. Norms of functions belonging to $H^s(\Omega)$ and $H^s(\Gamma)$ are denoted by $\|\cdot\|_s$ and $\|\cdot\|_{s,\Gamma}$, respectively. Of particular interest are the $L^2(\Omega)$-norm $\|\cdot\|_0$ and the $H^1(\Omega)$-norm

$$\|\phi\|_1^2 = \sum_{j=1}^{d} \left\| \frac{\partial \phi}{\partial x_j} \right\|_0^2 + \|\phi\|_0^2 .$$

Corresponding Sobolev spaces of real, vector-valued functions having r components will be denoted by $\mathbf{H}^s(\mathcal{D})$, e.g., $\mathbf{H}^1(\mathcal{D}) = [H^1(\mathcal{D})]^r$. Of particular interest will be the spaces $\mathbf{L}^2(\mathcal{D}) = \mathbf{H}^0(\mathcal{D}) = [L^2(\mathcal{D})]^r$ and

$$\mathbf{H}^1(\mathcal{D}) = \left\{ v_j \in L^2(\mathcal{D}) \;\middle|\; \frac{\partial v_j}{\partial x_k} \in L^2(\mathcal{D}) \quad \text{for } j = 1,\dots,r \text{ and } k = 1,\dots,d \right\},$$

where v_j, $j = 1,\dots,r$, denote the components of $\mathbf{v}$. Also of interest is the subspace

$$\mathbf{H}_0^1(\mathcal{D}) = \left\{ \mathbf{v} \in \mathbf{H}^1(\mathcal{D}) \;\middle|\; v_j = 0 \text{ on } \partial\mathcal{D},\ j = 1,\dots,r \right\}.$$

Norms for spaces of vector-valued functions will be denoted by the same notation as that used for their scalar counterparts. For example,

$$\|\mathbf{v}\|_s^2 = \sum_{j=1}^{r} \|v_j\|_s^2 \qquad \text{and} \qquad \|\mathbf{v}\|_{s,\Gamma}^2 = \sum_{j=1}^{r} \|v_j\|_{s,\Gamma}^2 .$$

We denote the $L^2(\Omega)$ and $\mathbf{L}^2(\Omega)$ inner products by $(\cdot,\cdot)$, i.e., for $p, q \in L^2(\Omega)$ and $\mathbf{u}, \mathbf{v} \in \mathbf{L}^2(\Omega)$,

$$(p, q) = \int_\Omega pq \, d\Omega \quad \text{and} \quad (\mathbf{u}, \mathbf{v}) = \int_\Omega \mathbf{u} \cdot \mathbf{v} \, d\Omega .$$

Similarly, we denote by $(\cdot,\cdot)_\Gamma$ the $L^2(\Gamma)$ and $\mathbf{L}^2(\Gamma)$ inner products, i.e., for $p, q \in L^2(\Gamma)$ and $\mathbf{u}, \mathbf{v} \in \mathbf{L}^2(\Gamma)$,

$$(p, q)_\Gamma = \int_\Gamma pq \, d\Gamma \quad \text{and} \quad (\mathbf{u}, \mathbf{v})_\Gamma = \int_\Gamma \mathbf{u} \cdot \mathbf{v} \, d\Gamma .$$

Since in all cases L^2-spaces will be used as pivot spaces, the above inner product notation can also be used to denote duality pairings between functions defined on H^s-spaces and their dual spaces.

For details concerning the notation employed, one may consult, e.g., [413].

We introduce the subspaces

$$L_0^2(\Omega) = \left\{ p \in L^2(\Omega) \;\middle|\; \int_\Omega p \, d\Omega = 0 \right\}$$

and

$$\mathbf{H}_n^1(\Gamma) = \left\{ \mathbf{g} \in \mathbf{H}^1(\Gamma) \;\middle|\; \int_\Gamma \mathbf{g} \cdot \mathbf{n} \, d\Gamma = 0 \right\}.$$

We also introduce the bilinear forms

$$a(\mathbf{u},\mathbf{v}) = \frac{1}{2}\int_\Omega \left((\nabla\mathbf{u}) + (\nabla\mathbf{u})^T\right) : \left((\nabla\mathbf{v}) + (\nabla\mathbf{v})^T\right) d\Omega \quad \forall\, \mathbf{u},\mathbf{v} \in \mathbf{H}^1(\Omega)$$

and

$$b(\mathbf{v},q) = -\int_\Omega q\,\nabla\cdot\mathbf{v}\,d\Omega \quad \forall\,\mathbf{v}\in\mathbf{H}^1(\Omega) \quad \text{and} \quad \forall\, q \in L^2(\Omega)$$

and the trilinear form

$$c(\mathbf{u},\mathbf{v},\mathbf{w}) = \int_\Omega \mathbf{u}\cdot\nabla\mathbf{v}\cdot\mathbf{w}\,d\Omega \quad \forall\,\mathbf{u},\mathbf{v},\mathbf{w}\in\mathbf{H}^1(\Omega)\,.$$

These forms are continuous over the spaces of definition indicated above. Moreover, we have the coercivity properties

$$a(\mathbf{v},\mathbf{v}) \geq C_a\|\mathbf{v}\|_1^2 \quad \forall\,\mathbf{v}\in\mathbf{H}_0^1(\Omega) \tag{6.51}$$

and

$$\sup_{0\neq\mathbf{v}\in\mathbf{H}_0^1(\Omega)} \frac{b(\mathbf{v},q)}{\|\mathbf{v}\|_1} \geq C_b\|q\|_0 \quad \forall\, q\in L_0^2(\Omega) \tag{6.52}$$

for some constants C_a and $C_b > 0$. For details concerning the notation employed for (6.51) and (6.52), one may consult [428, 430, 444].

The control problem

We will use Dirichlet boundary controls, i.e., control is effected through the data in a Dirichlet boundary condition. Let Ω denote a bounded domain in $\mathbb{R}^d$, $d = 2$ or 3, with a boundary denoted by Γ. Let $\mathbf{u}$ and p denote the velocity and pressure fields in Ω. The stationary Navier–Stokes equations for a viscous, incompressible flow are given by

$$-\nu\,\nabla\cdot\left((\nabla\mathbf{u}) + (\nabla\mathbf{u})^T\right) + \mathbf{u}\cdot\nabla\mathbf{u} + \nabla p = \mathbf{f} \quad \text{in } \Omega\,,$$

$$\nabla\cdot\mathbf{u} = 0 \quad \text{in } \Omega\,,$$

and

$$\mathbf{u} = \mathbf{b} + \mathbf{g} \quad \text{on } \Gamma\,,$$

where $\mathbf{f}$ is a given body force per unit mass, $\mathbf{b}$ and $\mathbf{g}$ are boundary velocity data with $\int_\Gamma \mathbf{b}\cdot\mathbf{n}\,d\Gamma = 0$ and $\int_\Gamma \mathbf{g}\cdot\mathbf{n}\,d\Gamma = 0$, and ν denotes the (constant) kinematic viscosity. We have absorbed the constant density into the pressure and the body force. If the variables in these equations are nondimensionalized, then ν is simply the inverse of the Reynolds number Re.

Setting $\lambda = 1/\nu = Re$ and replacing p with p/λ, $\mathbf{b}$ with $\lambda\mathbf{b}$, and $\mathbf{g}$ with $\lambda\mathbf{g}$, we may write the Navier–Stokes equations in the form

$$-\nabla\cdot\left((\nabla\mathbf{u}) + (\nabla\mathbf{u})^T\right) + \nabla p + \lambda\mathbf{u}\cdot\nabla\mathbf{u} = \lambda\mathbf{f} \quad \text{in } \Omega\,, \tag{6.53}$$

$$\nabla\cdot\mathbf{u} = 0 \quad \text{in } \Omega\,, \tag{6.54}$$

and

$$\mathbf{u} = \lambda(\mathbf{b} + \mathbf{g}) \quad \text{on } \Gamma . \tag{6.55}$$

We recast the Navier–Stokes equations (6.53)–(6.55) into the following particular weak form (see, e.g., [432]): seek $(\mathbf{u}, p, \mathbf{t}) \in \mathbf{H}^1(\Omega) \times L^2_0(\Omega) \times \mathbf{H}^{-1/2}(\Gamma)$ such that

$$a(\mathbf{u}, \mathbf{v}) + b(\mathbf{v}, p) - \langle \mathbf{t}, \mathbf{v} \rangle_\Gamma + \lambda\, c(\mathbf{u}, \mathbf{u}, \mathbf{v}) = \lambda\, \langle \mathbf{f}, \mathbf{v} \rangle \quad \forall\, \mathbf{v} \in \mathbf{H}^1(\Omega) , \tag{6.56}$$

$$b(\mathbf{u}, q) = 0 \quad \forall\, q \in L^2_0(\Omega) , \tag{6.57}$$

and

$$\langle \mathbf{s}, \mathbf{u} \rangle_\Gamma - \lambda\, \langle \mathbf{s}, \mathbf{g} \rangle_\Gamma = \lambda\, \langle \mathbf{s}, \mathbf{b} \rangle_\Gamma \quad \forall\, \mathbf{s} \in \mathbf{H}^{-1/2}(\Gamma) . \tag{6.58}$$

Formally we have

$$\mathbf{t} = \left[-p\mathbf{n} + \left(\nabla \mathbf{u} + (\nabla \mathbf{u})^T \right) \cdot \mathbf{n} \right]_\Gamma ,$$

i.e., $\mathbf{t}$ is the stress force on the boundary. The existence of a solution $(\mathbf{u}, p, \mathbf{t})$ for the system (6.56)–(6.58) was established in [432].

Given a desired velocity field $\mathbf{u}_0$, we define for any $(\mathbf{u}, p, \mathbf{t}) \in \mathbf{H}^1(\Omega) \times L^2_0(\Omega) \times \mathbf{H}^{-1/2}(\Gamma)$ and $\mathbf{g} \in \mathbf{H}^1_n(\Gamma)$ the functional

$$\mathcal{J}(\mathbf{u}, p, \mathbf{t}, \mathbf{g}) = \frac{\lambda}{4} \int_\Omega |\mathbf{u} - \mathbf{u}_0|^4 \, d\Omega + \frac{\lambda}{2} \int_\Gamma \left(|\nabla_s \mathbf{g}|^2 + |\mathbf{g}|^2 \right) d\Gamma , \tag{6.59}$$

where ∇_s denotes the surface gradient.

We define the spaces $X = \mathbf{H}^1(\Omega) \times L^2_0(\Omega) \times \mathbf{H}^{-1/2}(\Gamma)$, $Y = [\mathbf{H}^1(\Omega)]^* \times L^2_0(\Omega) \times \mathbf{H}^{1/2}(\Gamma)$, $G = \mathbf{H}^1_n(\Gamma)$, and $Z = \mathbf{L}^{3/2}(\Omega) \times \{0\} \times \mathbf{H}^1(\Gamma)$. By compact embedding results, Z is compactly embedded into Y. For the time being, we assume that the admissible set Θ for the control $\mathbf{g}$ is a closed, convex subset of $G = \mathbf{H}^1_n(\Gamma)$.

We then consider the following optimal control problem associated with the Navier–Stokes equations:

$$\min\{\mathcal{J}(\mathbf{u}, p, \mathbf{t}, \mathbf{g}) : (\mathbf{u}, p, \mathbf{t}) \in X, \mathbf{g} \in \Theta\} \quad \text{subject to} \quad (6.56)\text{–}(6.58) . \tag{6.60}$$

We define the continuous linear operator $T \in \mathcal{L}(Y; X)$ as follows: For each $(\boldsymbol{\zeta}, \eta, \boldsymbol{\kappa}) \in Y$, $T(\boldsymbol{\zeta}, \eta, \boldsymbol{\kappa}) = (\tilde{\mathbf{u}}, \tilde{p}, \tilde{\mathbf{t}}) \in X$ is the unique solution of

$$a(\tilde{\mathbf{u}}, \mathbf{v}) + b(\mathbf{v}, \tilde{p}) - \langle \tilde{\mathbf{t}}, \mathbf{v} \rangle_\Gamma = \langle \boldsymbol{\zeta}, \mathbf{v} \rangle \quad \forall\, \mathbf{v} \in \mathbf{H}^1(\Omega) ,$$

$$b(\tilde{\mathbf{u}}, q) = (\eta, q) \quad \forall\, q \in L^2_0(\Omega),$$

and

$$\langle \mathbf{s}, \tilde{\mathbf{u}} \rangle_\Gamma = \langle \mathbf{s}, \boldsymbol{\kappa} \rangle_\Gamma \quad \forall\, \mathbf{s} \in \mathbf{H}^{-1/2}(\Gamma) .$$

It can be easily verified that T is self-adjoint.

We define the (differentiable) nonlinear mapping $N : X \to Y$ by

$$N(\mathbf{u}, p, \mathbf{t}) = - \begin{pmatrix} \mathbf{f} - \mathbf{u} \cdot \nabla \mathbf{u} \\ 0 \\ \mathbf{b} \end{pmatrix}$$

or, equivalently,

$$\langle N(\mathbf{u}, p, \mathbf{t}), (\mathbf{v}, q, \mathbf{s})\rangle = -(\mathbf{f}, \mathbf{v}) + c(\mathbf{u}, \mathbf{u}, \mathbf{v}) - \langle \mathbf{s}, \mathbf{b}\rangle_\Gamma \quad \forall\, (\mathbf{v}, q, \mathbf{s}) \in X$$

and define $K : \mathbf{H}^{1/2}(\Gamma) \to Y$ by

$$K\mathbf{g} = -\begin{pmatrix} 0 \\ 0 \\ \mathbf{g} \end{pmatrix}$$

or, equivalently,

$$\langle K\mathbf{g}, (\mathbf{v}, q, \mathbf{s})\rangle = -\langle \mathbf{s}, \mathbf{g}\rangle_\Gamma \quad \forall\, \mathbf{g} \in \mathbf{H}^{1/2}(\Gamma)\,, \forall\, (\mathbf{v}, q, \mathbf{s}) \in X\,.$$

Clearly, the constraint equations (6.56)–(6.58) can be expressed as

$$(\mathbf{u}, p, \mathbf{t}) + \lambda\, T N(\mathbf{u}, p, \mathbf{t}) + \lambda\, T K \mathbf{g} = 0\,,$$

i.e., in the form (6.2). With the obvious definitions for $\mathcal{F}(\cdot)$ and $\mathcal{E}(\cdot)$, i.e.,

$$\mathcal{F}(\mathbf{u}, p, \mathbf{t}) = \frac{1}{4}\int_\Omega |\mathbf{u} - \mathbf{u}_0|^4\, d\Omega \quad \forall\, (\mathbf{u}, p, \mathbf{t}) \in X$$

and

$$\mathcal{E}(\mathbf{g}) = \frac{1}{2}\int_\Gamma \left(|\nabla_s \mathbf{g}|^2 + |\mathbf{g}|^2\right) d\Gamma\,,$$

the functional (6.59) can be expressed as

$$\mathcal{J}(\mathbf{u}, p, \mathbf{t}, \mathbf{g}) = \lambda\, \mathcal{F}(\mathbf{u}, p, \mathbf{t}) + \lambda\, \mathcal{E}(\mathbf{g})\,,$$

i.e., in the form (6.3).

For the minimization problem (6.60), all hypotheses (H1)–(H20) can be verified. Once these are verified, we can apply the results of Sections 6.1.1 and 6.1.2 for the abstract minimization problem (6.4) to derive the corresponding results for the minimization problem (6.60). We omit all details here; these may be found in [387].

Existence of optimal solutions and Lagrange multipliers

It can be shown that hypotheses (H1)–(H6) hold in the current setting. Then, it is just a matter of citing Theorem 6.1 to conclude the existence of an optimal solution that minimizes (6.59) subject to (6.56)–(6.58).

Theorem 6.11. *There exists a* $(\mathbf{u}, p, \mathbf{t}, \mathbf{g}) \in \mathbf{H}^1(\Omega) \times L^2_0(\Omega) \times \mathbf{H}^{-1/2}(\Omega) \times \Theta$ *such that* (6.59) *is minimized subject to* (6.56)–(6.58).

We now assume $(\mathbf{u}, p, \mathbf{t}, \mathbf{g})$ is an optimal solution. For any $(\mathbf{u}, p, \mathbf{t}) \in X$, the operator $N'(\mathbf{u}, p, \mathbf{t}) : X \to Y$ is given by

$$N'(\mathbf{u}, p, \mathbf{t}) \cdot (\mathbf{v}, q, \mathbf{s}) = -\begin{pmatrix} \mathbf{u} \cdot \nabla \mathbf{v} + \mathbf{v} \cdot \nabla \mathbf{u} \\ 0 \\ \mathbf{0} \end{pmatrix}$$

for all $(\mathbf{v}, q, \mathbf{s}) \in \mathbf{H}^1(\Omega) \times L^2_0(\Omega) \times \mathbf{H}^{-1/2}(\Gamma)$. The Lagrangian functional is given by

$$\begin{aligned}\mathcal{L}(\mathbf{u}, p, \mathbf{t}, \mathbf{g}, \boldsymbol{\mu}, \phi, \boldsymbol{\tau}, k) \\ = k\,\mathcal{J}(\mathbf{u}, \mathbf{g}) - \{a(\mathbf{u}, \boldsymbol{\mu}) + \lambda\, c(\mathbf{u}, \mathbf{u}, \boldsymbol{\mu}) + b(\boldsymbol{\mu}, p) + b(\mathbf{u}, \phi) - \langle \boldsymbol{\tau}, \mathbf{u}\rangle_\Gamma \\ - \langle \mathbf{t}, \boldsymbol{\mu}\rangle_\Gamma - \lambda\,\langle \mathbf{f}, \boldsymbol{\mu}\rangle_\Gamma + \lambda\,\langle \boldsymbol{\tau}, \mathbf{b}\rangle_\Gamma + \lambda\,\langle \boldsymbol{\tau}, \mathbf{g}\rangle_\Gamma\}\end{aligned}$$

for all $(\mathbf{u}, p, \mathbf{t}, \mathbf{g}, \boldsymbol{\mu}, \phi, \boldsymbol{\tau}, k) \in X \times G \times X \times \mathbb{R} = \mathbf{H}^1(\Omega) \times L^2_0(\Omega) \times \mathbf{H}^{-1/2}(\Gamma) \times \mathbf{H}^1_n(\Gamma) \times \mathbf{H}^1(\Omega) \times L^2_0(\Omega) \times \mathbf{H}^{-1/2}(\Gamma) \times \mathbb{R}$. Note that, in this form of the Lagrangian, the Lagrange multiplier $(\boldsymbol{\mu}, \phi, \boldsymbol{\tau}) \in X = Y^*$ so that we have already introduced the change of variables indicated between (6.20)–(6.22) and (6.23)–(6.25).

It can be shown that hypotheses (H7)–(H9) also hold in the current setting. We may then apply Theorem 6.3 to conclude that there exists a Lagrange multiplier $(\boldsymbol{\mu}, \phi, \boldsymbol{\tau}) \in X = \mathbf{H}^1(\Omega) \times L^2_0(\Omega) \times \mathbf{H}^{-1/2}(\Gamma)$ and a real number k such that

$$(\boldsymbol{\mu}, \phi, \boldsymbol{\tau}) + \lambda\, T^*\Big([N'(\mathbf{u}, p, \mathbf{t})]^* \cdot (\boldsymbol{\mu}, \phi, \boldsymbol{\tau}) - k\,\mathcal{J}_{(\mathbf{u},p,\mathbf{t})}(\mathbf{u}, p, \mathbf{t}, \mathbf{g})\Big) = 0 \tag{6.61}$$

and

$$\mathcal{L}(\mathbf{u}, p, \mathbf{t}, \mathbf{z}, \boldsymbol{\mu}, \phi, \boldsymbol{\tau}, k) \le \mathcal{L}(\mathbf{u}, p, \mathbf{t}, \mathbf{g}, \boldsymbol{\mu}, \phi, \boldsymbol{\tau}, k) \quad \forall\, \mathbf{z} \in \Theta \tag{6.62}$$

and that, for almost all values of λ, we may choose $k = 1$.

Recall that $T^* = T$. Also, note that, for $(\mathbf{u}, p, \mathbf{t}) \in X = \mathbf{H}^1(\Omega) \times L^2_0(\Omega) \times \mathbf{H}^{-1/2}(\Gamma)$, the operator $[N'(\mathbf{u}, p, \mathbf{t})]^* : X \to Y$ is given by

$$[N'(\mathbf{u}, p, \mathbf{t})]^* \cdot (\mathbf{v}, q, \mathbf{s}) = \begin{pmatrix} -\mathbf{u}\cdot\nabla\mathbf{v} + \mathbf{v}\cdot(\nabla\mathbf{u})^T \\ 0 \\ \mathbf{0} \end{pmatrix} \qquad \forall\,(\mathbf{v}, q, \mathbf{s}) \in X.$$

Thus, (6.61) with $k = 1$ can be rewritten as

$$\begin{aligned} a(\mathbf{w}, \boldsymbol{\mu}) + \lambda\, c(\mathbf{w}, \mathbf{u}, \boldsymbol{\mu}) + \lambda\, c(\mathbf{u}, \mathbf{w}, \boldsymbol{\mu}) + b(\mathbf{w}, \phi) - \langle \boldsymbol{\tau}, \mathbf{w}\rangle_\Gamma \\ = \lambda\,\big((\mathbf{u} - \mathbf{u}_0)^3, \mathbf{w}\big) \quad \forall\, \mathbf{w} \in \mathbf{H}^1(\Omega)\,, \end{aligned} \tag{6.63}$$

$$b(\boldsymbol{\mu}, r) = 0 \quad \forall\, r \in L^2_0(\Omega)\,, \tag{6.64}$$

and

$$\langle \mathbf{y}, \boldsymbol{\mu}\rangle_\Gamma = 0 \quad \forall\, \mathbf{y} \in \mathbf{H}^{-1/2}(\Gamma)\,. \tag{6.65}$$

In the right-hand side of (6.63), we use the notation $(\mathbf{v}^3, \mathbf{w}) = \sum_{j=1}^{d} (v_j^3, w_j)$.

Using the definition of the Lagrangian functional and (6.62) with $k = 1$, it can be shown that

$$\big(\nabla_s \mathbf{g}, \nabla_s(\mathbf{z} - \mathbf{g})\big)_\Gamma + (\mathbf{g}, \mathbf{z} - \mathbf{g})_\Gamma - \langle \boldsymbol{\tau}, \mathbf{z}\rangle_\Gamma \ge 0 \quad \forall\, \mathbf{z} \in \Theta\,. \tag{6.66}$$

We see that, for almost all values of λ, necessary conditions for an optimum are that (6.56)–(6.58), (6.63)–(6.65), and (6.66) are satisfied. The system formed by these equations is referred to as an *optimality system*.

We now specialize to the case $\Theta = \mathbf{H}^1_n(\Gamma)$. It can be shown that hypothesis (H10) is satisfied. Then, using Theorem 6.4, we see that inequality (6.66) becomes an equality and we now have, instead of (6.66),

$$(\nabla_s \mathbf{g}, \nabla_s \mathbf{z})_\Gamma + (\mathbf{g}, \mathbf{z})_\Gamma - \langle \boldsymbol{\tau}, \mathbf{z}\rangle_\Gamma = 0 \quad \forall\, \mathbf{z} \in \Theta = \mathbf{H}^1_n(\Gamma)\,. \tag{6.67}$$

Thus, according to that theorem, we have that, for almost all λ, an optimality system of equations is now given by (6.56)–(6.58), (6.63)–(6.65), and (6.67). However, we can go further and verify that hypothesis (H11) is valid, which in turn justifies the existence of a Lagrange multiplier satisfying the optimality system for *all* $\lambda \in \Lambda$.

Hence we conclude that, for *all* λ, the optimality system (6.56)–(6.58), (6.63)–(6.65), and (6.67) has a solution. Thus, we have Theorem 6.5 which, in the present context, is given as follows.

THEOREM 6.12. *Let* $(\mathbf{u}, p, \mathbf{t}, \mathbf{g}) \in \mathbf{H}^1(\Omega) \times L^2_0(\Omega) \times \mathbf{H}^{-1/2}(\Gamma) \times \mathbf{H}^1_n(\Gamma)$ *denote an optimal solution that minimizes* (6.59) *subject to* (6.56)–(6.58). *Then, for all* $\lambda \in \Lambda$, *there exists a nonzero Lagrange multiplier* $(\boldsymbol{\mu}, \phi, \boldsymbol{\tau}) \in \mathbf{H}^1(\Omega) \times L^2_0(\Omega) \times \mathbf{H}^{-1/2}(\Gamma)$ *satisfying the Euler–Lagrange equations* (6.63)–(6.65) *and* (6.67).

In the above, we have already employed hypothesis (H13), which in the current context is easily seen to be satisfied with $E : G \to G^*$ defined by

$$\langle E\mathbf{g}, \mathbf{z}\rangle = \int_\Gamma (\nabla_s \mathbf{g} \cdot \nabla_s \mathbf{z} + \mathbf{g} \cdot \mathbf{z})\, d\Gamma \quad \forall\, \mathbf{z} \in \mathbf{H}^1_n(\Gamma) = G\,.$$

We also note that, for each fixed $\boldsymbol{\tau}$, (6.67) with $\mathbf{g} \in \mathbf{H}^1_n(\Gamma)$ is equivalent to

$$(\nabla_s \mathbf{g}, \nabla_s \mathbf{k})_\Gamma + (\mathbf{g}, \mathbf{k})_\Gamma + \gamma \int_\Gamma \mathbf{k} \cdot \mathbf{n}\, d\Gamma = \langle \boldsymbol{\tau}, \mathbf{k}\rangle_\Gamma \quad \forall\, \mathbf{k} \in \mathbf{H}^1(\Gamma) \tag{6.68}$$

and

$$\int_\Gamma \mathbf{g} \cdot \mathbf{n}\, d\Gamma = 0\,, \tag{6.69}$$

where $\gamma \in \mathbb{R}$ is an additional unknown constant that accounts for the single integral constraint of (6.69). Although (6.67) and (6.68)–(6.69) are equivalent, the latter is more easily discretized.

Approximations of the optimality system and error estimates

A finite element discretization of the optimality system (6.56)–(6.58), (6.63)–(6.65), and (6.67) is defined as follows. First, one chooses families of finite-dimensional subspaces $\mathbf{V}^h \subset \mathbf{H}^1(\Omega)$ and $S^h \subset L^2(\Omega)$. These families are parameterized by the parameter h that tends to zero; commonly, this parameter is chosen to be some measure of the grid size in a subdivision of Ω into finite elements. We let $S^h_0 = S^h \cap L^2_0(\Omega)$ and $\mathbf{V}^h_0 = \mathbf{V}^h \cap \mathbf{H}^1_0(\Omega)$.

One may choose any pair of subspaces $\mathbf{V}^h$ and S^h that can be used for finding finite element approximations of solutions of the Navier–Stokes equations. Thus, concerning these subspaces, we make the following standard assumptions, which are exactly those employed in well-known finite element methods for the Navier–Stokes equations. First, we have the approximation properties: There exist an integer k and a constant C, independent of h, $\mathbf{v}$, and q, such that

$$\inf_{\mathbf{v}^h \in \mathbf{V}^h} \|\mathbf{v} - \mathbf{v}^h\|_1 \le C h^m \|\mathbf{v}\|_{m+1} \quad \forall\, \mathbf{v} \in \mathbf{H}^{m+1}(\Omega)\,,\ 1 \le m \le k, \tag{6.70}$$

and

$$\inf_{q^h \in S_0^h} \|q - q^h\|_0 \le C h^m \|q\|_m \quad \forall\, q \in H^m(\Omega) \cap L_0^2(\Omega)\,, \ 1 \le m \le k\,. \tag{6.71}$$

Next, we assume the *inf-sup condition* or *Ladyzhenskaya–Babuska–Brezzi condition*: there exists a constant C, independent of h, such that

$$\inf_{0 \ne q^h \in S_0^h} \sup_{\mathbf{0} \ne \mathbf{v}^h \in \mathbf{V}^h} \frac{b(\mathbf{v}^h, q^h)}{\|\mathbf{v}^h\|_1 \, \|q^h\|_0} \ge C\,. \tag{6.72}$$

This condition ensures the stability of finite element discretizations of the Navier–Stokes equations. For thorough discussions of the approximation properties (6.70)–(6.71), see, e.g., [415, 421], and for like discussions of the stability condition (6.72), see, e.g., [417, 418, 428, 430]. The latter references may also be consulted for a catalogue of finite element subspaces that meet the requirements of (6.70)–(6.72).

Next, let $\mathbf{P}^h = \mathbf{V}^h|_\Gamma$, i.e., $\mathbf{P}^h$ consists of the restriction, to the boundary Γ, of functions belonging to $\mathbf{V}^h$. For all choices of conforming finite element spaces $\mathbf{V}^h$, e.g., Lagrange type finite element spaces, we have that $\mathbf{P}^h \subset \mathbf{H}^{-1/2}(\Gamma)$. For the subspaces $\mathbf{P}^h = \mathbf{V}^h|_\Gamma$, we can show the following approximation property: There exist an integer k and a constant C, independent of h and $\mathbf{s}$, such that

$$\begin{aligned} &\inf_{\mathbf{s}^h \in \mathbf{P}^h} \|\mathbf{s} - \mathbf{s}^h\|_{-1/2,\Gamma} \\ &\quad \le C h^m \inf_{\mathbf{v} \in \mathbf{H}^m(\Omega), \mathbf{v}|_\Gamma = \mathbf{s}} \|\mathbf{v}\|_m \qquad \forall\, \mathbf{s} \in \mathbf{H}^m(\Omega)|_\Gamma\,, \ 1 \le m \le k\,. \end{aligned} \tag{6.73}$$

We also use the following inverse assumption: There exists a constant C, independent of h and $\mathbf{s}^h$, such that

$$\|\mathbf{s}^h\|_{s,\Gamma} \le C h^{s-q} \|\mathbf{s}^h\|_{q,\Gamma} \quad \forall\, \mathbf{s}^h \in \mathbf{P}^h\,, \ -1/2 \le q \le s \le 1/2\,. \tag{6.74}$$

See [415, 421] for details concerning (6.73) and (6.74). See also [387] for (6.73).

Now, let $\mathbf{Q}^h = \mathbf{V}^h|_\Gamma$, i.e., $\mathbf{Q}^h$ also consists of the restriction, to the boundary Γ, of functions belonging to $\mathbf{V}^h$. Again, for all choices of conforming finite element spaces $\mathbf{V}^h$ we have that $\mathbf{Q}^h \subset \mathbf{H}^1(\Gamma)$. We can show the approximation property: there exist an integer k and a constant C, independent of h and $\mathbf{k}$, such that for $1 \le m \le k$, $0 \le s \le 1$, and $\mathbf{k} \in \mathbf{H}^{m+1}(\Omega)|_\Gamma$,

$$\inf_{\mathbf{k}^h \in \mathbf{Q}^h} \|\mathbf{k} - \mathbf{k}^h\|_{s,\Gamma} \le C h^{m-s+\frac{1}{2}} \inf_{\mathbf{v} \in \mathbf{H}^{m+1}(\Omega), \mathbf{v}|_\Gamma = \mathbf{k}} \|\mathbf{v}\|_{m+1}\,. \tag{6.75}$$

This property follows from (6.70) once one notes that the same type of polynomials are used in $\mathbf{Q}^h$ as are used in $\mathbf{V}^h$.

We set $X^h = \mathbf{V}^h \times S_0^h \times \mathbf{P}^h$ and $G^h = \mathbf{Q}^h \cap \mathbf{H}_n^1(\Gamma)$.

Once the approximating subspaces have been chosen we seek $\mathbf{u}^h \in \mathbf{V}^h$, $p^h \in S_0^h$, $\mathbf{t}^h \in \mathbf{P}^h$, $\mathbf{g}^h \in \mathbf{Q}^h$, $\boldsymbol{\mu}^h \in \mathbf{V}^h$, $\phi^h \in S_0^h$, $\boldsymbol{\tau}^h \in \mathbf{P}^h$, and $\gamma^h \in \mathbb{R}$ such that

$$\begin{aligned} a(\mathbf{u}^h, \mathbf{v}^h) + \lambda\, c(\mathbf{u}^h, \mathbf{u}^h, \mathbf{v}^h) + b(\mathbf{v}^h, p^h) - \langle \mathbf{v}^h, \mathbf{t}^h \rangle_\Gamma& \\ = \lambda\, \langle \mathbf{f}, \mathbf{v}^h \rangle \quad \forall\, \mathbf{v}^h \in \mathbf{V}^h\,,& \end{aligned} \tag{6.76}$$

$$b(\mathbf{u}^h, q^h) = 0 \quad \forall\, q^h \in S_0^h\,, \tag{6.77}$$

$$\langle \mathbf{u}^h, \mathbf{s}^h \rangle_\Gamma - \lambda \langle \mathbf{g}^h, \mathbf{s}^h \rangle_\Gamma = \lambda \langle \mathbf{b}, \mathbf{s}^h \rangle_\Gamma \quad \forall\, \mathbf{s}^h \in \mathbf{P}^h\,, \tag{6.78}$$

$$(\nabla_s \mathbf{g}^h, \nabla_s \mathbf{k}^h)_\Gamma + \langle \mathbf{g}^h, \mathbf{k}^h \rangle_\Gamma + \gamma^h \int_\Gamma \mathbf{k}^h \cdot \mathbf{n}\, d\Gamma = \langle \boldsymbol{\tau}^h, \mathbf{k}^h \rangle_\Gamma \quad \forall\, \mathbf{k}^h \in \mathbf{Q}^h\,, \tag{6.79}$$

$$\int_\Gamma \mathbf{g}^h \cdot \mathbf{n}\, d\Gamma = 0\,, \tag{6.80}$$

$$\begin{aligned} a(\mathbf{w}^h, \boldsymbol{\mu}^h) + \lambda\, c(\mathbf{w}^h, \mathbf{u}^h, \boldsymbol{\mu}^h) + \lambda\, c(\mathbf{u}^h, \mathbf{w}^h, \boldsymbol{\mu}^h) + b(\mathbf{w}^h, \phi^h) - \langle \mathbf{w}^h, \boldsymbol{\tau}^h \rangle_\Gamma \\ = \lambda \left((\mathbf{u}^h - \mathbf{u}_0)^3, \mathbf{w}^h \right) \quad \forall\, \mathbf{w}^h \in \mathbf{V}^h\,, \end{aligned} \tag{6.81}$$

$$b(\boldsymbol{\mu}^h, r^h) = 0 \quad \forall\, r^h \in S_0^h\,, \tag{6.82}$$

and

$$\langle \boldsymbol{\mu}^h, \mathbf{y}^h \rangle = 0 \quad \forall\, \mathbf{y}^h \in \mathbf{P}^h\,. \tag{6.83}$$

Note that if (6.76)–(6.83) are satisfied, then necessarily $\mathbf{g}^h \in G^h$. Also, in the right-hand side of (6.81), we use notation similar to that used in the right-hand side of (6.63).

The operator $T^h \in \mathcal{L}(Y, X^h)$ is defined as the solution operator for

$$a(\mathbf{u}^h, \mathbf{v}^h) + b(\mathbf{v}^h, p^h) - \langle \mathbf{v}^h, \mathbf{t}^h \rangle_\Gamma = \langle \mathbf{f}, \mathbf{v}^h \rangle \quad \forall\, \mathbf{v}^h \in \mathbf{V}^h\,,$$

$$b(\mathbf{u}^h, q^h) = 0 \quad \forall\, q^h \in S_0^h\,,$$

and

$$\langle \mathbf{u}^h, \mathbf{s}^h \rangle_\Gamma = \langle \mathbf{b}, \mathbf{s}^h \rangle_\Gamma \quad \forall\, \mathbf{s}^h \in \mathbf{P}^h\,,$$

i.e, for each $\mathbf{f} \in Y$, $T^h\mathbf{f} = \boldsymbol{\psi}^h \in X^h$ is the solution of the above system of equations.

Since $T = T^*$, we define $(T^*)^h = T^h$.

We define the operator $E^h : G^* \to G^h$ as follows. For each $\boldsymbol{\tau} \in G^*$, $\mathbf{g}^h = E^h\boldsymbol{\tau}$ if and only if

$$(\nabla_s \mathbf{g}^h, \nabla_s \mathbf{z}^h)_\Gamma + \langle \mathbf{g}^h, \mathbf{z}^h \rangle_\Gamma + \gamma^h \int_\Gamma \mathbf{z}^h \cdot \mathbf{n}\, d\Gamma = \langle \boldsymbol{\tau}^h, \mathbf{z}^h \rangle_\Gamma \quad \forall\, \mathbf{z}^h \in \mathbf{Q}^h$$

and

$$\int_\Gamma \mathbf{g}^h \cdot \mathbf{n}\, d\Gamma = 0\,.$$

It can be shown that this system has a unique solution $(\mathbf{g}^h, \gamma^h) \in \mathbf{Q}^h \times \mathbb{R}$. The solution necessarily satisfies $\mathbf{g}^h \in G^h$. Thus the operator E^h is well defined.

With these definitions, we see that (6.76)–(6.83) can be written in the form (6.27)–(6.29).

With the help of results about the Navier–Stokes equations with inhomogeneous boundary conditions (see [432]), it can be shown that the hypotheses (H14)–(H20) hold in the current context. Hence, we can apply Theorem 6.10 to derive error estimates for the approximate solutions of the optimality system (6.56)–(6.58), (6.63)–(6.65), and (6.67). It should be noted that Lemma 6.9 implies that, for almost all values of λ, the solutions of the optimality system are regular.

Theorem 6.13. *Assume that Λ is a compact interval of $\mathbb{R}_+$ and that there exists a branch $\{(\lambda, \mathbf{u}(\lambda), p(\lambda), \mathbf{t}(\lambda), \mathbf{g}(\lambda), \boldsymbol{\mu}(\lambda), \phi(\lambda), \boldsymbol{\tau}(\lambda)) : \lambda \in \Lambda\}$ of regular solutions of the optimality system* (6.56)–(6.58), (6.63)–(6.65), *and* (6.67). *Assume that the finite element spaces X^h and G^h satisfy the hypotheses* (6.70)–(6.75). *Then, there exists a $\delta > 0$ and an $h_0 > 0$ such that for $h \le h_0$, the discrete optimality system* (6.76)–(6.83) *has a unique branch of solutions*

$$\{(\lambda, \mathbf{u}^h(\lambda), p^h(\lambda), \mathbf{t}^h(\lambda), \mathbf{g}^h(\lambda), \boldsymbol{\mu}^h(\lambda), \phi^h(\lambda), \boldsymbol{\tau}^h(\lambda)) : \lambda \in \Lambda\}$$

satisfying

$$\begin{aligned}\Big(&\|\mathbf{u}(\lambda) - \mathbf{u}^h(\lambda)\|_1 + \|p(\lambda) - p^h(\lambda)\|_0 + \|\mathbf{t}(\lambda) - \mathbf{t}^h(\lambda)\|_{-1/2,\Gamma}\\ &+\|\mathbf{g}(\lambda) - \mathbf{g}^h(\lambda)\|_{1,\Gamma} + \|\boldsymbol{\mu}(\lambda) - \boldsymbol{\mu}^h(\lambda)\|_1 + \|\phi(\lambda) - \phi^h(\lambda)\|_0\\ &+\|\boldsymbol{\tau}(\lambda) - \boldsymbol{\tau}^h(\lambda)\|_{-1/2,\Gamma}\Big) < \delta \qquad \textit{for all } \lambda \in \Lambda.\end{aligned}$$

Moreover,

$$\begin{aligned}\lim_{h\to 0}\Big(&\|\mathbf{u}(\lambda) - \mathbf{u}^h(\lambda)\|_1 + \|p(\lambda) - p^h(\lambda)\|_0 + \|\mathbf{t}(\lambda) - \mathbf{t}^h(\lambda)\|_{-1/2,\Gamma}\\ &+\|\mathbf{g}(\lambda) - \mathbf{g}^h(\lambda)\|_{1,\Gamma} + \|\boldsymbol{\mu}(\lambda) - \boldsymbol{\mu}^h(\lambda)\|_1\\ &+\|\phi(\lambda) - \phi^h(\lambda)\|_0 + \|\boldsymbol{\tau}(\lambda) - \boldsymbol{\tau}^h(\lambda)\|_{-1/2,\Gamma}\Big) = 0\end{aligned}$$

uniformly in $\lambda \in \Lambda$.

If, in addition, the solution satisfies $\big(\mathbf{u}(\lambda), p(\lambda), \mathbf{t}(\lambda), \mathbf{g}(\lambda), \boldsymbol{\mu}(\lambda), \phi(\lambda), \boldsymbol{\tau}(\lambda)\big) \in \mathbf{H}^{m+1}(\Omega) \times H^m(\Omega) \times \mathbf{H}^m(\Omega)|_\Gamma \times \mathbf{H}^{m+1}(\Omega)|_\Gamma \times \mathbf{H}^{m+1}(\Omega) \times H^m(\Omega) \times \mathbf{H}^m(\Omega)|_\Gamma$ for $\lambda \in \Lambda$, then there exists a constant C, independent of h, such that

$$\begin{aligned}\Big(&\|\mathbf{u}(\lambda) - \mathbf{u}^h(\lambda)\|_1 + \|p(\lambda) - p^h(\lambda)\|_0 + \|\mathbf{t}(\lambda) - \mathbf{t}^h(\lambda)\|_{-1/2,\Gamma}\\ &+\|\mathbf{g}(\lambda) - \mathbf{g}^h(\lambda)\|_{1,\Gamma} + \|\boldsymbol{\mu}(\lambda) - \boldsymbol{\mu}^h(\lambda)\|_1\\ &+\|\phi(\lambda) - \phi^h(\lambda)\|_0 + \|\boldsymbol{\tau}(\lambda) - \boldsymbol{\tau}^h(\lambda)\|_{-1/2,\Gamma}\Big)\\ \le Ch^{m-1/2}\Big(&\|\mathbf{u}(\lambda)\|_{m+1} + \|p(\lambda)\|_m + \inf_{\mathbf{v}\in\mathbf{H}^m(\Omega),\mathbf{v}|_\Gamma=\mathbf{t}} \|\mathbf{v}\|_m\\ &+ \inf_{\mathbf{v}\in\mathbf{H}^{m+1}(\Omega),\mathbf{v}|_\Gamma=\mathbf{g}} \|\mathbf{v}\|_{m+1} + \|\boldsymbol{\mu}(\lambda)\|_{m+1}\\ &+\|\phi(\lambda)\|_m + \inf_{\mathbf{w}\in\mathbf{H}^m(\Omega),\mathbf{w}|_\Gamma=\boldsymbol{\tau}} \|\mathbf{w}\|_m\Big)\end{aligned}$$

uniformly in $\lambda \in \Lambda$.

6.2 Analysis of a shape control problem for the stationary Navier–Stokes equations

We consider a two-dimensional shape optimization problem for the stationary Navier–Stokes equations for incompressible flow. We derive the first-order necessary conditions that can be used for developing a consistent numerical approach to the problem. The resulting optimality condition is a system of partial differential equations and variational inequalities. Details, including proofs, can be found in [190].

The model shape control problem

We consider the two-dimensional incompressible flow of a viscous fluid through the channel Ω shown in Figure 6.1. The velocity $\mathbf{u}$ and the pressure p satisfy the stationary Navier–Stokes system

$$-\nu\Delta\mathbf{u}+\mathbf{u}\cdot\nabla\mathbf{u}+\nabla p=\mathbf{f} \qquad \text{in } \Omega, \tag{6.84}$$

$$\nabla\cdot\mathbf{u}=0 \qquad \text{in } \Omega, \tag{6.85}$$

along with the Dirichlet boundary conditions

$$\mathbf{u}=\mathbf{g}=\begin{cases}\mathbf{g}_1 & \text{on } \Gamma_1,\\ \mathbf{g}_2 & \text{on } \Gamma_3,\\ \mathbf{0} & \text{on } \Gamma_2\cup\Gamma_4,\end{cases} \tag{6.86}$$

where $\mathbf{f}$ is the given body force per unit mass. In (6.84), ν denotes the inverse of the Reynolds number whenever the variables are appropriately nondimensionalized. The vectors $\mathbf{g}_1$ and $\mathbf{g}_2$ are the given velocities at the inflow Γ_1 and outflow Γ_3 of the channel, respectively. Along the bottom, Γ_4, and the top, Γ_2, of the channel the velocity vanishes. The function $\mathbf{g}$ must satisfy the compatibility condition

$$\int_\Gamma \mathbf{g}\cdot\mathbf{n}\,ds=0, \tag{6.87}$$

where $\mathbf{n}$ is the unit normal vector along the surface Γ.

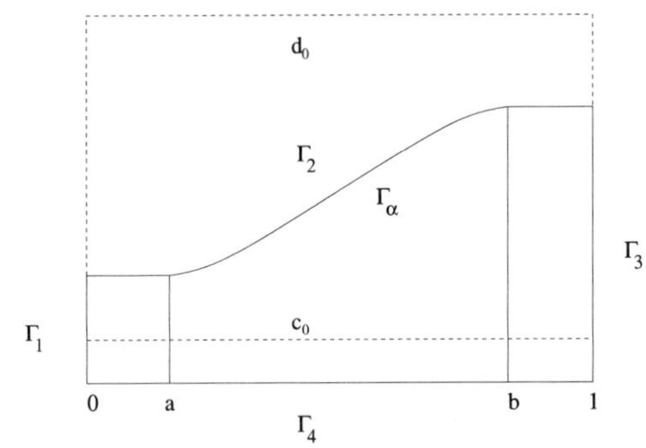

Figure 6.1. *The flow domain $\Omega(\alpha)$ and its boundary; Γ_α denotes the part of the boundary whose shape is to be determined by the optimization process.*

If some other types of boundary conditions, e.g., natural boundary conditions or outflow boundary conditions, are specified along the left, right, or bottom boundaries, the results given here are formally valid, but some technical details may differ and should be carefully revised. The set $\Gamma_\alpha=\{\mathbf{x}=(x,z)\in\mathbb{R}^2\,|\,x\in(a,b),\,z=\alpha(x)\}\subset\Gamma_2$ is the shape to be determined through the optimization process. We denote the interval (a,b) by I and the domain Ω by $\Omega(\alpha)$.

There is a substantial literature discussing the set of all possible shapes. Clearly, the function α must belong to $C^{0,1}(I)$ in order to be regular enough to suppress excessive oscillations; see, e.g., [249]. However, $\alpha \in C^{0,1}(I)$ may not be sufficient to enable one to explicitly derive a first-order necessary condition. Thus, here, the set of all admissible controls α is restricted to more regular shapes, namely, to belong to $C^{1,1}(I)$. In order to have a regular flow that is shape differentiable, we need a domain with piecewise $C^{1,1}$ boundary and convex corners (for details, see [188, 203]). Furthermore, there are some practical constraints that may be taken into account; for example, the first derivative at the points $x = a$ and $x = b$ should be specified so that Γ_α is connected smoothly to the rest of the boundary.

Specifically, we define a set of allowable shapes in the following way. Let c_0, d_0, c_2, and d_2 be positive constants and z_1 and z_2 be the location of the controlled surface Γ_α at $x = a$ and $x = b$, respectively. Then, the set

$$\{\alpha(x) \in C^1(I) \quad | \quad c_i \leq \alpha^{(i)} \leq d_i \text{ for } i = 0, 2,$$
$$\alpha(a) = z_1, \alpha(b) = z_2, \text{ and } \alpha^{(1)}(a) = \alpha^{(1)}(b) = 0\}$$

may be a suitable set of allowable shapes, where $\alpha^{(i)}$ denotes the ith derivative of α. We have fixed the values of the first derivatives at $x = a$ and $x = b$ to be zero but they can be adjusted if the boundary segments $\Gamma_2 \backslash \Gamma_\alpha$ are not horizontal when they meet Γ_α.

In order to enforce the regularity of the boundary, we take α to be the solution of the Poisson equation

$$\frac{d^2\alpha}{dx^2} = q \quad \text{on} \quad I, \qquad \alpha(a) = z_1, \quad \text{and} \quad \alpha(b) = z_2, \tag{6.88}$$

where $c_2 \leq q \leq d_2$ for all $x \in I$ is an unknown function. The other boundary conditions on α, i.e., $\alpha'(a) = \alpha'(b) = 0$, impose constraints on allowable functions q. In fact, if α and q are related by (6.88), then $\alpha'(a) = \alpha'(b) = 0$ if and only if

$$\int_a^b q(x)\,dx = 0 \qquad \text{and} \qquad \int_a^b \int_a^x q(\xi)\,d\xi dx = z_2 - z_1\,. \tag{6.89}$$

Since q is bounded, we have that $\alpha \in C^1(I)$ and Γ is piecewise $C^{1,1}$ with convex corners. We note that the lower bound for α, i.e., $\alpha(x) \geq c_0\, \forall x \in I$, is necessary to avoid the intersection of Γ_2 and Γ_4 (see Figure 6.1) and that the second derivative bounds are necessary to ensure that the curvature on Γ_α can be computed.

One could examine several practical objective functionals for determining the shape of the boundary, e.g., the reduction of the drag due to viscosity or the identification of the velocity at a fixed vertical slit downstream. To fix ideas, we focus on the minimization of the cost functional or, in the terminology of shape optimization, the design performance function,

$$\mathcal{J}(\mathbf{u}, q, \alpha) = 2\nu \int_{\Omega(\alpha)} D(\mathbf{u}) : D(\mathbf{u})\, d\mathbf{x} + \frac{\beta}{2} \int_I q^2\, dx\,, \tag{6.90}$$

where $\mathbf{u}$ is the velocity field defined on $\Omega(\alpha)$, $D(\mathbf{u}) = \frac{1}{2}(\nabla\mathbf{u} + (\nabla\mathbf{u})^T)$ is the deformation tensor for the flow $\mathbf{u}$, and β is a nonnegative constant. For $\beta = 0$, the functional (6.90) represents the rate of energy dissipation due to deformation and, except for an additive constant, can be physically identified with the viscous drag of the flow. We note that, for a solenoidal flow field $\mathbf{u}$, we have $2D(\mathbf{u}) : D(\mathbf{u}) = \nabla\mathbf{u} : \nabla\mathbf{u}$.

Formally speaking, the control problem we consider is to find $\mathbf{u}$, α, and q such that the functional (6.90) is minimized subject to the Navier–Stokes system (6.84)–(6.86) and the relations (6.87)–(6.89) being satisfied.

In Section 6.2.2, a first-order necessary condition is found through a direct sensitivity analysis, and the adjoint equation method is used to compute the shape gradient for the model problem; the derivations of this section assume that the standard condition guaranteeing the uniqueness of solutions of the stationary Navier–Stokes system holds. In Section 6.2.3, we use the Lagrange multiplier method to obtain similar results; although the derivations of this section are not as direct, they do not require that the "uniqueness condition" hold.

Notation

Depending on the context, C and K denote generic constants whose values also depend on the context. We denote by $H^s(\mathcal{O})$, $s \in \mathbb{R}$, the standard Sobolev space of order s with respect to the set $\mathcal{O}$, which is either the flow domain Ω, or its boundary Γ, or part of its boundary. Whenever m is a nonnegative integer, the inner product over $H^m(\mathcal{O})$ is denoted by $(f, g)_m$, and (f, g) denotes the inner product over $H^0(\mathcal{O}) = L^2(\mathcal{O})$. Hence, we associate with $H^m(\mathcal{O})$ its natural norm $\|f\|_{m,\mathcal{O}} = \sqrt{(f, f)_m}$. For $1 \le p < \infty$, the Sobolev space $W^{m,p}(\mathcal{O})$ is defined as the closure of $C^\infty(\mathcal{O})$ in the norm

$$\|f\|^p_{W^{m,p}(\mathcal{O})} = \sum_{|\alpha| \le m} \int_{\mathcal{O}} \left| \left(\frac{\partial}{\partial x}\right)^\alpha f(x) \right|^p dx \,.$$

The closure of $C_0^\infty(\mathcal{O})$ under the norm $\|\cdot\|_{W^{m,p}(\mathcal{O})}$ will be denoted by $W_0^{m,p}(\mathcal{O})$. Whenever possible, we will neglect the domain label in the norm.

For vector-valued functions and spaces, we use boldface notation. For example, $\mathbf{H}^s(\Omega) = [H^s(\Omega)]^n$ denotes the space of $\mathbb{R}^n$-valued functions such that each component belongs to $H^s(\Omega)$. Of special interest is the space

$$\mathbf{H}^1(\Omega) = \left\{ v_j \in L^2(\Omega) \;\middle|\; \frac{\partial v_j}{\partial x_k} \in L^2(\Omega) \quad \text{for } j, k = 1, 2 \right\}$$

equipped with the norm $\|\mathbf{v}\|_1 = (\sum_{k=1}^2 \|v_k\|_1^2)^{1/2}$. We define the space

$$\mathbf{V}(\Omega) = \{ \mathbf{u} \in \mathbf{H}^1(\Omega) \mid \nabla \cdot \mathbf{u} = 0 \}$$

and the space of infinite differentiable solenoidal functions by

$$\mathcal{V}(\Omega) = \{ \mathbf{u} \in \mathbf{C}_0^\infty(\bar{\Omega}) \mid \nabla \cdot \mathbf{u} = 0 \} \,.$$

For $\Gamma_s \subset \Gamma$ with nonzero measure, we also consider the subspace

$$\mathbf{H}^1_{\Gamma_s}(\Omega) = \{ \mathbf{v} \in \mathbf{H}^1(\Omega) \mid \mathbf{v} = \mathbf{0} \quad \text{on } \Gamma_s \} \,.$$

Also, we write $\mathbf{H}_0^1(\Omega) = \mathbf{H}^1_\Gamma(\Omega)$. For any $\mathbf{v} \in \mathbf{H}^1(\Omega)$, we write $\|\nabla \mathbf{v}\|$ for the seminorm. Let $(\mathbf{H}^1_{\Gamma_s})^*$ denote the dual space of $\mathbf{H}^1_{\Gamma_s}$. Note that $(\mathbf{H}^1_{\Gamma_s})^*$ is a subspace of $\mathbf{H}^{-1}(\Omega)$, where the latter is the dual space of $\mathbf{H}_0^1(\Omega)$. The duality pairing between $\mathbf{H}^{-1}(\Omega)$ and $\mathbf{H}_0^1(\Omega)$ is denoted by $\langle \cdot, \cdot \rangle$.

Let $\mathbf{g}$ be an element of $\mathbf{H}^{1/2}(\Gamma)$. It is well known that $\mathbf{H}^{1/2}(\Gamma)$ is a Hilbert space with norm

$$\|\mathbf{g}\|_{1/2,\Gamma} = \inf_{\mathbf{v}\in\mathbf{H}^1(\Omega);\ \gamma_\Gamma \mathbf{v}=\mathbf{g}} \|\mathbf{v}\|_1 \,,$$

where γ_Γ denotes the trace mapping $\gamma_\Gamma : \mathbf{H}^1(\Omega) \to \mathbf{H}^{1/2}(\Gamma)$. We let $(\mathbf{H}^{1/2}(\Gamma))^*$ denote the dual space of $\mathbf{H}^{1/2}(\Gamma)$ and $\langle\cdot,\cdot\rangle_\Gamma$ denote the duality pairing between $(\mathbf{H}^{1/2}(\Gamma))^*$ and $\mathbf{H}^{1/2}(\Gamma)$.

Let Γ_s be a smooth subset of Γ. Then, the trace mapping $\gamma_{\Gamma_s} : \mathbf{H}^1(\Omega) \to \mathbf{H}^{1/2}(\Gamma_s)$ is well defined and $\mathbf{H}^{1/2}(\Gamma_s) = \gamma_{\Gamma_s}(\mathbf{H}^1(\Omega))$.

Since the pressure is determined only up to an additive constant by the Navier–Stokes system with velocity boundary conditions, we define the space of square integrable functions having zero mean over Ω as

$$L_0^2(\Omega) = \left\{ p \in L^2(\Omega) \;\middle|\; \int_\Omega p\, d\mathbf{x} = 0 \right\} .$$

In order to define a weak form of the Navier–Stokes equations, we introduce the bilinear forms

$$a(\mathbf{u},\mathbf{v}) = 2\nu \int_\Omega D(\mathbf{u}) : D(\mathbf{v})\, d\mathbf{x} \qquad \forall\, \mathbf{u},\mathbf{v} \in \mathbf{H}^1(\Omega) \tag{6.91}$$

and

$$b(\mathbf{v},q) = -\int_\Omega q\, \nabla\cdot\mathbf{v}\, d\mathbf{x} \quad \forall q \in L_0^2(\Omega)\,, \quad \forall\, \mathbf{v} \in \mathbf{H}^1(\Omega), \tag{6.92}$$

and the trilinear form

$$\begin{aligned} c(\mathbf{w},\mathbf{u},\mathbf{v}) &= \int_\Omega \mathbf{w}\cdot\nabla\mathbf{u}\cdot\mathbf{v}\, d\mathbf{x} \\ &= \sum_{i,j=1}^{2} \int_\Omega w_j \left(\frac{\partial u_i}{\partial x_j}\right) v_i\, d\mathbf{x} \quad \forall\, \mathbf{w},\mathbf{u},\mathbf{v} \in \mathbf{H}^1(\Omega)\,. \end{aligned} \tag{6.93}$$

Obviously, $a(\cdot,\cdot)$ is a continuous bilinear form on $\mathbf{H}^1(\Omega)\times\mathbf{H}^1(\Omega)$, and $b(\cdot,\cdot)$ is a continuous bilinear form on $\mathbf{H}^1(\Omega) \times L_0^2(\Omega)$; also $c(\cdot,\cdot,\cdot)$ is a continuous trilinear form on $\mathbf{H}^1(\Omega) \times \mathbf{H}^1(\Omega) \times \mathbf{H}^1(\Omega)$, which can be verified by the Sobolev embedding of $\mathbf{H}^1(\Omega) \subset \mathbf{L}^4(\Omega)$ and Holder's inequality. We also have the coercivity property

$$a(\mathbf{v},\mathbf{v}) \geq C\|\mathbf{v}\|_1^2 \qquad \forall\, \mathbf{v} \in \mathbf{H}^1_{\Gamma_s}(\Omega)$$

whenever $\Gamma_s \subset \Gamma$ has positive measure and the inf-sup condition

$$\inf_{p\in L_0^2(\Omega)} \sup_{\mathbf{v}\in\mathbf{H}_0^1} \frac{b(\mathbf{v},p)}{\|\mathbf{v}\|_1\|p\|} \geq K\,.$$

For details concerning the function spaces we have introduced, one may consult [413, 428, 430, 444], and for details about the bilinear and trilinear forms and their properties, one may consult [428, 430, 444].

The associated boundary value problem

We consider the formulation of the direct problem for the Navier–Stokes system (6.84)–(6.86) for which the boundary and all the data functions are known. Let $\Gamma(\alpha)$ be the boundary that includes the segment Γ_α (see Figure 6.1) defined for a given $\alpha \in H^2(I)$. Given α, we can compute q by using (6.88).

A weak formulation of the Navier–Stokes system is given as follows:

Given $\mathbf{f} \in \mathbf{H}^{-1}(\Omega(\alpha))$ *and* $\mathbf{g} \in \mathbf{H}^{1/2}(\Gamma(\alpha))$*, find* $(\mathbf{u}, p) \in \mathbf{H}^1(\Omega(\alpha)) \times L_0^2(\Omega(\alpha))$ *satisfying*

$$\begin{cases} a(\mathbf{u}, \mathbf{v}) + c(\mathbf{u}, \mathbf{u}, \mathbf{v}) + b(\mathbf{v}, p) = \langle \mathbf{f}, \mathbf{v} \rangle \quad \forall \mathbf{v} \in \mathbf{H}_0^1(\Omega(\alpha)), \\ b(\mathbf{u}, q) = 0 \quad \forall q \in L_0^2(\Omega(\alpha)), \\ \langle \mathbf{u}, \mathbf{s} \rangle_{\Gamma(\alpha)} = \langle \mathbf{g}, \mathbf{s} \rangle_{\Gamma(\alpha)} \quad \forall \mathbf{s} \in \mathbf{H}^{-1/2}(\Gamma(\alpha)). \end{cases} \tag{6.94}$$

Existence and uniqueness results for solutions of the system (6.94) are contained in the following theorem; see, e.g., [444].

THEOREM 6.14. *Let* $\Omega(\alpha)$ *be an open, bounded set of* $\mathbb{R}^2$ *with Lipschitz-continuous boundary* $\Gamma(\alpha)$*. Let* $\mathbf{f} \in \mathbf{H}^{-1}(\Omega(\alpha))$ *and* $\mathbf{g} \in \mathbf{H}^{1/2}(\Gamma(\alpha))$ *and let* $\mathbf{g}$ *satisfy the compatibility condition* (6.87)*. Then,*

(i) *there exists at least one solution* $(\mathbf{u}, p) \in \mathbf{H}^1(\Omega(\alpha)) \times L^2(\Omega(\alpha))$ *of* (6.94)*;*

(ii) *the set of velocity fields that are solutions of* (6.94) *is closed in* $\mathbf{H}^1(\Omega(\alpha))$ *and is compact in* $\mathbf{L}^2(\Omega(\alpha))$*; and*

(iii) *if*

$$\nu > \nu_0(\Omega(\alpha), \mathbf{f}, \mathbf{g}) \tag{6.95}$$

for some positive ν_0 *whose value is determined by the given data, then the set of solutions of* (6.94) *consists of a single element.*

Note that solutions of (6.94) exist for any value of the Reynolds number. However, (iii) implies that uniqueness can be guaranteed only for "large enough" values of ν or for "small enough" data $\mathbf{f}$ and $\mathbf{g}$.

In order to write the first-order necessary condition that optimal solutions must satisfy, we shall need additional regularity for solutions of (6.94). To achieve the needed smoothness, we have to assume that the data of the problem are smoother than assumed in the hypotheses of Theorem 6.14. In that case, we have the following results; see [428, 444].

THEOREM 6.15. *Let the hypotheses of Theorem* 6.14 *hold. Let* $\Gamma(\alpha)$ *be piecewise* $C^{1,1}$ *with convex corners,* $\mathbf{g} \in \mathbf{H}^{3/2}(\Gamma(\alpha))$*, and* $\mathbf{f} \in \mathbf{L}^2(\Omega(\alpha))$*. Let* $(\mathbf{u}, p)$ *denote a solution of* (6.94)*. Then,*

(i) $(\mathbf{u}, p) \in \mathbf{H}^2(\Omega(\alpha)) \times H^1(\Omega(\alpha)) \cap L_0^2(\Omega(\alpha))$ *and*

(ii) *the set of solutions of* (6.94) *is closed in* $\mathbf{H}^2(\Omega(\alpha))$ *and compact in* $\mathbf{H}^1(\Omega(\alpha))$*.*

6.2.1 The model shape control problem and existence of solutions

In this section we give a precise formulation of the shape control problem we are considering and provide a result about the existence of solutions of that problem.

We define the closed convex set

$$\mathcal{Q}_{ad} = \big\{\alpha \in H^2(I) \mid 0 < c_0 \le \alpha \le d_0,\\ \alpha(a) = z_1, \alpha(b) = z_2, \text{ and } \alpha'(a) = \alpha'(b) = 0\big\}$$

and introduce the variable q belonging to the set

$$\mathcal{B}_{ad} = \big\{q \in L^2(I) \mid c_2 \le q \le d_2 \text{ almost everywhere (a.e.)}\big\}$$

defined by

$$\int_I qv\,dx = \int_I \frac{d^2\alpha}{dx^2}\,v\,dx \qquad \forall\, v \in L^2(I)\,. \tag{6.96}$$

The constants c_2 and d_2 are such that the set $\mathcal{Q}_{ad}$ is not empty. From the Sobolev embedding theorem, we have that $H^2(I) \subset C^1(\bar{I}) \subset C^{0,1}(\bar{I})$ and, therefore, if $\alpha \in \mathcal{Q}_{ad}$ and $q \in \mathcal{B}_{ad}$, then $\alpha \in C^{1,1}$ at least.

The optimal shape control problem can then be stated as

given $\mathbf{f} \in \mathbf{L}^2(\Omega(\alpha))$ *and* $\mathbf{g} \in \mathbf{H}^{3/2}(\Gamma(\alpha))$ *satisfying the compatibility condition* (6.87), *find* $(\widehat{\mathbf{u}}, \widehat{p}, \widehat{q}, \widehat{\alpha})$ *such that*

$$\mathcal{J}(\widehat{\mathbf{u}}, \widehat{q}, \widehat{\alpha}) \le \mathcal{J}(\mathbf{u}, q, \alpha) \tag{6.97}$$

for all $(\mathbf{u}, p, q, \alpha) \in \mathbf{H}^2(\Omega(\alpha)) \times \mathbf{H}^1(\Omega(\alpha)) \cap L^2_0(\Omega(\alpha)) \times \mathcal{B}_{ad} \times \mathcal{Q}_{ad}$ *satisfying* (6.94) *and* (6.96).

The extended domain $\widehat{\Omega}$ is defined to be the rectangle $[0, 1] \times [0, d_0]$. We also assume that $\mathbf{f}$ is defined over the extended domain $\widehat{\Omega}$ and write $\mathbf{f}(\Omega(\alpha))$ to denote its restriction to $\Omega(\alpha)$.

The admissible set of states and controls is given by

$$\mathcal{A}_{ad} = \{(\mathbf{u}, p, q, \alpha) \in \mathbf{H}^2(\Omega(\alpha)) \cap \mathbf{V}(\Omega(\alpha)) \times H^1(\Omega(\alpha)) \cap L^2_0(\Omega(\alpha)) \times \mathcal{B}_{ad} \times \mathcal{Q}_{ad}\\ \text{such that } \mathcal{J}(\mathbf{u}, q, \alpha) < \infty \text{ and } (\mathbf{u}, p, q, \alpha) \text{ satisfies (6.94) and (6.96)}\}\,.$$

Concerning the question of the existence of optimal solutions for the problem in (6.97), we have the following result.

Theorem 6.16. *There exists at least one optimal solution* $(\mathbf{u}, p, q, \alpha) \in \mathcal{A}_{ad}$ *of the optimal shape control problem* (6.97).

6.2.2 Direct determination of the optimality system and shape gradient

Generalized Liebniz rules

Let $\mathcal{C}_{ad}$ be the set of $\alpha \in \mathcal{Q}_{ad}$ such that $\alpha'' \in \mathcal{B}_{ad}$. Let $\alpha \in \mathcal{C}_{ad}$ be the function that determines $\Gamma_\alpha \subset \Gamma(\alpha)$. After deformation, the domain $\Omega(\alpha)$ takes a new shape $\Omega(\widetilde{\alpha})$ with boundary

$\Gamma(\widetilde{\alpha})$ corresponding to the function $\widetilde{\alpha} \in \mathcal{C}_{ad}$. The field, defined on Γ_α, transforming Γ_α into $\Gamma_{\widetilde{\alpha}}$ is indicated by $\mathbf{V}$ and the corresponding variation by $\delta\alpha = \widetilde{\alpha} - \alpha$. For example, in our specific case, if $\alpha, \widetilde{\alpha} \in \mathcal{Q}_{ad}$, then $\delta\alpha = \widetilde{\alpha} - \alpha$ and $\mathbf{V} = (0, \delta\alpha)$, where the vector $\mathbf{V}$ is defined on Γ_α. We note that, as proved in [408], the vector $\mathbf{V}$ can always be extended from Γ_α to every point $(x, z) \in \Omega(\alpha)$ or every point in $\mathbb{R}^2$, i.e., in our case we can simply take $\mathbf{V}(x, z) = (0, z\delta\alpha(x)/\alpha(x)) \in \mathbb{R}^2$.

We can generate a family of boundaries parameterized by t as

$$\Gamma_{\alpha+t\delta\alpha} = \{\mathbf{x}_\alpha + t\,\mathbf{V}(\mathbf{x}_\alpha) \,|\, \mathbf{x}_\alpha \in \Gamma_\alpha\} \tag{6.98}$$

for all $t \in [0, 1]$. Let $\mathcal{K}(\alpha)$ be a functional on the domain $\Omega(\alpha)$. We say that $\mathcal{K}(\alpha) : \mathcal{C}_{ad} \to \mathbb{R}$ is Gateaux differentiable at α in the direction $\delta\alpha$ if there is a real number $\mathcal{K}' = (D\mathcal{K}/D\alpha)\delta\alpha$ such that

$$\lim_{t\to 0+} \frac{|\mathcal{K}(\Omega(\alpha + t\delta\alpha)) - \mathcal{K}(\Omega(\alpha)) - t\,\mathcal{K}'|}{t} = 0\,. \tag{6.99}$$

In a similar way, when Y is a Banach space, the map $\mathbf{u}(\alpha) : \mathcal{C}_{ad} \to Y$ is said to be Gateaux differentiable at α in the direction $\delta\alpha$ ($\mathbf{V} = (0, \delta\alpha)$) if there exists a function $\mathbf{u}' \in Y(\Omega(\alpha))$ such that

$$\lim_{t\to 0+} \frac{\|\mathbf{u}(\Omega(\alpha + t\delta\alpha)) - \mathbf{u}(\Omega(\alpha)) - t\,\mathbf{u}'\|_{Y(\Omega(\alpha))}}{t} = 0\,. \tag{6.100}$$

Before proving differentiability, we need the following lemma from [408].

Lemma 6.17. *Given $\alpha, \widetilde{\alpha} \in \mathcal{C}_{ad}$ ($\delta\alpha = \widetilde{\alpha} - \alpha$), $\widehat{y}_1 \in W^{1,1}(\mathbb{R}^2)$, and $\widehat{y}_2 \in W^{2,1}(\mathbb{R}^2)$, let*

$$\mathcal{K}_{\Omega(\alpha)} = \int_{\Omega(\alpha)} \widehat{y}_1(\mathbf{x})\,d\mathbf{x} \qquad \text{and} \qquad \mathcal{K}_{\Gamma(\alpha)} = \int_{\Gamma(\alpha)} \widehat{y}_2(s)\,ds\,.$$

Then,

$$\frac{D\mathcal{K}_\Omega}{D\alpha}\cdot\delta\alpha = \int_{\Omega(\alpha)} \nabla\cdot(\mathbf{V}\,\widehat{y}_1(\mathbf{x}))\,d\mathbf{x} = \int_{\Gamma_\alpha} \widehat{y}_1(s)\,(\mathbf{V}\cdot\mathbf{n})\,ds, \tag{6.101}$$

$$\frac{D\mathcal{K}_\Gamma}{D\alpha}\cdot\delta\alpha = \int_{\Gamma_\alpha} \left(\frac{\partial\widehat{y}_2}{\partial n} + \kappa\widehat{y}_2(s)\right)(\mathbf{V}\cdot\mathbf{n})\,ds\,, \tag{6.102}$$

where $\mathbf{V}$ is the vector $(0, \delta\alpha)$ defined on Γ_α, and κ and $\mathbf{n}$ are the curvature and the unit vector normal to Γ_α, respectively.

This lemma has been stated in the context of our particular optimal control situation, but a more general framework can be found in [404, 408]. We remark that, in order to apply the above lemma, the functions $\widehat{y}_1$ and $\widehat{y}_2$ must be defined on $\mathbb{R}^2$ or extended outside $\Omega(\alpha)$, and their gradients must be integrable. This allows the limit to be represented as a boundary integral over Γ_α. The requirement $\widehat{y}_1 \in W^{1,1}(\mathbb{R}^2)$ implies a certain regularity for the solution of the Navier–Stokes system. If such a regularity is not present, we may extend the previous lemma to a function $\widehat{y}_1$ in $L^2(\mathbb{R}^2)$. In this case, the functional is weakly differentiable in H^{-2} [408], which may not imply regular solutions and its representation as a boundary integral on Γ_α.

Differentiability of candidate optimal solutions and the functional

We shall show that the optimal control solution must satisfy a first-order necessary condition that leads to a variational inequality. By studying this variational inequality, a possible candidate for the optimal control solution can be found. In order to obtain this result, we need to prove the differentiability of the functional and, before that, of functions satisfying the constraints.

Theorem 6.18. *Let (6.95) hold so that (6.94) has a unique solution. Let* $(\mathbf{u}, p, q, \alpha) \in \mathcal{A}_{ad}$. *The mapping* $(\mathbf{u}, p) = (\mathbf{u}(\alpha), p(\alpha))$ *has a Gateaux derivative* $(\mathbf{u}', p') = \left(\frac{d\mathbf{u}}{d\alpha} \cdot \delta\alpha, \frac{dp}{d\alpha} \cdot \delta\alpha\right)$ *in every direction* $\delta\alpha = \widetilde{\alpha} - \alpha$ *for all* $\widetilde{\alpha}$ *in* $\mathcal{A}_{ad}$. *Furthermore,* $(\mathbf{u}', p')$ *is the solution of the problem*

$$\left\{\begin{aligned} & a(\mathbf{u}', \mathbf{v}) + c(\mathbf{u}, \mathbf{u}', \mathbf{v}) + c(\mathbf{u}', \mathbf{u}, \mathbf{v}) \\ & \qquad + b(\mathbf{v}, p') = 0 \quad \forall\, \mathbf{v} \in \mathbf{H}^1_0(\Omega(\alpha)), \\ & b(\mathbf{u}', q) = 0 \quad \forall\, q \in L^2_0(\Omega(\alpha)), \\ & \int_{\Gamma(\alpha)} \mathbf{s} \cdot \left(\mathbf{u}' + \frac{\partial \mathbf{u}}{\partial n}(\mathbf{v} \cdot \mathbf{n})\right) ds = 0 \qquad \forall\, \mathbf{s} \in \mathbf{H}^{-1/2}(\Gamma(\alpha)), \end{aligned}\right. \tag{6.103}$$

where the vector $\mathbf{V}$, *defined by* $(0, \delta\alpha)$ *on* Γ_α, *vanishes on* $\Gamma(\alpha) - \Gamma_\alpha$. *The unit vector* $\mathbf{n}$ *is the unit normal to the boundary* $\Gamma(\alpha)$.

Under the hypotheses of Theorem 6.18, we have the existence of the Gateaux derivative of the map $(\mathbf{u}, p) = (\mathbf{u}(\alpha), p(\alpha))$. Clearly, the map $q = q(\alpha)$ from $\mathcal{Q}_{ad}$ to $L^2(I)$ is differentiable at $\alpha \in \mathcal{Q}_{ad}$ in the direction $\delta\alpha$, and its Gateaux derivative $q' = Dq/D\alpha \cdot \delta\alpha$ satisfies the following equation:

$$\int_I v\, q'\, dx + \int_I \frac{dv}{dx} \frac{d\delta\alpha}{dx}\, dx = 0 \quad \forall\, v \in H^1_0(I)\,. \tag{6.104}$$

Now, it is easy to show the differentiability of the functional $\mathcal{J}(\mathbf{u}, q, \alpha)$.

Theorem 6.19. *Let the hypotheses of Theorem* 6.18 *hold. Let* $(\mathbf{u}, p, q, \alpha) \in \mathcal{A}_{ad}$ *have Gateaux derivative* $\mathbf{u}' \in \mathbf{H}^1(\Omega(\alpha))$ *satisfying* (6.103). *Let* $q' \in L^2(I)$ *satisfy* (6.104). *Then, the functional in* (6.90) *defines a mapping*

$$\mathcal{J}(\mathbf{u}(\alpha), q(\alpha), \alpha) : \mathcal{Q}_{ad} \to \mathbb{R}\,. \tag{6.105}$$

This mapping has a Gateaux derivative $\mathcal{J}'(\mathbf{u}, q, \alpha)$ *in the direction* $\delta\alpha = \widetilde{\alpha} - \alpha$ *for all admissible* $\widetilde{\alpha}$. *Furthermore, we have*

$$\begin{aligned} \mathcal{J}'(\mathbf{u}(\alpha), q(\alpha), \alpha) = \frac{D\mathcal{J}(\mathbf{u}, q, \alpha)}{D\alpha} \cdot \delta\alpha = 2\nu \int_{\Omega(\alpha)} \nabla\mathbf{u} \cdot \nabla\mathbf{u}'\, d\mathbf{x} \\ + \nu \int_{\Gamma_\alpha} \frac{\partial \mathbf{u}}{\partial n} \frac{\partial \mathbf{u}}{\partial n} (\mathbf{V} \cdot \mathbf{n})\, ds + \beta \int_I q\, q'\, dx\,, \end{aligned} \tag{6.106}$$

where $\mathbf{V}$ *is defined by* $(0, \delta\alpha)$ *on* Γ_α *and* $\mathbf{n}$ *is the unit vector normal to the boundary.*

The optimality system

We show that the optimal control problem implies a first-order necessary condition. If $(\mathbf{u}, p, q, \alpha)$ is an optimal solution, then for every $\delta\alpha = \widetilde{\alpha} - \alpha$ such that $\widetilde{\alpha} \in \mathcal{Q}_{ad}$ and for every $\epsilon \in \mathbb{R}^+$, we have, from the definition of an optimal solution,

$$\mathcal{J}(\mathbf{u}(\alpha + \epsilon\delta\widetilde{\alpha}), q(\alpha + \epsilon\delta\widetilde{\alpha}), \alpha + \epsilon\delta\widetilde{\alpha}) \geq \mathcal{J}(\mathbf{u}, q, \alpha).$$

The above inequality implies

$$\frac{\mathcal{J}(\mathbf{u}(\alpha + \epsilon\delta\widetilde{\alpha}), q(\alpha + \epsilon\delta\widetilde{\alpha}), \alpha + \epsilon\delta\widetilde{\alpha}) - \mathcal{J}(\mathbf{u}, q, \alpha)}{\epsilon} \geq 0 \quad \text{if} \quad \epsilon \geq 0.$$

The limit must be nonnegative when ϵ tends to zero, which leads to the following first-order necessary condition.

Theorem 6.20. *If $(\mathbf{u}, p, q, \alpha)$ is an optimal pair for the problem in* (6.97) *and the functional in* (6.90) *is Gateaux differentiable, then the necessary condition for α to be a minimizer is*

$$\mathcal{J}'(\mathbf{u}, q, \widetilde{\alpha} - \alpha) \geq 0 \tag{6.107}$$

for all $\widetilde{\alpha} \in \mathcal{Q}_{ad}$ such that $\widetilde{\alpha}'' \in \mathcal{B}_{ad}$.

We recall that the set of all $\widetilde{\alpha} \in \mathcal{Q}_{ad}$ such that $\widetilde{\alpha}'' \in \mathcal{B}_{ad}$ is a convex closed set and is not empty from the definition of $\mathcal{B}_{ad}$.

In Theorem 6.23, we shall show that conditions (6.107) can be written explicitly and that the solution of the problem in (6.97) satisfies a system of equations and variational inequalities. In order to do this, we need the following interesting preliminary results.

Lemma 6.21. *Let α and $\widetilde{\alpha}$ be in $\mathcal{Q}_{ad}$ with $\delta\alpha = \widetilde{\alpha} - \alpha$, and let $\mathbf{u}'$ be defined by* (6.103). *Then, for every $\widetilde{\mathbf{h}}_2$ in $\mathbf{H}^1(\Omega(\alpha))$, we have*

$$\int_{\Omega(\alpha)} \nabla\widetilde{\mathbf{h}}_2 \cdot \nabla\mathbf{u}'\, d\mathbf{x} = -\nu \int_{\Gamma_\alpha} \frac{\partial \mathbf{u}}{\partial n}\frac{\partial \mathbf{w}}{\partial n} (\mathbf{V}\cdot\mathbf{n})\, ds,$$

where $\mathbf{V} = (0, \delta\alpha)$ is defined on Γ_α and $\mathbf{w}$ is the solution of the linear adjoint problem

$$\left\{\begin{aligned} & a(\mathbf{w}, \mathbf{v}) + c(\mathbf{v}, \mathbf{u}, \mathbf{w}) + c(\mathbf{u}, \mathbf{v}, \mathbf{w}) \\ & \qquad + b(\mathbf{v}, \sigma) = -\int_{\Omega(\alpha)} \nabla\widetilde{\mathbf{h}}_2 : \nabla\mathbf{v}\, d\mathbf{x} \qquad \forall\, \mathbf{v} \in \mathbf{H}_0^1(\Omega(\alpha)), \\ & b(\mathbf{w}, q) = 0 \quad \forall q \in L_0^2(\Omega(\alpha)), \\ & \langle \mathbf{w}, \mathbf{s}\rangle_{\Gamma(\alpha)} = 0 \quad \forall\, \mathbf{s} \in \mathbf{H}^{-1/2}(\Gamma(\alpha)). \end{aligned}\right. \tag{6.108}$$

Lemma 6.22. *Let α and $\widetilde{\alpha}$ be in $\mathcal{Q}_{ad}$ with $\delta\alpha = \widetilde{\alpha} - \alpha$, and let q' be defined by* (6.104). *Then, for every q in $L^2(I)$, we have*

$$\beta \int_I q\, q'\, dx = \int_I \frac{d\mu}{dx}\frac{d\delta\alpha}{dx}\, dx,$$

where μ is the solution of the linear equation

$$\int_I (\mu + \beta q)\,\phi\,dx = 0 \quad \forall\,\phi \in L^2(I)\ . \tag{6.109}$$

We now write out conditions (6.107) explicitly, using the solution of an adjoint system.

Theorem 6.23. *If* $(\mathbf{u}, p, q, \alpha)$ *is optimal for the problem* (6.97) *satisfying the hypotheses of Theorem* 6.18, *then* α *is the solution of*

$$\mathcal{J}'(\mathbf{u}, q, \alpha)(\mathbf{w}, \mu, \widetilde{\alpha} - \alpha) \geq 0 \quad \forall\,\widetilde{\alpha} \in \mathcal{Q}_{ad}\,, \tag{6.110}$$

where

$$\begin{aligned}&\mathcal{J}'(\mathbf{u}, q, \alpha)(\mathbf{w}, \mu, \zeta)\\ &\quad = \nu \int_{\Gamma_\alpha} \left(\frac{\partial \mathbf{u}}{\partial n}\frac{\partial \mathbf{u}}{\partial n} - \frac{\partial \mathbf{u}}{\partial n}\frac{\partial \mathbf{w}}{\partial n}\right)(\mathbf{V}(\zeta)\cdot\mathbf{n})\,ds + \int_I \frac{d\mu}{dx}\frac{d\zeta}{dx}dx\,.\end{aligned} \tag{6.111}$$

The vector $\mathbf{V}(\zeta)$ *is simply* $(0, \zeta)$, $\mathbf{n}$ *is the unit normal to* Γ_α, *and* $\mu + \beta q = 0$, *where* $q \in \mathcal{B}_{ad}$ *and the function* $\mathbf{w}$ *is the solution of the adjoint system*

$$\left\{\begin{aligned}&a(\mathbf{w}, \mathbf{v}) + c(\mathbf{v}, \mathbf{u}, \mathbf{w}) + c(\mathbf{u}, \mathbf{v}, \mathbf{w}) + b(\mathbf{v}, \sigma)\\ &\qquad\qquad = -2\nu \int_{\Omega(\alpha)} (\nabla\mathbf{u} : \nabla\mathbf{v})\,dx \quad \forall\,\mathbf{v} \in \mathbf{H}_0^1(\Omega(\alpha)),\\ &b(\mathbf{w}, q) = 0 \quad \forall\,q \in L_0^2(\Omega(\alpha)),\\ &\mathbf{w} = \mathbf{0} \quad \forall\,\mathbf{x} \in \Gamma(\alpha)\,.\end{aligned}\right. \tag{6.112}$$

To summarize, in order to solve the optimal problem for $(\mathbf{u}, \mathbf{w}, p, \sigma, q, \mu, \alpha)$ by using the adjoint equation method, we have to solve the Navier–Stokes system

$$\left\{\begin{aligned}&a(\mathbf{u}, \mathbf{v}) + c(\mathbf{u}, \mathbf{u}, \mathbf{v}) + b(\mathbf{v}, p) = \langle \mathbf{f}, \mathbf{v}\rangle \quad \forall\,\mathbf{v} \in \mathbf{H}_0^1(\Omega(\alpha)),\\ &b(\mathbf{u}, q) = 0 \quad \forall\,q \in L_0^2(\Omega(\alpha)),\\ &\langle \mathbf{u}, \mathbf{s}\rangle_{\Gamma(\alpha)} = \langle \mathbf{g}, \mathbf{s}\rangle_{\Gamma(\alpha)} \qquad \forall\,\mathbf{s} \in \mathbf{H}^{-1/2}(\Gamma(\alpha));\end{aligned}\right. \tag{6.113}$$

the adjoint system

$$\left\{\begin{aligned}&a(\mathbf{w}, \mathbf{v}) + c(\mathbf{w}, \mathbf{u}, \mathbf{v}) + c(\mathbf{u}, \mathbf{w}, \mathbf{v}) + b(\mathbf{v}, \sigma)\\ &\qquad\qquad = -2\nu \int_{\Omega(\alpha)} \nabla\mathbf{u}\cdot\nabla\mathbf{v}\,dx \quad \forall\,\mathbf{v} \in \mathbf{H}_0^1(\Omega(\alpha)),\\ &b(\mathbf{w}, q) = 0 \quad \forall\,q \in L_0^2(\Omega(\alpha)),\\ &\langle \mathbf{w}, \mathbf{s}\rangle_{\Gamma(\alpha)} = 0 \qquad \forall\,\mathbf{s} \in \mathbf{H}^{-1/2}(\Gamma(\alpha))\,;\end{aligned}\right. \tag{6.114}$$

and the inequality

$$\mathcal{L}'(\mathbf{u}, \alpha)(\mathbf{w}, \widetilde{\alpha} - \alpha) \geq 0 \tag{6.115}$$

for all $\widetilde{\alpha} \in \mathcal{C}_{ad}$. The function $\mathcal{L}'(\mathbf{u}, \alpha)(\mathbf{w}, \zeta)$ is defined by

$$\mathcal{L}'(\mathbf{u}, \alpha)(\mathbf{w}, \zeta) = \beta \int_I \frac{d^2\alpha}{dx^2} \frac{d^2\zeta}{dx^2} dx + \nu \int_{\Gamma_\alpha} \left(\frac{\partial \mathbf{u}}{\partial n} \frac{\partial \mathbf{u}}{\partial n} - \frac{\partial \mathbf{u}}{\partial n} \frac{\partial \mathbf{w}}{\partial n} \right) (\mathbf{V}(\zeta) \cdot \mathbf{n}) \, ds \,,$$

with boundary conditions $\alpha(a) = z_1$, $\alpha(b) = z_2$, and $\alpha'(a) = \alpha'(b) = 0$. The quantities q and μ can be computed through

$$\int_I q \, v \, dx = - \int_I \frac{d\alpha}{dx} \frac{dv}{dx} dx \quad \forall \, v \in H_0^1(I) \tag{6.116}$$

and $\mu + \beta q = 0$, respectively.

The shape gradient

We now have two methods for computing the shape gradient. Given a direction $\delta\alpha$ and a candidate optimizer $(\mathbf{u}, p, q, \alpha)$, we have that the shape gradient is given in terms of the *sensitivity variables* by (6.106), where $\mathbf{V} = (0, \delta\alpha)$ and where the flow sensitivity $\mathbf{u}'$ is determined from (6.103). Alternatively, the shape gradient is given in terms of *adjoint variables* by (6.111), where the adjoint variables $\mathbf{w}$ and μ are determined from (6.112) and $\mu + \beta q = 0$, respectively.

It is important to note that the sensitivity equations (6.103) involve $\delta\alpha$ through the appearance of $\mathbf{V}$. Thus, if one changes the direction $\delta\alpha$, one must recompute the sensitivities. On the other hand, the adjoint system (6.112) is independent of the choice of $\delta\alpha$. This can have a profound effect on the cost of determining optimal solutions through methods that use the shape gradient. For any candidate optimizer $(\mathbf{u}, p, q, \alpha)$, the shape gradient can be computed for multiple directions $\delta\alpha$ with a single linear adjoint system solution. On the other hand, the sensitivity equation must be solved for each distinct direction $\delta\alpha$.

6.2.3 The Lagrange multiplier method

In this section, we show that the Lagrange multiplier technique is well posed and can be used to obtain the first-order necessary condition. Further, the Lagrangian map can be shown to be strictly differentiable for all values of the external force, which allows us to apply the Lagrange multiplier method to a wider range of problems and completes the theoretical treatment of the problem for arbitrary values of the viscosity. Also, this method gives a different and better theoretical insight into the control process, allowing us to write the inequality constraints in a different form.

Preliminaries

First, we introduce auxiliary variables that allow us to transform the inequality constraints into equalities and then invoke well-known derivations for equality constrained minimization problems; see, e.g., [369, 410].

We begin by replacing

$$c_0 \le \alpha \le d_0 \qquad \text{and} \qquad c_2 \le q \le d_2 \qquad \forall\, x \in I \tag{6.117}$$

with

$$|\alpha - \alpha_0|^2 - \alpha_m^2 + s_0^2 = 0 \qquad \forall\, x \in I, \tag{6.118}$$

$$|q - q_0|^2 - q_m^2 + s_2^2 = 0 \qquad \forall\, x \in I \tag{6.119}$$

for some $s_2 \in L^2(I)$ and $s_0 \in H^2(I)$, where $\alpha_0 = (c_0 + d_0)/2$, $q_0 = (c_2 + d_2)/2$, $\alpha_m = (d_0 - c_0)/2$, and $q_m = (d_2 - c_2)/2$. Clearly, if (6.118)–(6.119) are satisfied, then so is (6.117). Also, note that if $(\mathbf{u}, p, q, \alpha)$ is a solution of the optimal control problem, then there exist s_0, s_2 such that α, q and s_0, s_2 satisfy (6.118)–(6.119).

In this section, we let $\Gamma(\alpha)$ be piecewise $C^{1,1}$ in agreement with the proposed model problem and $\mathbf{g} \in \mathbf{H}^{3/2}(\Gamma(\alpha))$, where $\int_{\Gamma(\alpha)} \mathbf{g} \cdot \mathbf{n} ds = 0$ with $\mathbf{g} = \mathbf{0}$ on $\Gamma_2 \cap \Gamma_4$, $\mathbf{g} = \mathbf{g}_1 \in \mathbf{H}^{3/2}(\Gamma_1)$ on Γ_1, and $\mathbf{g} = \mathbf{g}_2 \in \mathbf{H}^{3/2}(\Gamma_2)$ on Γ_2.

The Lagrange multiplier method proposed here is based on an embedded domain technique method. We write the Lagrangian on a larger domain and then choose the solution that fits our domain, forcing the boundary values as a constraint.

Let $\widehat{\Omega}$ be an open bounded domain with boundary $\widehat{\Gamma} \in C^{1,1}$ with convex corners, which contains $\Omega(\alpha)$ for all α. We use the "hat notation" for functions on $\widehat{\Omega}$. Also, let $\widehat{\mathbf{f}} \in \mathbf{L}^2(\widehat{\Omega})$ be the body force on the extended domain as previously discussed.

Let $\widehat{\mathbf{B}}_1 = (\mathbf{H}^2(\widehat{\Omega}) \cap \mathbf{H}_0^1(\widehat{\Omega})) \times (L_0^2(\widehat{\Omega}) \cap H^1(\widehat{\Omega})) \times \mathcal{B}_{ad} \times \mathcal{Q}_{ad} \times H^2(I) \times L^2(I)$, $\widehat{\mathbf{B}}_2 = \mathbf{H}^{-1}(\widehat{\Omega}) \times L_0^2(\widehat{\Omega}) \times \mathbf{H}^{1/2}(I) \times \mathbf{H}^{1/2}(\Gamma(\alpha) - \Gamma_\alpha) \times L^2(I) \times H^2(I) \times L^1(I)$ and $\widehat{\mathbf{B}}_3 = \mathbf{H}^{-1}(\widehat{\Omega}) \times L_0^2(\widehat{\Omega}) \times \mathbf{H}^{1/2}(I) \times \mathbf{H}^{1/2}(\Gamma(\alpha) - \Gamma_\alpha) \times L^2(I) \times W^{2,1}(I) \times L^1(I)$. We equip $\widehat{\mathbf{B}}_1$, $\widehat{\mathbf{B}}_2$, and $\widehat{\mathbf{B}}_3$ with the usual graph norms for the product spaces involved.

We define the nonlinear mapping $M : \widehat{\mathbf{B}}_1 \to \widehat{\mathbf{B}}_3$ by

$$M(\widehat{\mathbf{u}}, \widehat{p}, q, \alpha, s_0, s_2) = (\mathbf{l}_1, l_2, \mathbf{l}_3, \mathbf{l}_4, l_5, l_6, l_7)$$

for $(\widehat{\mathbf{u}}, \widehat{p}, q, \alpha, s_0, s_2) \in \widehat{\mathbf{B}}_1$ and $(\mathbf{l}_1, l_2, \mathbf{l}_3, \mathbf{l}_4, l_5, l_6, l_7) \in \widehat{\mathbf{B}}_3$ if and only if

$$\begin{cases} a(\widehat{\mathbf{u}}, \widehat{\mathbf{v}}) + c(\widehat{\mathbf{u}}, \widehat{\mathbf{u}}, \widehat{\mathbf{v}}) + b(\widehat{\mathbf{v}}, \widehat{p}), \\ \qquad - \displaystyle\int_{\widehat{\Omega}} \widehat{\mathbf{f}} \cdot \widehat{\mathbf{v}}\, d\mathbf{x} = \int_{\widehat{\Omega}} \mathbf{l}_1 \cdot \widehat{\mathbf{v}}\, d\mathbf{x} \quad \forall\, \widehat{\mathbf{v}} \in \mathbf{H}_0^1(\widehat{\Omega}), \\ b(\widehat{\mathbf{u}}, \widehat{z}) = \displaystyle\int_{\widehat{\Omega}} l_2 \widehat{z}\, d\mathbf{x} \quad \forall\, \widehat{z} \in L_0^2(\widehat{\Omega}), \\ \displaystyle\int_{\Gamma_\alpha} \widehat{\mathbf{u}} \cdot \mathbf{s}\, ds = \int_I \mathbf{l}_3 \cdot \mathbf{s}\, ds \qquad \forall\, \mathbf{s} \in \mathbf{H}^{-1/2}(I), \\ \displaystyle\int_{\Gamma(\alpha)-\Gamma_\alpha} (\widehat{\mathbf{u}} - \mathbf{g}) \cdot \mathbf{s}\, ds = \int_{\Gamma(\alpha)-\Gamma_\alpha} \mathbf{l}_4 \cdot \mathbf{s}\, ds \quad \forall\, \mathbf{s} \in \mathbf{H}^{-1/2}(\Gamma(\alpha) - \Gamma_\alpha), \\ \displaystyle\int_I v\, q\, dx + \int_I \frac{d\alpha}{dx} \frac{dv}{dx}\, dx = \int_I l_5\, v\, dx \quad \forall\, v \in H_0^1(I), \\ (\alpha - \alpha_0)^2 - \alpha_m^2 + s_0^2 = l_6 \quad \forall\, x \in I, \\ (q - q_0)^2 - q_m^2 + s_2^2 = l_7 \quad \forall\, x \in I, \end{cases} \tag{6.120}$$

with $\alpha(a) = z_1, \alpha(b) = z_2$, and $\alpha'(a) = \alpha'(b) = 0$. The set of constraint equations in the optimal shape control problem can be expressed as $M(\widehat{\mathbf{u}}, \widehat{p}, q, \alpha, s_0, s_2) = (\mathbf{0}, 0, \mathbf{0}, \mathbf{0}, 0, 0, 0)$ when the test functions $\widehat{\mathbf{v}}, \widehat{z}$ vanish outside $\widehat{\Omega}$.

Given $(\mathbf{u}_1, p_1, q_1, \alpha_1) \in \mathcal{A}_{ad}$, we define another nonlinear mapping $Q : \widehat{\mathbf{B}}_1 \to \mathbb{R} \times \widehat{\mathbf{B}}_3$ by $Q(\widehat{\mathbf{u}}, \widehat{p}, q, \alpha, s_0, s_2) = (a, \mathbf{l}_1, l_2, \mathbf{l}_3, \mathbf{l}_4, l_5, l_6, l_7)$ if and only if

$$\begin{pmatrix} \mathcal{J}(\widehat{\mathbf{u}}, q, \alpha) - \mathcal{J}(\mathbf{u}_1, q_1, \alpha_1) \\ M(\widehat{\mathbf{u}}, \widehat{p}, q, \alpha, s_0, s_2) \end{pmatrix} = \begin{pmatrix} a \\ (\mathbf{l}_1, l_2, \mathbf{l}_3, , \mathbf{l}_4, l_5, l_6, l_7) \end{pmatrix}. \tag{6.121}$$

Differentiability

These mappings are strictly differentiable, as shown in the following lemma. We recall the notion of strict differentiability (see [410]). Let X and Y denote Banach spaces. Then the mapping $\varphi : X \to Y$ is strictly differentiable at $x \in X$ if there exists a bounded, linear mapping D from X to Y such that, for any $\epsilon > 0$, there exists a $\delta > 0$ such that, whenever $\|x - x_1\|_X < \delta$ and $\|x - x_2\|_X < \delta$ for $x_1, x_2 \in X$, then

$$\|\varphi(x_1) - \varphi(x_2) - D(x_1 - x_2)\|_Y \le \epsilon \|x_1 - x_2\|_X \,.$$

The strict derivative D at the point $x \in X$, if it exists, will often be denoted by $D = \varphi'(x)$. The value of this mapping on an element $\widetilde{x} \in X$ will often be denoted by $\varphi'(x) \cdot \widetilde{x}$. In the next result we can identify $X = \widehat{\mathbf{B}}_1$ and $Y = \widehat{\mathbf{B}}_2$.

Lemma 6.24. *Let the nonlinear mappings $M : \widehat{\mathbf{B}}_1 \to \widehat{\mathbf{B}}_3$ and $Q : \widehat{\mathbf{B}}_1 \to \mathbb{R} \times \widehat{\mathbf{B}}_3$ be defined by* (6.120) *and* (6.121), *respectively. Then, these mappings are strictly differentiable at a point $(\widehat{\mathbf{u}}, \widehat{p}, q, \alpha, s_0, s_2) \in \widehat{\mathbf{B}}_1$ and their strict derivatives are given by the bounded linear operator $M'(\widehat{\mathbf{u}}, \widehat{p}, q, \alpha, s_0, s_2) : \widehat{\mathbf{B}}_1 \to \widehat{\mathbf{B}}_2$, where $M'(\widehat{\mathbf{u}}, \widehat{p}, q, \alpha, s_0, s_2) \cdot (\widetilde{\mathbf{u}}, \widetilde{p}, \widetilde{q}, \widetilde{\alpha}, \widetilde{s}_0, \widetilde{s}_2)$ $= (\bar{\mathbf{l}}_1, \bar{l}_2, \bar{\mathbf{l}}_3, \bar{\mathbf{l}}_4, \bar{l}_5, \bar{l}_6, \bar{l}_7)$ for $(\widetilde{\mathbf{u}}, \widetilde{p}, \widetilde{q}, \widetilde{\alpha}, \widetilde{s}_0, \widetilde{s}_2) \in \widehat{\mathbf{B}}_1$ and $(\bar{\mathbf{l}}_1, \bar{l}_2, \bar{\mathbf{l}}_3, \bar{\mathbf{l}}_4, \bar{l}_5, \bar{l}_6, \bar{l}_7) \in \widehat{\mathbf{B}}_2$ if and only if*

$$\left\{ \begin{aligned} & a(\widetilde{\mathbf{u}}, \widehat{\mathbf{v}}) + c(\widetilde{\mathbf{u}}, \widehat{\mathbf{u}}, \widehat{\mathbf{v}}) + c(\widehat{\mathbf{u}}, \widetilde{\mathbf{u}}, \widehat{\mathbf{v}}) + b(\widehat{\mathbf{v}}, \widetilde{p}) \\ & \qquad = \int_{\widehat{\Omega}} \bar{\mathbf{l}}_1 \cdot \widehat{\mathbf{v}}\, d\mathbf{x} \quad \forall\, \widehat{\mathbf{v}} \in \mathbf{H}_0^1(\widehat{\Omega}), \\ & b(\widetilde{\mathbf{u}}, \widehat{z}) = \int_{\widehat{\Omega}} \bar{l}_2 \widehat{z}\, d\mathbf{x} \quad \forall\, \widehat{z} \in L_0^2(\widehat{\Omega}), \\ & \int_{\Gamma_\alpha} \widetilde{\mathbf{u}} \cdot \mathbf{s}\, ds + \int_{\Gamma_\alpha} (\mathbf{V}(\widetilde{\alpha}) \cdot \mathbf{n}) \left(\kappa + \frac{\partial}{\partial n} \right) \mathbf{u} \cdot \mathbf{s}\, ds \\ & \qquad = \int_I \bar{\mathbf{l}}_3 \cdot \mathbf{s}\, ds \quad \forall\, \mathbf{s} \in \mathbf{H}^{-1/2}(I), \\ & \int_{\Gamma(\alpha)-\Gamma_\alpha} \widetilde{\mathbf{u}} \cdot \mathbf{s}\, ds = \int_{\Gamma(\alpha)-\Gamma_\alpha} \bar{\mathbf{l}}_4 \cdot \mathbf{s}\, ds \qquad \forall\, \mathbf{s} \in \mathbf{H}^{-1/2}(\Gamma(\alpha) - \Gamma_\alpha), \\ & \int_I \widetilde{q}\, v\, dx + \int_I \frac{d\widetilde{\alpha}}{dx} \frac{dv}{dx}\, dx = \int_I \bar{l}_5\, v\, dx \qquad \forall\, v \in H_0^1(I), \\ & 2\widetilde{\alpha}(\alpha - \alpha_0) + 2\widetilde{s}_0\, s_0 = \bar{l}_6 \quad \forall\, x \in I, \\ & 2\widetilde{q}\,(q - q_0) + 2\widetilde{s}_2\, s_2 = \bar{l}_7 \quad \forall\, x \in I, \\ & \widetilde{\alpha}(a) = \widetilde{\alpha}(b) = \widetilde{\alpha}'(a) = \widetilde{\alpha}'(b) = 0, \end{aligned} \right. \tag{6.122}$$

where $\mathbf{V}(\widetilde{\alpha}) = (0, \widetilde{\alpha})$, κ *denotes the curvature, and* $\mathbf{n}$ *is the normal vector to* Γ_α. *Moreover, the strict derivative of* Q *at a point* $(\widehat{\mathbf{u}}, \widehat{p}, q, \alpha, s_0, s_2) \in \widehat{\mathbf{B}}_1$ *is given by the bounded linear operator* $Q'(\widehat{\mathbf{u}}, \widehat{p}, q, \alpha, s_0, s_2) : \widehat{\mathbf{B}}_1 \to \mathbb{R} \times \widehat{\mathbf{B}}_2$, *where* $Q'(\widehat{\mathbf{u}}, \widehat{p}, q, \alpha, s_0, s_2) \cdot (\widetilde{\mathbf{u}}, \widetilde{p}, \widetilde{q}, \widetilde{\alpha}, \widetilde{s}_0, \widetilde{s}_2) = (\overline{a}, \overline{\mathbf{l}}_1, \overline{l}_2, \overline{\mathbf{l}}_3, \overline{\mathbf{l}}_4, \overline{l}_5, \overline{l}_6, \overline{l}_7)$ *for* $(\widetilde{\mathbf{u}}, \widetilde{p}, \widetilde{q}, \widetilde{\alpha}, \widetilde{s}_0, \widetilde{s}_2) \in \widehat{\mathbf{B}}_1$ *and* $(\overline{a}, \overline{\mathbf{l}}_1, \overline{l}_2, \overline{\mathbf{l}}_3, \overline{\mathbf{l}}_4, \overline{l}_5, \overline{l}_6, \overline{l}_7) \in \mathbb{R} \times \widehat{\mathbf{B}}_2$ *if and only if*

$$\begin{pmatrix} \mathcal{J}'(\widehat{\mathbf{u}}, q, \alpha) \cdot (\widetilde{\mathbf{u}}, \widetilde{p}, \widetilde{q}, \widetilde{\alpha}, \widetilde{s}_0, \widetilde{s}_2) \\ M'(\widehat{\mathbf{u}}, \widehat{p}, q, \alpha, s_0, s_2) \cdot (\widetilde{\mathbf{u}}, \widetilde{p}, \widetilde{q}, \widetilde{\alpha}, \widetilde{s}_0, \widetilde{s}_2) \end{pmatrix} = \begin{pmatrix} \overline{a} \\ (\overline{\mathbf{l}}_1, \overline{l}_2, \overline{\mathbf{l}}_3, \overline{\mathbf{l}}_4, \overline{l}_5, \overline{l}_6, \overline{l}_7) \end{pmatrix}, \tag{6.123}$$

where

$$\mathcal{J}'(\widehat{\mathbf{u}}, q, \alpha) \cdot (\widetilde{\mathbf{u}}, \widetilde{p}, \widetilde{q}, \widetilde{\alpha}, \widetilde{s}_0, \widetilde{s}_2) = \beta \int_I q \widetilde{q} \, dx + \nu \left(\int_{\Gamma_\alpha} \frac{\partial \widehat{\mathbf{u}}}{\partial n} \frac{\partial \widehat{\mathbf{u}}}{\partial n} (\mathbf{V}(\widetilde{\alpha}) \cdot \mathbf{n}) \, ds + 2 \int_{\Omega(\alpha)} (\nabla \widehat{\mathbf{u}} : \nabla \widetilde{\mathbf{u}}) \, d\mathbf{x} \right).$$

From the mapping (6.122) we can write the Gateaux derivative for our Navier–Stokes system and recover the results of the previous sections. In fact, the solution $(\mathbf{u}, p)$ of the Navier–Stokes system with boundary defined by $\Gamma(\alpha)$ can be seen as the restriction to $\Omega(\alpha)$ of the function $(\widehat{\mathbf{u}}, \widehat{p})$.

Let $\mathbf{B}_1$ and $\mathbf{B}_2$ be the spaces generated by all the restrictions from $\widehat{\Omega}$ to $\Omega(\alpha)$ of the functions in $\widehat{\mathbf{B}}_1$ and $\widehat{\mathbf{B}}_2$, respectively. With test functions in $\mathbf{H}_0^1(\Omega(\alpha))$ we write the mapping $M'(\mathbf{u}, p, q, \alpha, s_0, s_2)$ as

$$\left\{ \begin{aligned} & a(\widetilde{\mathbf{u}}, \mathbf{v}) + c(\widetilde{\mathbf{u}}, \mathbf{u}, \mathbf{v}) + c(\mathbf{u}, \widetilde{\mathbf{u}}, \mathbf{v}) \\ & \qquad + b(\mathbf{v}, \widetilde{p}) = \int_{\Omega(\alpha)} \overline{\mathbf{l}}_1 \cdot \mathbf{v} \, d\mathbf{x} \quad \forall \mathbf{v} \in H_0^1(\Omega(\alpha)), \\ & b(\widetilde{\mathbf{u}}, z) = \int_{\Omega(\alpha)} \overline{l}_2 \, z \, d\mathbf{x} \quad \forall z \in L_0^2(\Omega(\alpha)), \\ & \int_{\Gamma_\alpha} \widetilde{\mathbf{u}} \cdot \mathbf{s} \, ds + \int_{\Gamma_\alpha} (\mathbf{V} \cdot \mathbf{n}) \left(\kappa + \frac{\partial}{\partial n} \right) \mathbf{u} \cdot \mathbf{s} \, ds \\ & \qquad = \int_I \overline{\mathbf{l}}_3 \cdot \mathbf{s} \, ds \quad \forall \mathbf{s} \in \mathbf{H}^{-1/2}(I), \\ & \int_{\Gamma(\alpha) - \Gamma_\alpha} \widetilde{\mathbf{u}} \cdot \mathbf{s} \, ds = \int_{\Gamma(\alpha) - \Gamma_\alpha} \overline{\mathbf{l}}_4 \cdot \mathbf{s} \, ds \qquad \forall \mathbf{s} \in \mathbf{H}^{-1/2}(\Gamma(\alpha) - \Gamma_\alpha), \\ & \int_I v \widetilde{q} \, dx + \int_I \frac{d\widetilde{\alpha}}{dx} \frac{dv}{dx} \, dx = \int_I \overline{l}_5 \, v \, dx \qquad \forall v \in H_0^1(I), \\ & 2\widetilde{\alpha}(\alpha - \alpha_0) + 2\widetilde{s}_0 \, s_0 = \overline{l}_6 \quad \forall x \in I, \\ & 2\widetilde{q}\,(q - q_0) + 2\widetilde{s}_2 \, s_2 = \overline{l}_7 \quad \forall x \in I, \end{aligned} \right. \tag{6.124}$$

with $\widetilde{\alpha}(a) = 0$, $\widetilde{\alpha}(b) = 0$, and $\widetilde{\alpha}'(a) = \widetilde{\alpha}'(b) = 0$.

From (6.124), we note that the regularity of the Gateaux derivative cannot be the same as the solution of the Navier–Stokes system. In fact, the boundary conditions for the Gateaux derivative imply a different degree of regularity.

Also, we note that the map $M'(\mathbf{u}, p, q, \alpha, s_0, s_2)$ in (6.124) assumes exactly the same values of the restriction of $M'(\widehat{\mathbf{u}}, \widehat{p}, q, \alpha, s_0, s_2)$ to $\Omega(\alpha)$ since no further information is necessary to identify the function over the domain $\Omega(\alpha)$. The use of (6.124) or (6.122) over the domain $\Omega(\alpha)$ is equivalent and therefore, in the rest of this section, we always work with these operators on the domain $\Omega(\alpha)$ and use the notation $M'(\mathbf{u}, p, q, \alpha, s_0, s_2)$ and $Q'(\mathbf{u}, p, q, \alpha, s_0, s_2)$.

Some further properties of the derivatives of the mappings M and Q can be shown.

Lemma 6.25. *Let $(\mathbf{u}, p, q, \alpha, s_0, s_2) \in \mathbf{B}_1$ denote a solution of the optimal control problem. Then we have that*

(i) *the operator $M'(\mathbf{u}, p, q, \alpha, s_0, s_2)$ has closed range in $\mathbf{B}_2$,*

(ii) *the operator $Q'(\mathbf{u}, p, q, \alpha, s_0, s_2)$ has closed range in $\mathbb{R} \times \mathbf{B}_2$,*

(iii) *the operator $Q'(\mathbf{u}, p, q, \alpha, s_0, s_2)$ is not onto $\mathbb{R} \times \mathbf{B}_2$.*

The first-order necessary condition follows easily from the fact that the operator $Q'(\mathbf{u}, p, q, \alpha, s_0, s_2)$ is not onto $\mathbb{R} \times \mathbf{B}_2$; see, e.g., [71, 82].

Theorem 6.26. *Given $(\mathbf{u}, p, q, \alpha) \in \mathcal{A}_{ad}$, if $(\mathbf{u}, p, q, \alpha, s_0, s_2) \in \mathbf{B}_1$ is a solution of the optimal shape control problem, then there exists a nonzero Lagrange multiplier $(\lambda, \mathbf{w}, r, \boldsymbol{\theta}, \boldsymbol{\eta}, \mu, \tau_0, \tau_2) \in \mathbb{R} \times \mathbf{B}_2^*$ satisfying the Euler–Lagrange equations*

$$\begin{aligned}\lambda\, \mathcal{J}'(\mathbf{u}, q, \alpha) \cdot (\widetilde{\mathbf{u}}, \widetilde{r}, \widetilde{q}, \widetilde{\alpha}, \widetilde{s}_0, \widetilde{s}_2) \\ + \big\langle (\mathbf{w}, r, \boldsymbol{\theta}, \boldsymbol{\eta}, \mu, \tau_0, \tau_2), M'(\mathbf{u}, p, q, \alpha, s_0, s_2) \cdot (\widetilde{u}, \widetilde{r}, \widetilde{q}, \widetilde{\alpha}, \widetilde{s}_0, \widetilde{s}_2) \big\rangle = 0 \\ \forall\, (\widetilde{\mathbf{u}}, \widetilde{r}, \widetilde{q}, \widetilde{\alpha}, \widetilde{s}_0, \widetilde{s}_2) \in \mathbf{B}_1\,,\end{aligned} \tag{6.125}$$

where $\langle \cdot, \cdot \rangle$ denotes the duality pairing between $\mathbf{B}_2$ and $\mathbf{B}_2^*$.

The optimality system

Next, we examine the first-order necessary condition (6.125) to derive an optimality system from which optimal states and controls may be determined.

Theorem 6.27. *Let $(\mathbf{u}, p, q, \alpha, s_0, s_2) \in \mathbf{B}_1$ denote a solution of the optimal control problem. Then, if $s_0 \neq 0$ and $s_2 \neq 0$, we have that*

$$\int_I \frac{d\mu}{dx}\frac{d\zeta}{dx} dx + \nu \int_{\Gamma_\alpha} \left(\frac{\partial \mathbf{u}}{\partial n}\frac{\partial \mathbf{u}}{\partial n} - \frac{\partial \mathbf{u}}{\partial n}\frac{\partial \mathbf{w}}{\partial n} \right) (\mathbf{V}(\zeta) \cdot \mathbf{n})\, ds = 0 \quad \forall\, \zeta \in H_0^1(I),$$

$$\int_I (\mu + \beta q)\, v\, dx = 0 \quad \forall\, v \in L^2(I)$$

for all $\alpha \in H^2(I)$ *with* $\alpha(a) = z_1$, $\alpha(b) = z_2$, *and* $\alpha'(a) = \alpha'(b) = 0$, *where* $(\mathbf{w}, r) \in \mathbf{H}^1_0(\Omega(\alpha)) \times L^2_0(\Omega(\alpha))$ *satisfies the adjoint problem*

$$\begin{cases} a(\mathbf{w}, \mathbf{v}) + c(\mathbf{v}, \mathbf{u}, \mathbf{w}) + c(\mathbf{u}, \mathbf{v}, \mathbf{w}) + b(\mathbf{v}, r) \\ \qquad = -2\nu \displaystyle\int_{\Omega(\alpha)} \nabla \mathbf{u} : \nabla \mathbf{v} \, d\mathbf{x} \quad \forall \mathbf{v} \in \mathbf{H}^1_0(\Omega(\alpha)), \\ b(\mathbf{w}, q) = 0 \quad \forall q \in L^2_0(\Omega(\alpha)), \\ \mathbf{w} \in \mathbf{H}^1_0(\Omega(\alpha)) . \end{cases} \tag{6.126}$$

If $s_0 = 0$ *we have* $\alpha = c_0$ *or* $\alpha = d_0$. *If* $s_2 = 0$ *we have* $q = c_2$ *or* $q = d_2$, *which gives* α *through* (6.96) *and the appropriate boundary conditions.*

It is obvious that the first-order necessary condition obtained in Section 6.2.2, when applicable, is the same as that obtained by the Lagrangian formulation. The introduction of the variables s_0 and s_2 gives us insight into the structure of the optimal solution, but it is clear that it is not convenient in the numerical computations. Numerically it is difficult to evaluate the variables s_0 and s_2, and many local extremal points can be introduced in such an attempt.

As a consequence, the optimality system we have to solve is given by

$$\begin{cases} a(\mathbf{u}, \mathbf{v}) + c(\mathbf{u}, \mathbf{u}, \mathbf{v}) + b(\mathbf{v}, p) = \langle \mathbf{f}, \mathbf{v} \rangle \quad \forall \mathbf{v} \in \mathbf{H}^1_0(\Omega(\alpha)), \\ b(\mathbf{u}, r) = 0 \quad \forall r \in L^2_0(\Omega(\alpha)), \\ \langle \mathbf{u}, \mathbf{s} \rangle_{\Gamma(\alpha)} = \langle \mathbf{g}, \mathbf{s} \rangle_{\Gamma(\alpha)} \qquad \forall \mathbf{s} \in \mathbf{H}^{1/2}(\Gamma(\alpha)); \end{cases}$$

$$\begin{cases} a(\mathbf{w}, \mathbf{v}) + c(\mathbf{v}, \mathbf{u}, \mathbf{w}) + c(\mathbf{u}, \mathbf{v}, \mathbf{w}) + b(\mathbf{v}, r) \\ \qquad = -2\nu \displaystyle\int_{\Omega(\alpha)} \nabla \mathbf{u} : \nabla \mathbf{v} \, d\mathbf{x} \quad \forall \mathbf{v} \in \mathbf{H}^1_0(\Omega(\alpha)), \\ b(\mathbf{w}, q) = 0 \quad \forall q \in L^2_0(\Omega(\alpha)), \\ \mathbf{w} = \mathbf{0} \quad on \quad \Gamma(\alpha); \end{cases}$$

$$\int_I \frac{d\mu}{dx}\frac{d\zeta}{dx}\,dx + \nu \int_{\Gamma_\alpha} \Big(\frac{\partial \mathbf{u}}{\partial n}\frac{\partial \mathbf{u}}{\partial n} - \frac{\partial \mathbf{u}}{\partial n}\frac{\partial \mathbf{w}}{\partial n} \Big) (\mathbf{V}(\zeta) \cdot \mathbf{n}) \, ds = 0 \quad \forall \zeta \in H^1_0(I);$$

$$-\int_I \frac{d\alpha}{dx}\frac{dv}{dx}\,dx = \int_I q\, v \, dx \qquad \forall v \in H^1_0(I)$$

with $\alpha(a) = z_0$ and $\alpha(a) = z_1$; and

$$\int_I (\beta q + \mu)(\widetilde{q} - q) \geq 0$$

for all $\widetilde{\alpha} \in \mathcal{Q}_{ad}$ and for all $\widetilde{q} \in \mathcal{B}_{ad}$.

6.3 The velocity tracking problem for unsteady Navier–Stokes flows with boundary control

We now consider the analysis and numerical approximation of an optimal boundary control problem of velocity tracking type for time-dependent Navier–Stokes flows in a bounded, two-dimensional domain. We examine the existence of optimal solutions and provide an optimality system from which optimal solutions may be determined. We also define and analyze semidiscrete-in-time and full space-time discrete approximations of the optimality system and of a gradient method for the solution of the fully discrete system. Details and proofs may be found in [84]. Computational experiments with this problem are given in Section 3.3.3.

The control problem

The velocity tracking problem we consider reflects the desire to steer, over time, a candidate velocity field $\mathbf{u}$ to a given target velocity field $\mathbf{U}$ by appropriately controlling the velocity along a portion of the boundary of the flow domain. We consider a two-dimensional flow over the time interval $[0, T]$ in the physical domain Ω with boundary Γ with control effected over $\Gamma_c \subset \Gamma$. The equations considered here are the nondimensional incompressible Navier–Stokes equations

$$\begin{cases} \mathbf{u}_t + \mathbf{u} \cdot \nabla \mathbf{u} - \nu \Delta \mathbf{u} + \nabla p = \mathbf{0} & \text{in } (0, T) \times \Omega, \\ \nabla \cdot \mathbf{u} = 0 & \text{in } (0, T) \times \Omega, \\ \mathbf{u} = \mathbf{g} & \text{on } (0, T) \times \Gamma_c, \\ \mathbf{u} = \mathbf{0} & \text{on } (0, T) \times (\Gamma \setminus \Gamma_c) \end{cases} \tag{6.127}$$

with initial velocity $\mathbf{u}(0, \mathbf{x}) = \mathbf{u}_0(\mathbf{x})$. The vector $\mathbf{u}$ denotes the velocity, p the pressure, and ν the constant kinematic viscosity coefficient. We note that for appropriate nondimensionalizations, the Reynolds number is equal to $1/\nu$. The boundary velocity control is denoted by $\mathbf{g}$ and is required to satisfy the compatibility conditions

$$\int_{\Gamma_c} \mathbf{g} \cdot \mathbf{n} \, d\mathbf{x} = 0, \tag{6.128}$$

where $\mathbf{n}$ denotes the unit outward normal vector along Γ, and

$$\mathbf{g}|_{t=0} = \mathbf{u}_0|_{\Gamma_c}. \tag{6.129}$$

Thus, the control is required to effect zero mass flow across the boundary and to match, at the initial time, the initial flow $\mathbf{u}_0$ on the boundary. The first of these is necessary in view of the incompressibility condition and the second in order to obtain the appropriate regularity for the solution of the Navier–Stokes system.

The optimal control problem is formulated as

find a boundary control $\mathbf{g}$ *and a velocity field* $\mathbf{u}$ *such that the cost functional*

$$
\begin{aligned}
\mathcal{J}(\mathbf{u},\mathbf{g}) = {} & \frac{\alpha}{2}\int_0^T\int_\Omega |\mathbf{u}-\mathbf{U}|^2\,d\mathbf{x}dt \\
& +\frac{\beta}{2}\int_0^T\int_{\Gamma_c} (|\mathbf{g}|^2+\beta_1|\mathbf{g}_t|^2+\beta_2|\mathbf{g}_x|^2)\,d\mathbf{x}dt
\end{aligned}
\tag{6.130}
$$

is minimized subject to $(\mathbf{u},\mathbf{g})$ *satisfying* (6.127)–(6.129).

The minimization of the first term involving $(\mathbf{u}-\mathbf{U})$ in (6.130) is the real goal of the velocity tracking problem; the other terms have been introduced in order to bound the control function and to prove the existence of an optimal control. We can effectively limit the size of the control through an appropriate choice of the positive coefficients β, β_1, and β_2.

Notation and preliminary results

We introduce the following standard notation over a bounded, connected, open set Ω in $\mathbb{R}^2$ with boundary $\Gamma\in C^2$. Let $\mathbf{n}$ and $\boldsymbol{\tau}$ denote the unit normal and tangent vectors, respectively. Let $I=(0,T)$, $Q=I\times\Omega$, $S=I\times\Gamma$, and $S_c=I\times\Gamma_c$, where Γ_c denotes the part of the boundary on which control is applied. Also, we denote $\Omega_0=\Omega\times\{0\}$ and $\Gamma_0=\Gamma\times\{0\}$.

We shall use the standard notation for the Sobolev spaces $H^m(\Omega)$ with norm $\|\cdot\|_m$ (their vector-valued, i.e., $\mathbb{R}^2$-valued, counterparts will be denoted by boldface letters); we also use the notation $L^2(\Omega)=H^0(\Omega)$ with $\|\cdot\|=\|\cdot\|_0$ and $\mathcal{D}(\Omega)$ for the space of distributions. Let $H_0^m(\Omega)$ denote the closure of $C_0^\infty(\Omega)$ under the norm $\|\cdot\|_m$, and let $H_0^{-m}(\Omega)$ denote the dual space of $H_0^m(\Omega)$. We introduce the solenoidal spaces $\mathcal{V}(\Omega)$, $\mathbf{V}(\Omega)$, and $\mathbf{W}(\Omega)$ as

$$
\begin{aligned}
\mathcal{V}(\Omega) &= \{\mathbf{u}\in\mathbf{C}_0^\infty(\Omega) : \nabla\cdot\mathbf{u}=0\}\,,\\
\mathbf{V}(\Omega) &= \{\mathbf{u}\in\mathbf{H}_0^1(\Omega) : \nabla\cdot\mathbf{u}=0\}\,,\\
\mathbf{W}(\Omega) &= \{\mathbf{u}\in\mathbf{L}^2(\Omega) : \nabla\cdot\mathbf{u}=0\}\,.
\end{aligned}
$$

The dual space of $\mathbf{V}(\Omega)$ is denoted by $\mathbf{V}(\Omega)^*$. Also, we define

$$
L_0^2(\Omega)=\left\{p\in L^2(\Omega) : \int_\Omega p\,d\mathbf{x}=0\right\}.
$$

Let X be a Banach space and (a,b) an open set of $\mathbb{R}$. We denote by $L^p((a,b);X)$ $(1\le p<\infty)$ the space of functions $f(t):(a,b)\to X$ such that f is measurable and

$$
\|f\|_{L^p((a,b);X)}=\left(\int_a^b \|f(t)\|_X^p\right)^{1/p}
$$

is finite. We also denote by $L^\infty((a,b);X)$ the space of functions f from (a,b) into X such that f is measurable and bounded almost everywhere over (a,b), and we set

$$
\|f\|_{L^\infty((a,b);X)}=\inf_{\|f(t)\|\le M\ a.e.}(M)\,.
$$

We define the following anisotropic Sobolev spaces. Let r and $s\ge 0$ and $Q=(a,b)\times\Omega$. We let

$$
H^{r,s}(Q)=L^2((a,b);H^r(\Omega))\cap H^s((a,b);L^2(\Omega))
\tag{6.131}
$$

with the norm

$$\|u\|_{H^{r,s}} = \left(\|u\|^2_{L^2((a,b);H^r)} + \|u\|^2_{H^s((a,b);L^2)}\right)^{1/2}.$$

For details about these spaces, see, e.g., [413, 424, 428, 429].

In order to define a weak form of the Navier–Stokes equations, we introduce two continuous bilinear forms

$$a(\mathbf{u}, \mathbf{v}) = 2\nu \sum_{i,j=1}^{n} \int_\Omega D_{ij}(\mathbf{u}) D_{ij}(\mathbf{v})\, d\mathbf{x} \qquad \forall \mathbf{u},\ \mathbf{v} \in \mathbf{H}^1(\Omega), \tag{6.132}$$

$$b(\mathbf{v}, q) = -\int_\Omega q \nabla \cdot \mathbf{v}\, d\mathbf{x} \qquad \forall q \in L^2(\Omega), \quad \forall \mathbf{v} \in \mathbf{H}^1(\Omega), \tag{6.133}$$

where $D_{ij}(\mathbf{v}) = \frac{1}{2}(\partial v_i/\partial x_j + \partial v_j/\partial x_i)$, and the continuous trilinear form

$$c(\mathbf{w}, \mathbf{u}, \mathbf{v}) = \sum_{i,j=1}^{n} \int_\Omega w_j \left(\frac{\partial u_i}{\partial x_j}\right) v_i\, d\mathbf{x} \qquad \forall \mathbf{w},\ \mathbf{u},\ \mathbf{v} \in \mathbf{H}^1(\Omega).$$

We restrict the domain $\Omega \subset \mathbb{R}^2$ to be an open bounded set with simply connected boundary $\Gamma \in C^2$. We define

$$curl(H^2)(\Omega) = \left\{\mathbf{v} \in \mathbf{H}^1(\Omega)\ :\ \nabla \cdot \mathbf{v} = 0, \quad \int_\Gamma \mathbf{v} \cdot \mathbf{n}\, d\mathbf{x} = 0\right\},$$

$$\mathbf{H}^1_n(\Gamma) = \left\{\mathbf{g} \in \mathbf{H}^1(\Gamma)\ :\ \int_\Gamma \mathbf{g} \cdot \mathbf{n}\, d\mathbf{x} = 0\right\},$$

$$\mathbf{H}^1_{n0}(\Gamma_c) = \mathbf{H}^1_0(\Gamma_c) \cap \mathbf{H}^1_n(\Gamma_c),$$

where Γ_c is part of the boundary Γ. The set $curl(H^2)(\Omega)$ is a closed subspace of $\mathbf{H}^1(\Omega)$; $\mathbf{H}^1_n(\Gamma)$ and $\mathbf{H}^1_{n0}(\Gamma)$ are closed subspaces of $\mathbf{H}^1(\Gamma)$. For details concerning these subspaces, see, e.g., [424]. We remark that the space $\mathbf{H}^1(\Gamma)$ can be decomposed into $\mathbf{H}^1_n(\Gamma) \oplus (\mathbf{H}^1_n(\Gamma))^\perp$, where $(\mathbf{H}^1_n(\Gamma))^\perp$ is the space of vectors normal to the surface with constant length. If $\mathbf{g} \in \mathbf{H}^1(\Gamma)$, one can write $\mathbf{g} = \mathbf{g}_1 + \mathbf{g}_2$, where

$$\mathbf{g}_2 = d\mathbf{n}, \qquad d = \frac{\int_\Gamma \mathbf{g} \cdot \mathbf{n}\,, d\mathbf{x}}{\mu(\Gamma)},$$

$$\mathbf{g}_1 = \boldsymbol{\tau}(\mathbf{g} \cdot \boldsymbol{\tau}) + \mathbf{n}(\mathbf{g} \cdot \mathbf{n} - d),$$

where $\mathbf{g}_1 \in \mathbf{H}^1_n(\Gamma)$ and $\mathbf{g}_2 \in (\mathbf{H}^1_n(\Gamma))^\perp$.

In the rest of this section, we shall use γ and γ_k to denote trace operators, i.e., $\gamma\mathbf{f} = \gamma_0\mathbf{f} = \mathbf{f}_\Gamma$ and $\gamma_0 \partial_n^k \mathbf{f} = \gamma_k \mathbf{f}$, where $\partial_n \mathbf{f} = n_1 \partial_1 \mathbf{f} + n_2 \partial_2 \mathbf{f}$ with $\partial_j \mathbf{f} = \partial \mathbf{f}/\partial x_j$.

6.3.1 Formulation and analysis of the optimal control problem

Weak formulation of the optimal control problem

We consider an open bounded set $\Omega \subset \mathbb{R}^2$ with a boundary $\Gamma \in C^2$. $\mathbf{U}(t, \mathbf{x})$ is said to be in the set of *admissible target velocities* U_{ad} if

$$\left\{\begin{array}{l} \mathbf{U} = \mathbf{U}(t, \mathbf{x}) \in C([0, T]; \mathbf{H}^1(\Omega)), \\ \mathbf{F}_{\mathbf{U}}(t, \mathbf{x}) \in L^\infty((0, T); \mathbf{L}^2(\Omega)), \end{array}\right. \tag{6.134}$$

where $\mathbf{F_U} = \mathbf{U}_t - \nu\nabla^2\mathbf{U} + (\mathbf{U}\cdot\nabla)\mathbf{U}$.

Let $\mathbf{u} \in L^2((0,T);\mathbf{H}^1(\Omega))$ and $p \in L^2((0,T);L_0^2(\Omega))$ denote the state variables, i.e., the velocity and pressure fields, respectively. Let the boundary control $\mathbf{g}$ belong to $L^2((0,T);\mathbf{H}_{n0}^1(\Gamma_c))$ with $\mathbf{g}_t \in \mathbf{L}^2((0,T);L^2(\Gamma_c))$. The state variables are constrained to satisfy the weak form of the Navier–Stokes system (6.127) for almost all t in $(0,T)$, i.e.,

$$\begin{cases} \langle \mathbf{u}_t, \mathbf{v}\rangle + a(\mathbf{u},\mathbf{v}) + c(\mathbf{u},\mathbf{u},\mathbf{v}) + b(\mathbf{v},p) = 0 \quad \forall\, \mathbf{v}\in\mathbf{H}_0^1(\Omega), \\ b(\mathbf{u},q) = 0 \quad \forall q \in L_0^2(\Omega), \\ (\mathbf{u},\mathbf{s})_\Gamma = (\mathbf{g}(t,\mathbf{x}),\mathbf{s})_{\Gamma_c} \quad \forall\,\mathbf{s}\in\mathbf{H}^{-1/2}(\Gamma), \\ \mathbf{u}(0,\mathbf{x}) = \mathbf{u}_0(\mathbf{x}) \in curl(H^2)(\Omega)\,. \end{cases} \tag{6.135}$$

More precisely, let $\mathbf{g} \in \mathbf{H}^{1,1}(S_c) \cap L^2((0,T);\mathbf{H}_{n0}^1(\Gamma_c))$ and $\mathbf{u}_0 \in curl(H^2)(\Omega)$. Then, $(\mathbf{u},p) \in L^2((0,T);\mathbf{H}^1(\Omega)) \times L^2((0,T);L_0^2(\Omega))$ is called a *weak solution for the Navier–Stokes equations* if it satisfies (6.135).

If $\mathbf{u}$ is a solution of (6.127), then it is also a solution of the weak formulation (6.135). If $\mathbf{u}$ is a solution of (6.135), then it satisfies (6.127) in the sense of distributions on $(0,T)$ so that solutions of (6.135) generalize the notion of solutions of (6.127). If $\mathbf{g}$ and $\mathbf{u}_0$ are given as above, then we can show that there exists a unique admissible weak solution $(\mathbf{u},p)$ of (6.135) such that $\mathbf{u} \in L^\infty((0,T);\mathbf{W}(\Omega)) \cap L^2((0,T);\mathbf{H}^1(\Omega))$ and $\mathbf{u}_t \in L^2((0,T);\mathbf{H}^{-1}(\Omega))$, i.e, it is a.e. equal to a continuous function [424].

Theorem 6.28. *Let $\Omega \subset \mathbb{R}^2$ be an open, bounded domain with boundary Γ of class C^2. Let $\mathbf{g}(t,\mathbf{x})$ be a function belonging to $\mathbf{H}^{1/2,1}(S)$ satisfying the compatibility conditions*

$$\int_\Gamma \mathbf{g}\cdot\mathbf{n}\,d\Gamma = 0\,, \tag{6.136}$$

$$\mathbf{g}(0,\mathbf{x}) = \mathbf{u}_0|_\Gamma\,. \tag{6.137}$$

Then, there exists a unique $\mathbf{u} \in L^2((0,T);\mathbf{H}^1(\Omega)) \cap L^\infty((0,T);\mathbf{L}^2(\Omega))$ and $p \in L^2((0,T);L_0^2(\Omega))$ that are the solution of the nonhomogeneous Navier–Stokes problem

$$\begin{cases} \langle \mathbf{u}_t, \mathbf{v}\rangle + a(\mathbf{u},\mathbf{v}) + c(\mathbf{u},\mathbf{u},\mathbf{v}) + b(\mathbf{v},p) = 0 \qquad \forall\, \mathbf{v}\in\mathbf{H}_0^1(\Omega), \\ b(\mathbf{u},q) = 0 \quad \forall q \in L_0^2(\Omega), \\ \mathbf{u} = \mathbf{g}(t,\mathbf{x}) \quad \forall\,\mathbf{x}\in\Gamma, \\ \mathbf{u}(0,\mathbf{x}) = \mathbf{u}_0(\mathbf{x}) \in curl(H^2)(\Omega) \end{cases} \tag{6.138}$$

for almost all $t \in (0,T)$. Moreover,

$$\|\mathbf{u}\|^2_{L^2((0,T);\mathbf{H}^1)} + \|\mathbf{u}\|^2_{L^\infty((0,T);\mathbf{L}^2)} \le K(\|\mathbf{g}\|^2_{\mathbf{H}^{1/2,1}(S)} + \|\mathbf{u}_0\|^2_{\mathbf{H}^{1/2}(\Gamma)})\,, \tag{6.139}$$

where K depends on $\mathbf{g}$.

A proof can be found in, e.g., [60]. It is worthwhile to recall that if $\|\mathbf{g}\|^2_{\mathbf{H}^{1/2,1}(S)}$ is uniformly bounded, then the left-hand side of (6.139) is also uniformly bounded.

From the previous discussion we can define more precisely the set of admissible solutions, which we denote by A_d:

Given $T > 0$, $\mathbf{u}_0 \in curl(H^2)(\Omega)$, and $\mathbf{U} \in U_{ad}$, then $(\mathbf{u}, p, \mathbf{g})$ is called an admissible solution for the optimal control problem if $(\mathbf{u}, p, \mathbf{g}) \in L^2((0,T); \mathbf{H}^1(\Omega)) \times L^2((0,T); L^2_0(\Omega)) \times \mathbf{H}^{1,1}(S_c) \cap L^2(0,T; \mathbf{H}^1_{n0}(\Gamma_c))$ is a solution of (6.135), the control $\mathbf{g}$ satisfies the compatibility conditions (6.136)–(6.137), and the functional $\mathcal{J}(\mathbf{u}, \mathbf{g})$ is bounded.

The optimal control problem can then be formulated as

given $\mathbf{u}_0 \in curl(H^2)(\Omega)$ and $\mathbf{U} \in U_{ad}$, find $(\mathbf{u}, p, \mathbf{g}) \in A_d$ such that the control $\mathbf{g}$ minimizes the cost functional

$$\mathcal{J}(\mathbf{u}, \mathbf{g}) = \frac{\alpha}{2} \int_0^T \int_\Omega (\mathbf{u} - \mathbf{U})^2 \, d\mathbf{x}dt + \frac{\beta}{2} \int_0^T \int_{\Gamma_c} (\mathbf{g}^2 + \beta_1 \mathbf{g}_x^2 + \beta_2 \mathbf{g}_t^2) \, d\mathbf{x}dt \tag{6.140}$$

with $\alpha, \beta, \beta_1, \beta_2 > 0$.

The first term represents the goal of our optimization and the second term is the penalty term necessary to regularize the solution. The requirement that β, β_1, β_2 be different from zero is necessary if we want $\mathbf{g} \in \mathbf{H}^{1,1}(S_c)$.

Existence of an optimal solution and the first-order necessary conditions

The optimal control problem (6.140) is well posed and has at least one solution.

Theorem 6.29. *Given $T > 0$ and $\mathbf{u}_0 \in curl(H^2)(\Omega)$, there exists a solution $(\mathbf{u}, p, \mathbf{g}) \in A_d$ of the optimal control problem* (6.140).

We next derive the first-order necessary conditions. Let $\mathbf{G}$ be the set of all $\mathbf{g} \in \mathbf{H}^{1,1}(S) \cap L^2(0,T; \mathbf{H}^1_{n0}(\Gamma))$ satisfying the compatibility conditions in (6.136)–(6.137). For all $\mathbf{g} \in \mathbf{G}$, the first-order necessary condition is available if the map

$$\mathbf{u}(\mathbf{g}) : \mathbf{G} \to L^2((0,T); \mathbf{H}^1(\Omega))$$

is Gateaux differentiable. The following theorem provides for the existence of the Gateaux derivative for directions $\widetilde{\mathbf{h}}$ in $\mathbf{H}^{1,1}(S) \cap L^2(0,T; \mathbf{H}^1_{n0}(\Gamma))$.

Theorem 6.30. *Given $\Omega \in C^2$, $\mathbf{u}_0 \in curl(H^2)$, and $\mathbf{g} \in \mathbf{G}$, the mapping*

$$\mathbf{u}(\mathbf{g}) : \mathbf{G} \to L^2((0,T); \mathbf{H}^1(\Omega))$$

has a Gateaux derivative $\frac{d\mathbf{u}}{d\mathbf{g}} \cdot \widetilde{\mathbf{h}}$ in every direction $\widetilde{\mathbf{h}} \in \mathbf{H}^{1,1}(S) \cap L^2(0,T; \mathbf{H}^1_{n0}(\Gamma))$ with $\widetilde{\mathbf{h}} = \mathbf{0}$ at $t = 0$. Furthermore, $\widetilde{\mathbf{w}}(h) = \frac{d\mathbf{u}}{d\mathbf{g}} \cdot \widetilde{\mathbf{h}}$ is the solution of the problem

$$\begin{cases} \langle \widetilde{\mathbf{w}}_t, \mathbf{v} \rangle + a(\widetilde{\mathbf{w}}, \mathbf{v}) + c(\mathbf{u}, \widetilde{\mathbf{w}}, \mathbf{v}) + c(\widetilde{\mathbf{w}}, \mathbf{u}, \mathbf{v}) + b(\mathbf{v}, \widetilde{p}) = 0 \quad \forall \mathbf{v} \in \mathbf{H}^1_0(\Omega), \\ b(\widetilde{\mathbf{w}}, q) = 0 \quad \forall q \in L^2_0(\Omega), \\ (\widetilde{\mathbf{w}}(t, \mathbf{x}), \mathbf{s}) = (\widetilde{\mathbf{h}}(t, \mathbf{x}), \mathbf{s}) \quad \forall \mathbf{s} \in \mathbf{H}^{-1/2}(\Gamma), \\ \widetilde{\mathbf{w}}(0, \mathbf{x}) = \mathbf{0} \quad \mathbf{x} \in \Omega, \end{cases} \tag{6.141}$$

where $\widetilde{\mathbf{w}} \in L^\infty((0,T); \mathbf{L}^2(\Omega)) \cap L^2((0,T); \mathbf{H}^1(\Omega))$.

The canonical extension $\widetilde{\mathbf{h}}_c \to \widetilde{\mathbf{h}}$ from $\mathbf{H}^1_{n0}(\Gamma_c)$ to $\mathbf{H}^1(\Gamma)$, where

$$\widetilde{\mathbf{h}} = \begin{cases} \widetilde{\mathbf{h}}_c & \mathbf{x} \in \Gamma_c, \\ 0 & \mathbf{x} \in \Gamma \setminus \Gamma_c, \end{cases}$$

is a continuous mapping; see [424]. This allows us to take variations in subdomains of the boundary, i.e., $\mathbf{H}^1(\Gamma_c)$, and to claim the existence of the Gateaux derivative for such configurations. For a variation $\widetilde{\mathbf{h}}_c \in \mathbf{H}^{1,1}(S_c) \cap L^2(0,T; \mathbf{H}^1_{n0}(\Gamma_c))$, i.e., $\widetilde{\mathbf{h}}_c \in L^2((0,T); \mathbf{H}^1_{n0}(\Gamma_c))$ and $\widetilde{\mathbf{h}}_{ct} \in L^2((0,T); \mathbf{L}^2(\Gamma_c))$, of the control $\mathbf{g}$, the Gateaux derivative of the Navier–Stokes system can be written in the following form:

$$\begin{cases} \langle \widetilde{\mathbf{w}}_t, \mathbf{v}\rangle + a(\widetilde{\mathbf{w}}, \mathbf{v}) + c(\widetilde{\mathbf{w}}, \mathbf{u}, \mathbf{v}) + c(\mathbf{u}, \widetilde{\mathbf{w}}, \mathbf{v}) \\ \qquad\qquad\qquad + b(\mathbf{v}, \widetilde{p}_1) = 0 \quad \forall \mathbf{v} \in \mathbf{H}^1_0(\Omega), \\ b(\widetilde{\mathbf{w}}, q) = 0 \quad \forall q \in L^2_0(\Omega), \\ (\widetilde{\mathbf{w}}, \mathbf{s})_\Gamma = (\widetilde{\mathbf{h}}_c(t, \mathbf{x}), \mathbf{s})_{\Gamma_c} \quad \forall \mathbf{s} \in \mathbf{H}^{-1/2}(\Gamma), \\ \widetilde{\mathbf{w}}(0, \mathbf{x}) = \mathbf{0} \quad \mathbf{x} \in \Omega. \end{cases} \tag{6.142}$$

Now, it can be shown that that the optimal solution must satisfy a first-order necessary condition. If $(\mathbf{u}, \mathbf{g})$ is an optimal pair, then for every $\widetilde{\mathbf{h}} \in \mathbf{H}^{1,1}(S_c) \cap L^2(0,T; \mathbf{H}^1_{n0}(\Gamma_c))$ and for every $\epsilon \in \mathbb{R}$, we have, from the definition of an optimal solution,

$$\mathcal{J}(\mathbf{g} + \epsilon \widetilde{\mathbf{h}}) \geq \mathcal{J}(\mathbf{g}).$$

The above inequality implies

$$\frac{\mathcal{J}(\mathbf{g} + \epsilon\widetilde{\mathbf{h}}) - \mathcal{J}(\mathbf{g})}{\epsilon} \geq 0 \quad \text{if} \quad \epsilon \geq 0 \quad \text{and} \quad \frac{\mathcal{J}(\mathbf{g} + \epsilon\widetilde{\mathbf{h}}) - \mathcal{J}(\mathbf{g})}{\epsilon} \leq 0 \quad \text{if} \quad \epsilon \leq 0.$$

The limit must vanish when ϵ tends to zero, which leads to the following first-order necessary condition.

THEOREM 6.31. *If* $(\mathbf{u}, p, \mathbf{g})$ *is an optimal pair for the problem in* (6.140), *then the Gateaux derivative of* $\mathcal{J}(\cdot,\cdot)$ *vanishes at* $(\mathbf{u}, p, \mathbf{g})$.

We would like to write the first-order necessary condition in a more explicit form. In order to do this, we need this interesting preliminary result.

LEMMA 6.32. *Given* $\Omega \in C^2$ *and* $\mathbf{u}_0 \in curl(H^2)(\Omega)$, *let* $\widetilde{\mathbf{h}}_c$ *be given in* $\mathbf{H}^{1,1}(S_c) \cap L^2(0,T; \mathbf{H}^1_{n0}(\Gamma_c))$ *and let* $\widetilde{\mathbf{w}}(\widetilde{\mathbf{h}}_c)$ *be defined by* (6.142). *Then, for every* $\widetilde{\mathbf{h}}_2$ *belonging to* $L^2((0,T); \mathbf{H}^1(\Omega))$, *we have*

$$\int_0^T \int_\Omega \widetilde{\mathbf{h}}_2 \cdot \widetilde{\mathbf{w}}(\widetilde{\mathbf{h}}_c)\, d\mathbf{x} dt = -\int_0^T \int_{\Gamma_c} \boldsymbol{\xi} \cdot \mathbf{w}\, d\mathbf{x} dt,$$

where $\mathbf{w}$ *is the solution of the adjoint linearized problem*

$$\begin{cases} -(\mathbf{w}_t, \mathbf{v}) + a(\mathbf{w}, \mathbf{v}) + c(\mathbf{v}, \mathbf{u}, \mathbf{w}) + c(\mathbf{u}, \mathbf{v}, \mathbf{w}) \\ \qquad\qquad + b(\mathbf{v}, \sigma) = (\widetilde{\mathbf{h}}_2, \mathbf{v}) \quad \forall \mathbf{v} \in \mathbf{H}_0^1(\Omega), \\ b(\mathbf{w}, q) = 0 \quad \forall q \in L_0^2(\Omega), \\ \mathbf{w} = \mathbf{0} \quad \forall \mathbf{x} \in \Gamma, \\ \mathbf{w}(T, \mathbf{x}) = \mathbf{0} \quad \forall \mathbf{x} \in \Omega. \end{cases} \tag{6.143}$$

The function $\boldsymbol{\xi} = (\nu\gamma_1\mathbf{w} - \sigma\mathbf{n}) \in L^2((0,T); \mathbf{H}^{-1/2}(\Gamma_c))$ *is defined by*

$$\begin{aligned} \int_{\Gamma_c} \boldsymbol{\xi} \cdot \mathbf{v}\, d\mathbf{x} = {} & -(\mathbf{w}_t, \mathbf{v}) + a(\mathbf{w}, \mathbf{v}) + c(\mathbf{v}, \mathbf{u}, \mathbf{w}) + c(\mathbf{u}, \mathbf{v}, \mathbf{w}) \\ & + b(\mathbf{v}, \sigma) - (\widetilde{\mathbf{h}}_2, \mathbf{v}) \quad \forall \mathbf{v} \in \mathbf{H}^1(\Omega). \end{aligned} \tag{6.144}$$

The next theorem shows that if the Gateaux derivative vanishes, then $\mathbf{g}$ must be a solution of a differential equation.

THEOREM 6.33. *If* $(\mathbf{u}, \mathbf{g})$ *is an optimal pair for the problem in* (6.140), *then* $\mathbf{g} \in \mathbf{H}^{1,1}(S_c) \cap L^2(0, T; \mathbf{H}_{n0}^1(\Gamma_c))$ *with* $\mathbf{g}(0, \mathbf{x}) = \gamma\mathbf{u}_0$ *is a solution of*

$$\int_0^T \int_{\Gamma_c} \left[\mathbf{g} \cdot \widetilde{\mathbf{h}} + \beta_1 \mathbf{g}_t \cdot \widetilde{\mathbf{h}}_t + \beta_2 \partial_s \mathbf{g} \cdot \partial_s \widetilde{\mathbf{h}} - \frac{1}{\beta}(\boldsymbol{\xi} \cdot \widetilde{\mathbf{h}}) \right] d\mathbf{x}dt = 0 \tag{6.145}$$

for all $\widetilde{\mathbf{h}} \in \mathbf{H}^{1,1}(S_c) \cap L^2(0, T; \mathbf{H}_{n0}^1(\Gamma_c))$ *with* $\widetilde{\mathbf{h}} = 0$ *at* $t = 0$. *The function* $\mathbf{w} \in L^\infty((0,T); \mathbf{L}^2(\Omega)) \cap L^2((0,T); \mathbf{V}(\Omega))$ *and is the solution of the adjoint linearized problem*

$$\begin{cases} -(\mathbf{w}_t, \mathbf{v}) + a(\mathbf{w}, \mathbf{v}) + c(\mathbf{u}, \mathbf{v}, \mathbf{w}) + c(\mathbf{v}, \mathbf{u}, \mathbf{w}) \\ \qquad\qquad + b(\mathbf{v}, \sigma) = \alpha(\mathbf{u} - \mathbf{U}, \mathbf{v}) \quad \forall \mathbf{w} \in \mathbf{H}_0^1(\Omega), \\ b(\mathbf{w}, q) = 0 \quad \forall q \in L_0^2(\Omega), \\ \mathbf{w} = \mathbf{0} \quad \forall \mathbf{x} \in \Gamma, \\ \mathbf{w}(T, \mathbf{x}) = \mathbf{0} \quad \forall \mathbf{x} \in \Omega, \end{cases} \tag{6.146}$$

and $\boldsymbol{\xi} = (\nu\gamma_1\mathbf{w} - \sigma\mathbf{n})$ *on* Γ_c *is defined by*

$$\begin{aligned} \int_{\Gamma_c} \boldsymbol{\xi} \cdot \mathbf{v}\, d\mathbf{x} &= \int_{\Gamma_c} (\nu\gamma_1\mathbf{w} \cdot \gamma_0\mathbf{v} - \sigma\gamma_0\mathbf{v} \cdot \mathbf{n})\, d\mathbf{x} \\ &= -(\mathbf{w}_t, \mathbf{v}) + a(\mathbf{w}, \mathbf{v}) + c(\mathbf{u}, \mathbf{v}, \mathbf{w}) \\ &\qquad + c(\mathbf{v}, \mathbf{u}, \mathbf{w}) + b(\mathbf{v}, \sigma) - \alpha(\mathbf{u} - \mathbf{U}, \mathbf{v}) \quad \forall \mathbf{v} \in \mathbf{H}^1(\Omega). \end{aligned} \tag{6.147}$$

Equation (6.145) provides the solution for the boundary control. Since $\widetilde{\mathbf{h}} \in \mathbf{H}^{1,1}(S_c) \cap L^2(0, T; \mathbf{H}_{n0}^1(\Gamma_c))$ with $\widetilde{\mathbf{h}}(0, \mathbf{x}) = 0$, we can take $\widetilde{\mathbf{h}} = \psi(t)\mathbf{r}(\mathbf{x})$, where $\psi \in \mathcal{D}((0,T))$ with

$\psi(0) = 0$ and $\mathbf{r}(\mathbf{x})$ in $\mathbf{H}^1_{n0}(\Gamma_c)$. After integration by parts, we have

$$\beta_1(\mathbf{g}_t(T), \mathbf{r})\psi(T) + \int_0^T \psi(t)\left[(\mathbf{g}, \mathbf{r}) - \beta_1(\mathbf{g}_{tt}, \mathbf{r}) + \beta_2(\partial_s \mathbf{g}, \partial_s \mathbf{r}) - \frac{1}{\beta}(\boldsymbol{\xi}, \mathbf{r})\right] dt = 0 \tag{6.148}$$

for all $\psi \in \mathcal{D}((0, T))$ with $\psi(0) = 0$ and for all $\mathbf{r} \in \mathbf{H}^1_{n0}(\Gamma_c)$. In this way, a necessary condition to satisfy (6.145) is to satisfy the differential equation

$$(\mathbf{g}, \mathbf{r}) - \beta_1(\mathbf{g}_{tt}, \mathbf{r}) + \beta_2(\partial_s \mathbf{g}, \partial_s \mathbf{r}) - \frac{1}{\beta}(\boldsymbol{\xi}, \mathbf{r}) = 0 \quad \forall \mathbf{r} \in \mathbf{H}^1_{n0}(\Gamma_c) \tag{6.149}$$

with $\mathbf{g}(0, \mathbf{x}) = \gamma_0 \mathbf{u}_0(\mathbf{x})$ and $\mathbf{g}_t(T, \mathbf{x}) = \mathbf{0}$. The first of these boundary conditions is imposed on candidate minimizers in order to ensure the regularity of solutions; see (6.137). The second boundary condition is a result of the minimization process. Now we can use the space $\mathbf{H}^1_n(\Gamma_c)$ for the test functions. Because of the orthogonality between $\mathbf{H}^1_n(\Gamma_c)$ and $(\mathbf{H}^1_n)^\perp(\Gamma_c)$, we can write a weak formulation of (6.149) with test functions in $\mathbf{H}^1(\Gamma_c)$ by adding an arbitrary constant vector in the normal direction. We recall that $\mathbf{H}^1(\Gamma_c) = \mathbf{H}^1_n(\Gamma_c) \oplus (\mathbf{H}^1_n(\Gamma_c))^\perp$ and thus

$$\mathbf{r} = \mathbf{r}_1 - \mathbf{n}\,\frac{\int_{\Gamma_c} \mathbf{r}_1 \cdot \mathbf{n}\, d\mathbf{x}}{\mu(\Gamma_c)},$$

where $\mathbf{r}_1 \in \mathbf{H}^1(\Gamma_c)$. Now, the equation can be tested against $\mathbf{r}_1 \in \mathbf{H}^1_0(\Gamma_c)$ in the following weak form:

$$(\mathbf{g}, \mathbf{r}_1) - \beta_1(\mathbf{g}_{tt}, \mathbf{r}_1) + \beta_2(\partial_s \mathbf{g}, \partial_s \mathbf{r}_1) + k(t)(\mathbf{n}, \mathbf{r}_1) = \frac{1}{\beta}(\boldsymbol{\xi}, \mathbf{r}_1) \quad \forall \mathbf{r}_1 \in \mathbf{H}^1_0(\Gamma_c), \tag{6.150}$$

where $k(t)$ is specified by the constraint

$$\int_\Gamma \mathbf{g} \cdot \mathbf{n} d\mathbf{x} = 0.$$

Finally, in order to obtain the solution of the optimal control problem, we have to solve the Navier–Stokes system

$$\begin{cases} \langle \mathbf{u}_t, \mathbf{v} \rangle + a(\mathbf{u}, \mathbf{v}) + c(\mathbf{u}, \mathbf{u}, \mathbf{v}) + b(\mathbf{v}, p) = 0 \quad \forall \mathbf{v} \in \mathbf{H}^1_0(\Omega), \\ b(\mathbf{u}, q) = 0 \quad \forall q \in L^2_0(\Omega), \\ (\mathbf{u}, \mathbf{s})_\Gamma = (g(t, \mathbf{x}), \mathbf{s})_\Gamma \quad \forall \mathbf{s} \in \mathbf{H}^{-1/2}(\Gamma), \\ \mathbf{u}(0, \mathbf{x}) = \mathbf{u}_0(\mathbf{x}) \in curl(H^2)(\Omega); \end{cases} \tag{6.151}$$

the adjoint system

$$\begin{cases} -\langle \mathbf{w}_t, \mathbf{v}\rangle + \; a(\mathbf{w}, \mathbf{v}) + c(\mathbf{w}, \mathbf{u}, \mathbf{v}) + c(\mathbf{u}, \mathbf{w}, \mathbf{v}) \\ \qquad\qquad + b(\mathbf{v}, \sigma) = \alpha(\mathbf{u} - \mathbf{U}, \mathbf{v}) \quad \forall\, \mathbf{v} \in \mathbf{H}_0^1(\Omega), \\ b(\mathbf{w}, q) = 0 \quad \forall\, q \in L_0^2(\Omega), \\ \mathbf{w} = \mathbf{0} \quad \forall\, \mathbf{x} \in \Gamma, \\ \mathbf{w}(T, \mathbf{x}) = \mathbf{0} \quad \forall\, \mathbf{x} \in \Omega; \end{cases} \tag{6.152}$$

the optimality condition

$$\begin{cases} (\mathbf{g}, \mathbf{r}) - \beta_1(\mathbf{g}_{tt}, \mathbf{r}) + \beta_2(\partial_s \mathbf{g}, \partial_s \mathbf{r}) + k(t)(\mathbf{n}, \mathbf{r}) \\ \qquad\qquad = \dfrac{1}{\beta}[(\gamma_1 \mathbf{w}, \mathbf{r}) - (\mathbf{n}\sigma, \mathbf{r})] \quad \forall\, \mathbf{r} \in \mathbf{H}_0^1(\Gamma_c), \\ \mathbf{g}(0, \mathbf{x}) = \gamma_0 \mathbf{u}_0 \quad \forall\, \mathbf{x} \in \Gamma_c, \\ \mathbf{g}_t(T, \mathbf{x}) = \mathbf{0} \quad \forall\, \mathbf{x} \in \Gamma_c, \\ \mathbf{g} = 0 \quad \forall\, \mathbf{x} \in \partial\Gamma_c; \end{cases} \tag{6.153}$$

and the compatibility condition for the boundary control

$$\int_\Gamma \mathbf{g} \cdot \mathbf{n}\, d\Gamma = 0. \tag{6.154}$$

Equation (6.154) is needed in order to calculate the variable $k(t)$. If the control is a tangential control, then terms containing the adjoint pressure σ and $k(t)$ can be neglected in (6.153).

6.3.2 Semidiscrete-in-time approximations

Formulation of the semidiscrete-in-time optimal control problem

Let $\sigma_N = \{t_j\}_{j=0}^N$ be a partition of $[0, T]$ into equal intervals $\Delta t = T/N$ with $t_0 = 0$ and $t_N = T$. For each fixed Δt (or N) and for every quantity $q(t, \mathbf{x})$, we associate the corresponding set $\{q^{(j)}(\mathbf{x})\}_{j=0}^N$ and a continuous piecewise linear function $q^N = q^N(t, \mathbf{x})$ such as $q^N(t_j, \mathbf{x}) = q^{(j)}(\mathbf{x})$ for all $j = 0, 1, \dots, N$. We will denote with vector notation $\vec{q}$ the vector $(q^{(1)}, q^{(2)}, \dots, q^{(N)})$ of the discrete time components; for vector-valued functions, we have the notation $\vec{\mathbf{v}}$ for $(\mathbf{v}^{(1)}, \mathbf{v}^{(2)}, \dots, \mathbf{v}^{(N)})$. Also, the space X^N will be denoted as $\vec{X}$. On this partition we define the discrete target velocity as $\mathbf{U}^{(j)}(\mathbf{x}) = \mathbf{U}(t_j, \mathbf{x})$ for $j = 0, 1, \dots, N$ when $\mathbf{U} \in U_{ad}$.

The state variables $\mathbf{u}^{(j)} \in \mathbf{H}_0^1(\Omega)$ and $p^{(j)} \in L_0^2(\Omega)$ are constrained to satisfy the semidiscrete Navier–Stokes system

$$\begin{cases} \dfrac{1}{\Delta t}(\mathbf{u}^{(j)} - \mathbf{u}^{(j-1)}, \mathbf{v}) + a(\mathbf{u}^{(j)}, \mathbf{v}) \\ \qquad\qquad + c(\mathbf{u}^{(j)}, \mathbf{u}^{(j)}, \mathbf{v}) + b(\mathbf{v}, p^{(j)}) = 0 \quad \forall\, \mathbf{v} \in \mathbf{H}_0^1(\Omega), \\ b(\mathbf{u}^{(j)}, q) = 0 \quad \forall\, q \in L_0^2(\Omega), \\ (\mathbf{u}^{(j)}(\mathbf{x}), \mathbf{s})_\Gamma = (\mathbf{g}^{(j)}(\mathbf{x}), \mathbf{s})_{\Gamma_c} \quad \forall\, \mathbf{s} \in \mathbf{H}^{-1/2}(\Gamma) \end{cases} \tag{6.155}$$

for $j = 1, 2, \ldots, N$ with $\mathbf{u}^{(0)} = \mathbf{u}_0(\mathbf{x}) \in curl(H^2)(\Omega)$.

Optimization is achieved by means of the minimization of the discretized functional

$$\begin{aligned}\mathcal{J}^N(\mathbf{u}, \mathbf{g}) &= \frac{\alpha}{2} \sum_{j=1}^{N} \|\mathbf{u}^{(j)} - \mathbf{U}^{(j)}\|^2 \Delta t \\ &+ \frac{\beta}{2} \sum_{j=1}^{N} \left[\|\mathbf{g}^{(j)}\|^2_{\Gamma_c} \Delta t + \beta_1 \|\partial_s \mathbf{g}^{(j)}\|^2_{\Gamma_c} \Delta t + \beta_2 \| \mathbf{g}^{(j)} - \mathbf{g}^{(j-1)} \|^2_{\Gamma_c} \right].\end{aligned} \tag{6.156}$$

Of course, if Δt tends to zero, this functional tends to the corresponding continuous functional (6.130).

The *admissibility set* A_{ad} is defined by

$$A_{ad} = \{(\vec{\mathbf{u}}, \vec{p}, \vec{\mathbf{g}}) \in \vec{\mathbf{H}}^1(\Omega) \times \vec{L}^2_0(\Omega) \times \vec{\mathbf{H}}^1_{n0}(\Gamma_c) \quad \textit{such that } (\vec{\mathbf{u}}, \vec{p}, \vec{\mathbf{g}}) \textit{ is a solution of } (6.155),\ \mathbf{g}^{(0)} = \gamma \mathbf{u}_0, \textit{ and the functional in } (6.156) \textit{ is bounded} \}.$$

The formulation of the optimal control problem in the semidiscrete approximation is

given $\Delta t = T/N$, $\mathbf{u}_0 \in curl(H^2)(\Omega)$, *and* $\mathbf{U} \in U_{ad}$, $(\vec{\mathbf{u}}, \vec{p}, \vec{\mathbf{g}}) \in A_{ad}$ *is called an optimal solution if there exists* $\epsilon > 0$ *such that*

$$\mathcal{J}^N(\vec{\mathbf{u}}, \vec{\mathbf{g}}) \le \mathcal{J}^N(\widetilde{\mathbf{u}}, \widetilde{\mathbf{h}}) \quad \forall\, \widetilde{\mathbf{h}} \in \vec{\mathbf{H}}^1_{n0}$$

whenever $\|\mathbf{g}^{(j)} - \widetilde{\mathbf{h}}^{(j)}\|_{\Gamma_c} \le \epsilon$ *with* $j = 1, 2, \ldots, N$.

For the semidiscrete Navier–Stokes nonhomogeneous boundary problem, one can prove the following theorem.

Theorem 6.34. *Let* $\Delta t = T/N$ *and* $\mathbf{u}_0 \in curl(H^2)(\Omega)$. *Let* $\epsilon > 0$, $\vec{\mathbf{g}} \in \vec{\mathbf{H}}^1_{n0}(\Omega)$ *such that* $\sum_{i=1}^{N}(\|\mathbf{g}^{(j)}\|^2_1 \Delta t + \|\mathbf{g}^{(j)} - \mathbf{g}^{(j-1)}\|^2) \le \epsilon$, *i.e.,* $\mathbf{g}^N$ *and* $\mathbf{g}'^N$ *are uniformly bounded by* ϵ *in* $L^2((0,T); \mathbf{H}^1(\Gamma_c))$ *and in* $L^2((0,T); \mathbf{L}^2(\Gamma_c))$, *respectively. Then, there exists a function* $\vec{\mathbf{u}} \in \vec{\mathbf{H}}^1(\Omega)$ *that is a solution of the system*

$$\left\{\begin{aligned} &\frac{1}{\Delta t}(\mathbf{u}^{(j)} - \mathbf{u}^{(j-1)}, \mathbf{v}) + a(\mathbf{u}^{(j)}, \mathbf{v}) + c(\mathbf{u}^{(j)}, \mathbf{u}^{(j)}, \mathbf{v}) \\ &\qquad + b(\mathbf{v}, p^{(j)}) = 0 \quad \forall \mathbf{v} \in \mathbf{H}^1_0(\Omega) \quad \textit{for} \quad j = 1, \ldots, N, \\ &b(\mathbf{u}^{(j)}, q) = 0 \quad \forall q \in L^2_0(\Omega) \quad \textit{for} \quad j = 1, \ldots, N, \\ &\mathbf{u}^{(j)}(\mathbf{x}) = \mathbf{g}^{(j)} \quad \textit{for } \mathbf{x} \in \Gamma_c \quad \textit{for} \quad j = 1, \ldots, N, \\ &\mathbf{u}^{(j)}(\mathbf{x}) = 0 \quad \textit{for } \mathbf{x} \in \Gamma \setminus \Gamma_c \quad \textit{for} \quad j = 1, \ldots, N, \\ &\mathbf{u}^{(0)} = \mathbf{u}_0(\mathbf{x}) \in curl(H^2)(\Omega) \end{aligned}\right. \tag{6.157}$$

with the following estimates:

$$\|\mathbf{u}^{(j)}\|^2_1 \le K \quad n = 1, 2, \ldots, N, \tag{6.158}$$

$$\sum_{j=1}^{N} \|\nabla \mathbf{u}^{(j)}\|^2 \Delta t \le K, \tag{6.159}$$

$$\sum_{j=1}^{N} \|\mathbf{u}^{(j)} - \mathbf{u}^{(j-1)}\|_{-1}^2 \leq K, \tag{6.160}$$

where the constant K is independent of Δt.

If $\mathbf{g}$ and its time derivative are uniformly bounded, then the existence of solutions of the semidiscrete-in-time optimal control problem can be proved. This fact is an easy consequence of the definition of the optimal control problem and the boundedness of the functional.

Lemma 6.35. *Let $\Delta t = T/N$, $\mathbf{u}_0 \in curl(H^2)(\Omega)$, and $\mathbf{U} \in U_{ad}$. If $(\vec{\mathbf{u}}, \vec{\mathbf{g}})$ is the solution of the semidiscrete optimal control problem, then for all β_1 and $\beta_2 > 0$ there exists a constant C independent of Δt such that*

$$\sum_{j=1}^{N} \|\mathbf{g}^{(j)}\|_{1,\Gamma}^2 \Delta t \leq C, \tag{6.161}$$

$$\sum_{j=1}^{N} \|\mathbf{g}^{(j)} - \mathbf{g}^{(j-1)}\|^2 \leq C, \tag{6.162}$$

$$\sum_{j=1}^{N} \|\mathbf{u}^{(j)}\|^2 \Delta t \leq C\,. \tag{6.163}$$

Hence, we have that $\mathbf{g}^N \in L^2((0,T); \mathbf{v}^1(\Gamma))$, $\mathbf{g}'^N \in L^2((0,T); \mathbf{L}^2(\Gamma))$, and $\mathbf{u}^N \in L^2((0,T); \mathbf{W}(\Omega))$ for all N.

We can recall that if the norm of $\mathbf{g}^N \in L^2((0,T); \mathbf{H}^1(\Gamma))$ and the norm of $\mathbf{g}'^N \in \mathbf{L}^2((0,T); L^2(\Gamma))$ are uniformly bounded for all N, then $\mathbf{g}^N$ is uniformly bounded in $L^2((0,T); \mathbf{L}^2(\Gamma))$ for all N. Now, one can prove the existence of solutions for the semidiscrete optimal control problem in an open bounded domain Ω with boundary Γ in C^2.

Theorem 6.36. *Given $\Delta t = T/N$, $\mathbf{u}_0 \in curl(H^2)(\Omega)$, and $\mathbf{U} \in U_{ad}$, there exists a solution $(\vec{\mathbf{u}}, \vec{p}, \vec{\mathbf{g}}) \in A_{ad}$ of* (6.155) *such that $\vec{\mathbf{g}}$ minimizes the cost functional* (6.156).

First-order necessary condition

We now examine the first-order necessary condition for the semidiscrete optimal control problem. Denote by $\vec{\mathbf{B}}_1$ and $\vec{\mathbf{B}}_2$ the following sets:

$$\begin{cases} \vec{\mathbf{B}}_1 = \vec{\mathbf{H}}^1(\Omega) \times \vec{L}_0^2(\Omega) \times \vec{\mathbf{H}}_{n0}^1(\Gamma_c), \\ \vec{\mathbf{B}}_2 = \vec{\mathbf{H}}^{-1}(\Omega) \times \vec{L}_0^2(\Omega) \times \vec{\mathbf{H}}^{1/2}(\Gamma_c)\,. \end{cases} \tag{6.164}$$

We define the nonlinear map

$$M(\vec{\mathbf{u}}, \vec{p}, \vec{\mathbf{g}}) : \vec{\mathbf{B}}_1 \to \vec{\mathbf{B}}_2$$

as $M(\vec{\mathbf{u}}, \vec{p}, \vec{\mathbf{g}}) = (\vec{\mathbf{f}}, \vec{z}, \vec{\mathbf{b}})$ if and only if

$$\begin{cases} \dfrac{1}{\Delta t}(\mathbf{u}^{(j)} - \mathbf{u}^{(j-1)}, \mathbf{v}) + a(\mathbf{u}^{(j)}, \mathbf{v}) + c(\mathbf{u}^{(j)}, \mathbf{u}^{(j)}, \mathbf{v}) \\ \qquad + b(\mathbf{v}, p^{(j)}) = (\mathbf{f}^{(j)}, \mathbf{v}) \quad \forall \mathbf{v} \in \mathbf{H}_0^1(\Omega) \quad \text{for} \quad j = 1, \dots, N, \\ b(\mathbf{u}^{(j)}, q) = (z^{(j)}, q) \quad \forall q \in L_0^2(\Omega) \quad \text{for} \quad j = 1, \dots, N, \\ (\mathbf{u}^{(j)}, \mathbf{s})_\Gamma - (\mathbf{g}^{(j)}, \mathbf{s})_{\Gamma_c} = (\mathbf{b}^{(j)}, \mathbf{s})_\Gamma \quad \forall \mathbf{s} \in \mathbf{H}^{-1/2}(\Gamma) \quad \text{for} \quad j = 1, \dots, N, \\ \mathbf{u}^{(0)} = \mathbf{u}_0(\mathbf{x}) \in curl(H^2)(\Omega) . \end{cases} \tag{6.165}$$

In the same manner, let $\widehat{\mathbf{g}}$ be an optimal solution and define

$$N(\vec{\mathbf{u}}, \vec{p}, \vec{\mathbf{g}}) : \vec{\mathbf{B}}_1 \to \mathbb{R} \times \vec{\mathbf{B}}_2$$

as $N(\vec{\mathbf{u}}, \vec{p}, \vec{\mathbf{g}}) = (a, \vec{\mathbf{f}}, \vec{z}, \vec{\mathbf{b}})$ if and only if

$$\begin{pmatrix} \mathcal{J}^N(\vec{\mathbf{u}}, \vec{\mathbf{g}}) - \mathcal{J}^N(\widehat{\mathbf{u}}, \widehat{\mathbf{g}}) \\ M(\vec{\mathbf{u}}, \vec{p}, \vec{\mathbf{g}}) \end{pmatrix} = \begin{pmatrix} a \\ (\vec{\mathbf{f}}, \vec{z}, \vec{\mathbf{b}}) \end{pmatrix} . \tag{6.166}$$

Thus, the constraints can be expressed as $M(\vec{\mathbf{u}}, \vec{p}, \vec{\mathbf{g}}) = (\vec{\mathbf{0}}, \vec{0}, \vec{\mathbf{0}})$, and the optimal control problem can be reformulated as follows:

> *Find* $(\widehat{\mathbf{u}}, \widehat{\mathbf{p}}, \widehat{\mathbf{g}})$ *and* $a \le 0$ *such that the equation* $N(\widehat{\mathbf{u}}, \widehat{p}, \widehat{\mathbf{g}}) = (a, \vec{\mathbf{0}}, \vec{0}, \vec{\mathbf{0}})$ *is satisfied for all* $\vec{\mathbf{g}}$ *such that* $\|\mathbf{g}^{(j)} - \widehat{\mathbf{g}}^{(j)}\| \le \epsilon$ *for* $j = 1, 2, \dots, N$ *for some* $\epsilon > 0$.

Since we are looking for local minimum points, it is natural to define the operators $M'(\vec{\mathbf{u}}, \vec{p}, \vec{\mathbf{g}})$ and $N'(\vec{\mathbf{u}}, \vec{p}, \vec{\mathbf{g}})$. Given a $(\vec{\mathbf{u}}, \vec{p}, \vec{\mathbf{g}})$, we define the linear operator

$$M'(\vec{\mathbf{u}}, \vec{p}, \vec{\mathbf{g}}) : \vec{\mathbf{B}}_1 \to \vec{\mathbf{B}}_2$$

as $M'(\vec{\mathbf{u}}, \vec{p}, \vec{\mathbf{g}}) \cdot (\widetilde{\mathbf{w}}, \widetilde{r}, \widetilde{\mathbf{h}}) = (\bar{\mathbf{f}}, \bar{z}, \bar{\mathbf{b}})$ if and only if

$$\begin{cases} \dfrac{1}{\Delta t}(\widetilde{\mathbf{w}}^{(j)} - \widetilde{\mathbf{w}}^{(j-1)}, \mathbf{v}) + a(\widetilde{\mathbf{w}}^{(j)}, \mathbf{v}) \\ \qquad + c(\widetilde{\mathbf{w}}^{(j)}, \mathbf{u}^{(j)}, \mathbf{v}) + c(\mathbf{u}^{(j)}, \widetilde{\mathbf{w}}^{(j)}, \mathbf{v}) \\ \qquad + b(\mathbf{v}, \widetilde{r}^{(j)}) = (\bar{\mathbf{f}}^{(j)}, \mathbf{v}) \quad \forall \mathbf{v} \in \mathbf{v}_0^1(\Omega) \quad \text{for} \quad j = 1, \dots, N, \\ b(\widetilde{\mathbf{w}}^{(j)}, q) = (\bar{z}^{(j)}, q) \quad \forall q \in L_0^2(\Omega) \quad \text{for} \quad j = 1, \dots, N, \\ (\widetilde{\mathbf{w}}^{(j)}, \mathbf{s})_\Gamma - (\widetilde{\mathbf{h}}^{(j)}, \mathbf{s})_{\Gamma_c} = (\bar{\mathbf{b}}^{(j)}, \mathbf{s})_\Gamma \\ \qquad \forall \mathbf{s} \in \mathbf{v}^{-1/2}(\Gamma) \quad \text{for} \quad j = 1, \dots, N, \\ \mathbf{u}^{(0)} = \mathbf{u}_0(\mathbf{x}) \in curl(H^2)(\Omega) . \end{cases} \tag{6.167}$$

Let

$$N'(\vec{\mathbf{u}}, \vec{p}, \vec{\mathbf{g}}) : \vec{\mathbf{B}}_1 \to \mathbb{R} \times \vec{\mathbf{B}}_2$$

be defined as $N'(\vec{\mathbf{u}}, \vec{p}, \vec{\mathbf{g}}) \cdot (\widetilde{a}, \widetilde{\mathbf{w}}, \widetilde{r}, \widetilde{\mathbf{h}}) = (\bar{a}, \bar{\mathbf{f}}, \bar{z}, \bar{\mathbf{b}})$ if and only if

$$\begin{pmatrix} \mathcal{J}^{N'}(\vec{\mathbf{u}}, \vec{\mathbf{g}}) \cdot (\widetilde{a}, \widetilde{\mathbf{w}}, \widetilde{r}, \widetilde{\mathbf{h}}) \\ M'(\vec{\mathbf{u}}, \vec{p}, \vec{\mathbf{g}}) \cdot (\widetilde{a}, \widetilde{\mathbf{w}}, \widetilde{r}, \widetilde{\mathbf{h}}) \end{pmatrix} = \begin{pmatrix} \bar{a} \\ (\bar{\mathbf{f}}, \bar{z}, \bar{\mathbf{b}}) \end{pmatrix}. \tag{6.168}$$

Now we have to show that these operators are well defined, i.e., the equations for the Gateaux derivatives are well posed and have solutions.

Lemma 6.37. *Given $\Delta t = T/N$ and $\mathbf{u}_0 \in curl(H^2)(\Omega)$, let $(\widehat{\mathbf{u}}, \widehat{p}, \widehat{\mathbf{g}}) \in \vec{\mathbf{B}}_1$ denote a solution of the semidiscrete optimal control problem. Then, we have that*

(i) *the operator $M'(\widehat{\mathbf{u}}, \widehat{p}, \widehat{\mathbf{g}})$ has closed range and is onto $\vec{\mathbf{B}}_2$;*

(ii) *the operator $N'(\widehat{\mathbf{u}}, \widehat{p}, \widehat{\mathbf{g}})$ has closed range in $\mathbb{R} \times \vec{\mathbf{B}}_2$.*

The operator $N'(\widehat{\mathbf{u}}, \widehat{p}, \widehat{\mathbf{g}})$ cannot be onto. If it were, by the implicit function theorem, we would have that there exists a solution, which is different from the optimal solution, that minimizes the functional for every small neighborhood of $(\widehat{\mathbf{u}}, \widehat{p}, \widehat{\mathbf{g}})$. This contradicts the hypothesis that $(\widehat{\mathbf{u}}, \widehat{p}, \widehat{\mathbf{g}})$ is an optimal solution. Therefore, the optimality condition implies the following theorem.

Theorem 6.38. *Given $\Delta t = T/N$ and $\mathbf{u}_0 \in curl(H^2)(\Omega)$, if $(\widehat{\mathbf{u}}, \widehat{p}, \widehat{\mathbf{g}}) \in (\vec{\mathbf{H}}^1(\Omega) \times \vec{L}^2_0(\Omega) \times \vec{\mathbf{H}}^1_{n0}(\Gamma_c))$ is a solution of the semidiscrete optimal control problem, then the operator $N'(\widehat{\mathbf{u}}, \widehat{p}, \widehat{\mathbf{g}})$ is not onto $\mathbb{R} \times \vec{\mathbf{B}}_2$.*

In the next theorem, we write the first-order necessary condition and characterize the optimal control solution as a solution of the corresponding Euler–Lagrange system of equations.

Theorem 6.39. *Given $\Delta t = T/N$ and $\mathbf{u}_0 \in curl(H^2)(\Omega)$, if $(\widehat{\mathbf{u}}, \widehat{p}, \widehat{\mathbf{g}}) \in \vec{\mathbf{H}}^1(\Omega) \times \vec{L}^2_0(\Omega) \times \vec{\mathbf{H}}^1_{n0}(\Gamma_c)$ is a solution of the semidiscrete optimal control problem, i.e., the operator $N'(\widehat{\mathbf{u}}, \widehat{p}, \widehat{\mathbf{g}})$ is not onto, then there exists a nonzero Lagrangian multiplier $(\vec{\mathbf{w}}, \vec{\sigma}, \vec{\boldsymbol{\xi}}) \in \vec{\mathbf{H}}^1_0(\Omega) \times \vec{L}^2_0(\Omega) \times \vec{\mathbf{H}}^{-1/2}(\Gamma)$ satisfying the Euler equations*

$$\mathcal{J}^{N'}(\widehat{\mathbf{u}}, \widehat{\mathbf{g}}) \cdot (\widetilde{\mathbf{w}}, \widetilde{r}, \widetilde{\mathbf{h}}) + \Big\langle (\vec{\mathbf{w}}, \vec{\sigma}, \vec{\boldsymbol{\xi}}), M'(\widehat{\mathbf{u}}, \widehat{p}, \widehat{\mathbf{g}}) \cdot (\widetilde{\mathbf{w}}, \widetilde{r}, \widetilde{\mathbf{h}}) \Big\rangle = 0$$
$$\forall\, (\widetilde{\mathbf{w}}, \widetilde{r}, \widetilde{\mathbf{h}}) \in \vec{\mathbf{H}}^1(\Omega) \times \vec{L}^2_0(\Omega) \times \vec{\mathbf{H}}^1_{n0}(\Gamma_c),$$

where $\langle \cdot, \cdot \rangle$ denotes the duality pairing between $\mathbb{R} \times \vec{\mathbf{B}}_2$ and $\mathbb{R} \times \vec{\mathbf{B}}_2^*$.

The optimality system

From the first-order necessary condition, we can further characterize the optimal control as a solution of a differential equation.

Theorem 6.40. *Given $\Delta t = T/N$ and $\mathbf{u}_0 \in curl(H^2)(\Omega)$, let $(\vec{\mathbf{u}}, \vec{p}, \vec{\mathbf{g}}) \in (\vec{\mathbf{H}}^1(\Omega) \times \vec{L}^2_0(\Omega) \times \vec{\mathbf{H}}^1_{n0}(\Gamma_c))$ denote an optimal control solution. Then, the control $\mathbf{g}^{(j)}$ satisfying the compatibility conditions*

$$\mathbf{g}^{(0)} = \gamma_0 \mathbf{u}_0\,, \qquad \int_\Gamma \mathbf{g}^{(j)} \cdot \mathbf{n}\, d\mathbf{x} = 0 \quad \textit{for } j = 0, 1, \ldots, N$$

is a solution of the system

$$\int_{\Gamma_c}\left[\mathbf{g}^{(j)}\cdot\widetilde{\mathbf{h}}-\frac{\beta_1}{\Delta t^2}(\mathbf{g}^{(j+1)}-2\mathbf{g}^{(j)}+\mathbf{g}^{(j-1)})\cdot\widetilde{\mathbf{h}}+\beta_2\partial_s\mathbf{g}^{(j)}\cdot\partial_s\widetilde{\mathbf{h}}+k^{(j)}\mathbf{n}\cdot\widetilde{\mathbf{h}}\right]d\mathbf{x}$$

$$=\frac{1}{\beta}\int_{\Gamma_c}(\boldsymbol{\xi}^{(j)}\cdot\widetilde{\mathbf{h}})\,d\mathbf{x}\qquad\forall\,\widetilde{\mathbf{h}}\in\mathbf{v}_0^1(\Gamma_c)$$

for $j=1,\dots,N-1$ ($\mathbf{g}^N=\mathbf{g}^{N-1}$). *The function* $\vec{\boldsymbol{\xi}}\in\vec{\mathbf{H}}^{-1/2}(\Gamma_c)$ *is defined by*

$$\int_{\Gamma_c}\boldsymbol{\xi}^{(j-1)}\cdot\widetilde{\mathbf{v}}\,d\mathbf{x}=\int_{\Gamma_c}(\gamma_1\mathbf{w}^{(j-1)}-\sigma^{(j-1)}\mathbf{n})\cdot\widetilde{\mathbf{v}}\,d\mathbf{x} \tag{6.169}$$

$$=-\alpha(\mathbf{u}^{(j)}-\mathbf{U}^{(j)},\widetilde{\mathbf{v}})-\frac{1}{\Delta t}(\mathbf{w}^{(j)}-\mathbf{w}^{(j-1)},\widetilde{\mathbf{v}})_\Omega+a(\mathbf{w}^{(j-1)},\widetilde{\mathbf{v}})$$

$$+c(\widetilde{\mathbf{v}},\mathbf{u}^{(j)},\mathbf{w}^{(j-1)})+c(\mathbf{u}^{(j)},\widetilde{\mathbf{v}},\mathbf{w}^{(j-1)})+b(\widetilde{\mathbf{v}},\sigma^{(j-1)})$$

$$\forall\,\widetilde{\mathbf{v}}\in\mathbf{H}^1(\Omega)\quad\text{for}\quad j=1,\dots,N,$$

where $\vec{\mathbf{w}}$ *and* $\vec{\sigma}$ *satisfy*

$$\begin{cases}-\dfrac{1}{\Delta t}(\mathbf{w}^{(j)}-\mathbf{w}^{(j-1)},\mathbf{v})+a(\mathbf{w}^{(j-1)},\mathbf{v})+c(\mathbf{w}^{(j-1)},\mathbf{u}^{(j)},\mathbf{v})\\ \qquad+\,c(\mathbf{u}^{(j)},\mathbf{w}^{(j-1)},\mathbf{v})+b(\mathbf{v},\sigma^{(j-1)})=\alpha(\mathbf{u}^{(j)}-\mathbf{U}^{(j)},\mathbf{v})\\ \qquad\qquad\forall\,\mathbf{v}\in\mathbf{H}_0^1(\Omega)\quad \textit{for}\quad j=1,\dots,N,\\ b(\mathbf{w}^{(j-1)},q)=0\quad\forall\,q\in L_0^2(\Omega)\quad\textit{for}\quad j=1,\dots,N,\\ (\mathbf{w}^{(j-1)},\mathbf{s})=0\quad\forall\,\mathbf{s}\in\mathbf{H}^{-1/2}(\Gamma)\quad\textit{for}\quad j=1,\dots,N,\\ \mathbf{w}^{(N)}=0.\end{cases} \tag{6.170}$$

Now, in order to obtain the solution of the semidiscrete-in-time optimal control problem, we have to solve the semidiscrete Navier–Stokes system

$$\begin{cases}\dfrac{1}{\Delta t}(\mathbf{u}^{(j)}-\mathbf{u}^{(j-1)},\mathbf{v})+a(\mathbf{u}^{(j)},\mathbf{v})+c(\mathbf{u}^{(j)},\mathbf{u}^{(j)},\mathbf{v})\\ \qquad+\,b(\mathbf{v},p^{(j)})=0\quad\forall\,\mathbf{v}\in\mathbf{H}_0^1(\Omega)\quad\text{for}\quad j=1,\dots,N,\\ b(\mathbf{u}^{(j)},q)=0\quad\forall\,q\in L_0^2(\Omega)\quad\text{for}\quad j=1,\dots,N,\\ (\mathbf{u}^{(j)},\mathbf{s})_\Gamma=(\mathbf{g}^{(j)},\mathbf{s})_{\Gamma_c}\quad\forall\,\mathbf{s}\in\mathbf{H}^{-1/2}(\Gamma)\quad\text{for}\quad j=1,\dots,N,\\ \mathbf{w}^{(0)}=\mathbf{u}_0\in curl(H^2)(\Omega);\end{cases} \tag{6.171}$$

the semidiscrete adjoint system

$$\left\{\begin{array}{l} -\dfrac{1}{\Delta t}(\mathbf{w}^{(j)}-\mathbf{w}^{(j-1)},\mathbf{v})+a(\mathbf{w}^{(j-1)},\mathbf{v})+c(\mathbf{u}^{(j)},\mathbf{v},\mathbf{w}^{(j-1)}) \\ \qquad +\,c(\mathbf{v},\mathbf{u}^{(j)},\mathbf{w}^{(j-1)})+b(\mathbf{v},\sigma^{(j-1)})=\alpha(\mathbf{u}^{(j)}-\mathbf{U}^{(j)},\mathbf{v}) \\ \qquad\qquad \forall\,\mathbf{v}\in\mathbf{H}_0^1(\Omega)\quad\text{for}\quad j=1,\dots,N, \\ b(\mathbf{w}^{(j)},q)=0\quad\forall q\in L_0^2(\Omega)\quad\text{for}\quad j=1,\dots,N, \\ \mathbf{w}^{(j-1)}=0\quad\text{on}\quad\Gamma\quad\text{for}\quad j=1,\dots,N, \\ \mathbf{w}^{(N)}=0\quad\text{in}\quad\Omega; \end{array}\right. \tag{6.172}$$

and the optimality condition

$$\left\{\begin{array}{l} -\dfrac{\beta_1}{\Delta t^2}(\mathbf{g}^{(j+1)}-2\mathbf{g}^{(j)}+\mathbf{g}^{(j-1)},\widetilde{\mathbf{h}})+\beta_2(\partial_s\mathbf{g}^{(j)},\partial_s\widetilde{\mathbf{h}}) \\ \qquad +\,(\mathbf{g}^{(j)},\widetilde{\mathbf{h}})+k^{(j)}(\mathbf{n},\widetilde{\mathbf{h}})=\dfrac{1}{\beta}(\gamma_1\omega^{(j)}-\mathbf{n}\sigma^{(j)},\widetilde{\mathbf{h}}) \\ \qquad\qquad \forall\widetilde{\mathbf{h}}\in\mathbf{H}_0^1(\Gamma_c)\quad\text{for}\quad j=1,\dots,N-1, \\ (\mathbf{g}^{(j)},\mathbf{n})_\Gamma=0\quad\text{for}\quad j=1,\dots,N-1, \\ \mathbf{g}^{(j)}=0\quad\text{on}\quad\Gamma\setminus\Gamma_c\quad\text{for}\quad j=1,\dots,N-1, \\ \mathbf{g}^{(0)}=\gamma_0\mathbf{u}_0\quad\text{on}\quad\Gamma_c, \\ \mathbf{g}^{(N)}=\mathbf{g}^{(N-1)}\quad\text{on}\quad\Gamma_c. \end{array}\right. \tag{6.173}$$

6.3.3 Fully discrete space-time approximations

Assumptions on the finite element spaces

We consider only conforming finite element approximations. Let $\mathbf{X}^h\subset\mathbf{H}^1(\Omega)$ and $S^h\subset L^2(\Omega)$ be two families of finite-dimensional subspaces parameterized by an h that tends to zero. We also denote $\mathbf{X}_0^h=\mathbf{X}^h\cap\mathbf{H}_0^1(\Omega)$ and $S_0^h=S^h\cap L_0^2(\Omega)$. We make the following assumptions on $\mathbf{X}^h$ and S^h.

(a) *The approximation hypotheses.*

There exists an integer l and a constant C, independent of h, $\mathbf{u}$, and p, such that for $1\le k\le l$ we have

$$\inf_{\mathbf{u}_h\in\mathbf{X}^h}\|\mathbf{u}_h-\mathbf{u}\|_1\le Ch^k\|\mathbf{u}\|_{k+1}\quad\forall\,\mathbf{u}\in\mathbf{H}^{k+1}(\Omega)\cap\mathbf{H}_0^1(\Omega), \tag{6.174}$$

$$\inf_{p_h\in S^h}\|p-p_h\|\le Ch^k\|p\|_k\qquad\forall\,p\in H^k(\Omega)\cap L_0^2(\Omega). \tag{6.175}$$

(b) *The inf-sup condition or LBB condition.*

There exists a constant C', independent of h, such that

$$\inf_{0\neq q_h\in S^h}\ \sup_{0\neq\mathbf{u}_h\in\mathbf{X}^h}\frac{\int_\Omega q_h\nabla\cdot\mathbf{u}_h\,dx}{\|\mathbf{u}_h\|_1\|q_h\|}\ge C'>0. \tag{6.176}$$

This condition ensures the stability of the discrete Navier–Stokes solutions.

Next, let $\mathbf{P}^h = \mathbf{X}^h|_\Gamma$, i.e., $\mathbf{P}^h$ consists of the restriction, to the boundary Γ, of functions belonging to $\mathbf{X}^h$. For all choices of conforming finite element space $\mathbf{X}^h$ we then have that $\mathbf{P}^h \subset \mathbf{H}^{-1/2}(\Gamma)$. For the subspaces $\mathbf{P}^h = \mathbf{X}^h|_\Gamma$, we assume the approximation property: there exists an integer l and a constant C, independent of h and $\mathbf{s}$, such that for $1 \le k \le l$ we have

$$\inf_{\mathbf{s}_h \in \mathbf{P}_h} \|\mathbf{s}_h - \mathbf{s}\|_{-1/2,\Gamma} \le Ch^k \|\mathbf{u}\|_{k-1/2} \quad \forall\, \mathbf{s} \in \mathbf{H}^{k-1/2}(\Gamma)\,. \tag{6.177}$$

Now, let $\mathbf{Q}^h = \mathbf{X}^h|_{\Gamma_c}$, i.e., $\mathbf{Q}^h$ also consists of the restriction, to the boundary segment Γ_c, of the functions belonging to $\mathbf{X}^h$. For all choices of conforming finite element spaces $\mathbf{X}^h$, we have that $\mathbf{Q}^h \subset \mathbf{H}^1(\Gamma_c)$. We define $\mathbf{Q}_0^h = \mathbf{Q}^h \cap \mathbf{H}^1_{n0}(\Gamma_c)$. If the same type of polynomials are used in $\mathbf{Q}_0^h$, we have the following property.

(c) *Boundary approximating property.*

There exist an integer k and a constant C, independent of h and $\mathbf{s}$, such that for $1 \le m \le k$ we have

$$\begin{aligned} \inf_{\mathbf{s}_h \in \mathbf{Q}_0^h} \|\mathbf{s}_h - \mathbf{s}\|_{s,\Gamma_c} &\le Ch^{m-s+1/2} \|\mathbf{s}\|_{m+1/2} \\ &\forall\, \mathbf{s} \in \mathbf{H}^1_{n0}(\Gamma_c), \quad 0 \le s \le 1\,. \end{aligned} \tag{6.178}$$

See [414, 432] for details concerning the approximation on the boundary.

Formulation of the fully discrete optimal control approximation

Let $\sigma_N = \{t_j\}_{j=0}^N$ be a partition of $[0, T]$ into equal intervals $\Delta t = T/N$ with $t_0 = 0$ and $t_N = T$. For each fixed Δt (or N) and for every quantity $q(t, \mathbf{x})$, we associate the corresponding set $\{q_h^{(j)}\}_{j=1}^N$. We will denote the vector $(q_h^{(1)}, q_h^{(2)}, \ldots, q_h^{(N)})$ by $\vec{q}_h$ and $(\mathbf{v}_h^{(1)}, \mathbf{v}_h^{(2)}, \ldots, \mathbf{v}_h^{(N)})$ by $\vec{\mathbf{v}}_h$ and the space Y^N as $\vec{Y}$. The continuous linear function $\mathbf{q}_h^N(t, \mathbf{x})$ is defined by $\mathbf{q}_h^N(t_j, \mathbf{x}) = q_h(t_j, \mathbf{x})$ for all $j = 0, 1, 2, \ldots, N$.

Given $\Delta t = T/N$, $\vec{\mathbf{g}} \in \vec{\mathbf{H}}^{1/2}(\Gamma)$, and $\mathbf{u}_0 \in curl(H^2)(\Omega)$, $(\vec{\mathbf{u}}_h, \vec{p}_h)$ is called a *generalized solution* of the fully discrete time-space approximate Navier–Stokes equations if $\mathbf{u}_h^{(j)} \in \mathbf{X}^h$, $p_h^{(j)} \in S_0^h$, and $(\mathbf{u}_h^{(j)}, p_h^{(j)})$ satisfies the following system of equations:

$$\left\{ \begin{aligned} & \frac{1}{\Delta t}(\mathbf{u}_h^{(j)} - \mathbf{u}_h^{(j-1)}, \mathbf{v}_h) + a(\mathbf{u}_h^{(j)}, \mathbf{v}_h) \\ & \qquad + c(\mathbf{u}_h^{(j)}, \mathbf{u}_h^{(j)}, \mathbf{v}_h) + b(\mathbf{v}_h, p^{(j)}) = 0 \quad \forall\, \mathbf{v}_h \in \mathbf{X}_0^h, \\ & b(\mathbf{v}_h^{(j)}, q_h) = 0 \quad \forall\, q_h \in S_0^h, \\ & (\mathbf{u}_h^{(j)}, \mathbf{s})_\Gamma = (\mathbf{g}_h^{(j)}, \mathbf{s})_{\Gamma_c} \quad \forall\, \mathbf{s} \in \mathbf{Q}_0^h \end{aligned} \right. \tag{6.179}$$

for $j = 1, 2, \ldots, N$ with initial velocity $\mathbf{u}_h^{(0)} = \pi^h \mathbf{u}_0(\mathbf{x})$.

The formulation of the optimal control problem in the fully discrete approximation becomes

given $\Delta t = T/N$, $\mathbf{u}_0 \in curl(H^2)(\Omega)$, *and* $\mathbf{U} \in U_{ad}$, *find* $(\vec{\mathbf{u}}_h, \vec{p}_h, \vec{\mathbf{g}}_h)$ *in* $\vec{\mathbf{X}}^h \times \vec{S}_0^h \times \vec{\mathbf{Q}}_0^h$ *such that* $(\mathbf{u}_h^{(j)}, p_h^{(j)}, \mathbf{g}_h^{(j)})$ *is a solution of* (6.179) *and minimizes the cost function*

$$\mathcal{J}_h^N(\vec{\mathbf{u}}_h, \vec{\mathbf{g}}_h) = \frac{\alpha}{2}\sum_{j=1}^{N} \|\mathbf{u}_h^{(j)} - \mathbf{U}^{(j)}\|^2 \Delta t$$
$$+\frac{\beta}{2}\sum_{j=1}^{N}\Big[\|\mathbf{g}_h^{(j)}\|_{\Gamma_c}^2 \Delta t + \beta_1 \|\partial_s \mathbf{g}_h^{(j)}\|_{\Gamma_c}^2 \Delta t \qquad (6.180)$$
$$+\beta_2\|(\mathbf{g}_h^{(j)} - \mathbf{g}_h^{(j-1)})\|_{\Gamma_c}^2\Big]$$

with $\mathbf{g}^{(0)} = \pi_h \gamma \mathbf{u}_0$.

In the above definition, the operator π_h approximates the trace of a function in the corresponding finite element space. In the framework of conforming finite element methods, the existence of solutions of the fully discrete optimal control problem can be shown.

Theorem 6.41. *Given* $\Delta t = T/N$, $\mathbf{u}_0 \in curl(H^2)(\Omega)$, *and* $\mathbf{U} \in U_{ad}$, *there exists a solution* $(\vec{\mathbf{u}}_h, \vec{p}_h, \vec{\mathbf{g}}_h)$ *in* $\vec{\mathbf{X}}^h \times \vec{S}_0^h \times \vec{\mathbf{Q}}_0^h$ *of the fully discrete optimal control problem.*

First-order necessary condition and the optimality system

We can derive the first-order necessary condition, the corresponding Euler–Lagrange equation, and the final characterization for the optimal control. For conforming finite elements we can state the following theorem.

Theorem 6.42. *Given* $\Delta t = T/N$ *and* $\mathbf{u}_0 \in curl(H^2)(\Omega)$, *let* $(\vec{\mathbf{u}}_h, \vec{\sigma}_h, \vec{\mathbf{g}}_h) \in \vec{\mathbf{X}}^h \times \vec{S}_0^h \times \vec{\mathbf{Q}}_0^h$ *denote a solution of the fully discrete optimal control problem. Then, the control* $\vec{\mathbf{g}}_h$ *satisfies the following system:*

$$\int_{\Gamma_c}\Big[\mathbf{g}_h^{(j)}\cdot\widetilde{r}_h - \frac{\beta_1}{\Delta t^2}(\mathbf{g}_h^{(j+1)} - 2\mathbf{g}_h^{(j)} + \mathbf{g}_h^{(j-1)})\cdot\widetilde{r}_h + \beta_2 \partial_s \mathbf{g}_h^{(j)}\cdot\partial_s\widetilde{r}_h$$
$$+ k^{(j)}\mathbf{n}\cdot\widetilde{r}_h - \frac{1}{\beta}(\boldsymbol{\xi}_h^{(j)}\cdot\widetilde{r}_h)\Big]\, d\mathbf{x}dt = 0, \quad \textit{for} \quad j = 1,\dots,N-1$$

with

$$\int_{\Gamma}\mathbf{g}_h^{(j)}\cdot\mathbf{n}\, d\mathbf{x} = 0 \quad \textit{for} \quad j = 1,\dots,N-1,$$
$$\mathbf{g}^{(0)} = \gamma_0\mathbf{u}_0 \quad \textit{and} \quad \mathbf{g}^{(N)} = \mathbf{g}^{(N-1)} \quad \textit{on} \quad \Gamma_c,$$

where $\vec{\boldsymbol{\xi}}_h \in \vec{\mathbf{P}}^h(\Gamma)$ *is defined by*

$$(\boldsymbol{\xi}_h^{(j-1)}, \widetilde{\mathbf{v}}_h)_{\Gamma_c} = \left(\frac{\partial \mathbf{w}_h^{(j-1)}}{\partial n} - \sigma_h^{(j-1)}\mathbf{n}, \widetilde{\mathbf{v}}_h\right)_{\Gamma_c}$$
$$= -\alpha(\mathbf{u}_h^{(j)} - \mathbf{U}^{(j)}, \widetilde{\mathbf{v}}_h)_\Omega - \frac{1}{\Delta t}(\mathbf{w}_h^{(j)} - \mathbf{w}_h^{(j-1)}, \widetilde{\mathbf{v}}_h)_\Omega$$
$$+a(\mathbf{w}_h^{(j-1)}, \widetilde{\mathbf{v}}_h) + c(\widetilde{\mathbf{v}}_h, \mathbf{u}_h^{(j)}, \mathbf{w}_h^{(j-1)}) + c(\mathbf{u}_h^{(j)}, \widetilde{\mathbf{v}}_h, \mathbf{w}_h^{(j-1)})$$
$$+b(\widetilde{\mathbf{v}}_h, \sigma_h^{(j-1)}) \quad \forall\widetilde{\mathbf{v}}_h \in \mathbf{X}^h \quad \textit{for} \quad j = 1,\dots,N,$$

and $\vec{\mathbf{w}}_h$ and $\vec{\sigma}_h$ satisfy

$$\begin{cases} -\dfrac{1}{\Delta t}(\mathbf{w}_h^{(j)} - \mathbf{w}_h^{(j-1)}, \mathbf{v}_h) + a(\mathbf{w}_h^{(j-1)}, \mathbf{v}_h) \\ \qquad + c(\mathbf{v}_h, \mathbf{u}_h^{(j)}, \mathbf{w}_h^{(j-1)}) + c(\mathbf{u}_h^{(j)}, \mathbf{v}_h, \mathbf{w}_h^{(j-1)}) \\ \qquad + b(\mathbf{v}_h, \sigma_h^{(j-1)}) = \alpha(\mathbf{u}_h^{(j)} - \mathbf{U}^{(j)}, \mathbf{v}_h) \\ \qquad\qquad \forall\, \mathbf{v}_h \in \mathbf{X}_0^h \quad \textit{for} \quad j = 1, \dots, N, \\ b(\mathbf{w}_h^{(j-1)}, q_h) = 0 \quad \forall\, q_h \in S_0^h(\Omega) \quad \textit{for} \quad j = 1, \dots, N, \\ \mathbf{w}_h^{(j-1)} = 0 \quad \textit{on} \quad \Gamma \quad \textit{for} \quad j = 1, \dots, N, \\ \mathbf{w}_h^{(N)} = 0\,. \end{cases}$$

We remark that an optimal solution is a solution of the above system, but among the solutions of this system there may be solutions that are not optimal.

6.3.4 Implementation of fully discrete space-time approximations

For simplicity, we consider the case of tangential control, i.e., for $j = 0, 1, 2, \dots$, $\mathbf{g}^{(j)} = \lambda^{(j)}\boldsymbol{\tau}$, where $\boldsymbol{\tau}$ denotes the unit tangent vector to Γ. This enables us to dispense with the unknown function $k^{(j)}$ in the optimality system. The optimality system that determines the optimal tangential control consists of

(a) the fully discrete Navier–Stokes system

$$\begin{cases} \dfrac{1}{\Delta t}(\mathbf{u}_h^{(j)} - \mathbf{u}_h^{(j-1)}, \mathbf{v}_h) + a(\mathbf{u}_h^{(j)}, \mathbf{v}_h) + c(\mathbf{u}_h^{(j)}, \mathbf{u}_h^{(j)}, \mathbf{v}_h) \\ \qquad + b(\mathbf{v}_h, p_h^{(j)}) = 0 \quad \forall\, \mathbf{v}_h \in \mathbf{X}_0^h \quad \text{for} \quad j = 1, \dots, N, \\ b(\mathbf{u}_h^{(j)}, q_h) = 0 \quad \forall\, q_h \in S_0^h(\Omega) \quad \text{for} \quad j = 1, \dots, N, \\ (\mathbf{u}_h^{(j)}, \mathbf{s}_h)_\Gamma = (\lambda_h^{(j)}\boldsymbol{\tau}, \mathbf{s}_h)_{\Gamma_c} \quad \forall\, \mathbf{s}_h \in \mathbf{P}^h \quad \text{for} \quad j = 1, \dots, N, \\ \mathbf{w}_h^{(0)} = \pi^h \mathbf{u}_0 \quad \text{in} \quad \Omega; \end{cases} \tag{6.181}$$

(b) the fully discrete adjoint system

$$\begin{cases} -\dfrac{1}{\Delta t}(\mathbf{w}_h^{(j)} - \mathbf{w}_h^{(j-1)}, \mathbf{v}_h) + a(\mathbf{w}_h^{(j-1)}, \mathbf{v}_h) + c(\mathbf{u}_h^{(j)}, \mathbf{v}_h, \mathbf{w}_h^{(j-1)}) \\ \qquad + c(\mathbf{v}_h, \mathbf{u}_h^{(j)}, \mathbf{w}_h^{(j-1)}) + b(\mathbf{v}_h, \sigma_h^{(j-1)}) = \alpha(\mathbf{u}_h^{(j)} - \mathbf{U}^{(j)}, \mathbf{v}_h) \\ \qquad\qquad \forall\, \mathbf{v}_h \in \mathbf{X}_0^h \quad \text{for} \quad j = 1, \dots, N, \\ b(\mathbf{w}_h^{(j-1)}, q_h) = 0 \quad \forall\, q_h \in S_0^h(\Omega) \quad \text{for} \quad j = 1, \dots, N, \\ \mathbf{w}_h^{(j-1)} = 0 \quad \text{on} \quad \Gamma \quad \text{for} \quad j = 1, \dots, N, \\ \mathbf{w}_h^{(N)} = 0 \quad \text{in} \quad \Omega; \end{cases} \tag{6.182}$$

(c) the fully discrete optimality condition

$$
\begin{cases}
-\dfrac{\beta_1}{\Delta t^2}(\lambda_h^{(j+1)} - 2\lambda_h^{(j)} + \lambda_h^{(j-1)}, \mathbf{r}_h) + \beta_2(\partial_s \lambda_h^{(j)}, \partial_s \mathbf{r}_h) + (\lambda_h^{(j)}, \mathbf{r}_h) \\
\qquad = \dfrac{1}{\beta}\left(\dfrac{\partial \mathbf{w}^{(j)}}{\partial n}, \mathbf{r}_h\right) \quad \forall\, \mathbf{r}_h \in \mathbf{Q}_0^h(\Gamma_c) \quad \text{for} \quad j = 1, \dots, N-1, \\
\lambda_h^{(j)} = 0 \quad \text{on} \quad \Gamma \setminus \Gamma_c \quad \text{for} \quad j = 1, \dots, N-1, \\
\lambda_h^{(0)} = \pi_h \gamma_0 \mathbf{u}_0 \quad \text{on} \quad \Gamma_c \\
\lambda_h^{(N)} = \lambda_h^{(N-1)} \quad \text{on} \quad \Gamma_c\,.
\end{cases}
\tag{6.183}
$$

In order to solve this discrete system, let us consider a gradient method for the optimal control problem. We have to split the system into three parts in order to apply the algorithm: the Navier–Stokes system (6.181), the adjoint system (6.182), and the optimality condition (6.183). Let $\mathcal{J}_h^N(m) = \mathcal{J}_h^N(\vec{\mathbf{u}}_h(m), \vec{\lambda}_h(m))$ and τ be the tolerance required for the convergence of the functional. The *gradient algorithm* proceeds as follows.

Algorithm 6.1. (Gradient method for solution of discrete optimality system)

(a) initialization:

(i) given $\vec{\lambda}_h(0)$, τ, and $s = 1$;
(ii) solve (6.181) with $\vec{\lambda}_h(0)$ for $\vec{\mathbf{u}}_h(0)$;
(iii) evaluate $\mathcal{J}_h^N(0)$;

(b) main loop: for $m = 1, 2, \dots,$

(iv) solve (6.182) with $\vec{\mathbf{u}}_h(m-1)$ for $\vec{\mathbf{w}}_h(m)$;
(v) solve (6.183) with $\vec{\mathbf{w}}_h(m)$ for $\vec{\lambda}_h(m)$;
(vi) for $j = 1, \dots, N$, set $\lambda_h^{(j)}(m) = \lambda_h^{(j)}(m-1) - s\left(\lambda_h^{(j)}(m-1) - \lambda_h^{(j)}(m)\right)$
(vii) solve (6.181) for $\vec{\mathbf{u}}_h(m)$;
(viii) evaluate $\mathcal{J}_h^N(m)$;
(ix) if $\mathcal{J}_h^N(m) \le \mathcal{J}_h^N(m-1)$, set $s = 1.5s$ and go to (iv);
if $\mathcal{J}_h^N(m) > \mathcal{J}_h^N(m-1)$, set $s = 0.5s$ and go to (vi).

The algorithm stops when $|\mathcal{J}_h^N(m) - \mathcal{J}_h^N(m-1)|/\mathcal{J}_h^N(m) \le \tau$. It can be shown that this gradient algorithm converges to a solution of the discrete optimality system.

Theorem 6.43. *Let $(\vec{\mathbf{u}}_h(m), \vec{p}_h(m), \vec{\mathbf{w}}_h(m), \vec{\sigma}_h(m), \vec{\lambda}_h(m))$ be the mth iterate of Algorithm* 6.1 *and let $(\vec{\mathbf{u}}_h, \vec{p}_h, \vec{\mathbf{w}}_h, \vec{\sigma}_h, \vec{\lambda}_h)$ denote a solution of the fully discrete optimality system* (6.181)–(6.183)*. Then, for Δt sufficiently small, there exists a ball $\mathbf{B} \in \mathbf{Q}_0^h$ whose radius depends on the ratio α/β such that, if $\lambda_h^{(j)}(0)\boldsymbol{\tau} \in \mathbf{B}$ for $j = 0, 1, \dots,$ then $(\vec{\mathbf{u}}_h(m), \vec{p}_h(m), \vec{\mathbf{w}}_h(m), \vec{\sigma}_h(m), \vec{\lambda}_h(m))$ converges to $(\vec{\mathbf{u}}_h, \vec{p}_h, \vec{\mathbf{w}}_h, \vec{\sigma}_h, \vec{\lambda}_h)$ as $m \to \infty$.*

Chapter 7

A Brief Look at the Feedback Control of Fluids Flows

In this chapter, we discuss, in very brief terms, *feedback control problems* for fluid flows. We will look at the set-up of feedback control problems in an abstract setting to contrast it with our previous discussion, in Section 2.1, of optimal control problems. We will also look at three simple concrete examples: linear feedback control of Navier–Stokes flows, feedback control of oscillations in the lift for flow past a cylinder, and suppression of instabilities, through feedback, in boundary layer flows.

An abstract feedback control problem

The ingredients in a feedback control problem are as follows:

- state variables ϕ: the variables that describe the flow;
- control variables g: what actuators do;
- observations y: what sensors measure;
- constraints $F(\phi, g) = 0$: the relation between the state and controls;
- observation relation $y = f(\phi)$: the relation between the observations and state;
- feedback law $g = G(y)$: the relation between the controls and observations, which may be combined with the observation relation to yield $g = G(f(\phi)) = g(\phi)$, the feedback law as a relation between the controls and state.

Abstractly, if $G(y)$ and $f(\phi)$ are known, or if $g(\phi)$ is known, we can then substitute in the constraint equations to obtain

$$F(\phi, g(\phi)) = 0,$$

from which the state may be obtained.

What to observe is usually determined by what sensors are available, or cheap, or effective; the observation law is then deduced from what the sensor measures.

The feedback law is chosen so that some objective is met. Ideally, we would like to choose a feedback law so that an objective functional is minimized. This is very difficult to do in the nonlinear fluids setting. For example, even for *linear ordinary differential equations* with *quadratic cost functionals*, finding the optimal feedback law requires the solution of a Riccati equation, i.e., a *quadratic (and therefore nonlinear) matrix differential equation*. The difficulty with fluid models is that they involve partial differential equations and, more seriously, that the partial differential equations are nonlinear. Thus, the usual approaches to designing feedback laws in the fluid flow setting are

- to use ad hoc (and often semi-empirically designed) feedback laws or
- to design feedback laws for the linearized flow equations.

Both approaches have been successfully employed, but neither approach can be guaranteed to work. Here, we give simple illustrations of the first approach.

7.1 Linear feedback control of Navier–Stokes flows

Given a domain Ω with boundary Γ and a terminal time T, we consider solutions of the Navier–Stokes system

$$\begin{cases} \mathbf{u}_t + \mathbf{u}\cdot\nabla\mathbf{u} - \nu\Delta\mathbf{u} + \nabla p = \mathbf{f} & \text{in } (0,T)\times\Omega\,, \\ \nabla\cdot\mathbf{u} = 0 & \text{in } (0,T)\times\Omega\,, \\ \mathbf{u} = 0 & \text{on } (0,T)\times\Gamma\,, \\ \mathbf{u}|_{t=0} = \mathbf{u}_0 & \text{in } \Omega\,, \end{cases} \tag{7.1}$$

where $\mathbf{u}_0$ is a given initial flow. Given a solenoidal target flow $\mathbf{U}$ defined on $(0, T)\times\Omega$, we want to choose the forcing function $\mathbf{f}$ so that the solution $\mathbf{u}$ of (7.1) is "close" to the target flow $\mathbf{U}$.[59] Details about this example and the approach used here may be found in [324].

Here, we examine how well we can make the solution $\mathbf{u}$ of (7.1) match the given target flow $\mathbf{U}$ by a simple, linear feedback law. To define the feedback law, we let $\mathbf{F}$ denote the forcing function generated by the target flow, i.e.,

$$\mathbf{F} = \mathbf{U}_t - \nu\Delta\mathbf{U} + \mathbf{U}\cdot\nabla\mathbf{U}\,. \tag{7.2}$$

Then, the feedback law is given by

$$\begin{aligned} \mathbf{f} &= \mathbf{F} - \gamma(\mathbf{u}-\mathbf{U}) \\ &= \mathbf{U}_t - \nu\Delta\mathbf{U} + \mathbf{U}\cdot\nabla\mathbf{U} - \gamma(\mathbf{u}-\mathbf{U}) \end{aligned} \qquad \text{on } (0,T)\times\Omega \tag{7.3}$$

for some constant $\gamma > 0$. For (7.3), we observe the state at all points in Ω for all times in $(0, T)$ and we also apply control over all $(0, T)\times\Omega$. This simple feedback law cannot be expected to work, i.e., to drive the flow $\mathbf{u}$ to the target flow $\mathbf{U}$ in general situations. However, we will see in two computational examples that there are problems for which the law does indeed work.

[59] In Sections 3.3.1 and 3.3.2, we considered a similar problem in optimal control settings.

Velocity matching by feedback for a stationary target flow

We consider a unit square domain $(0, 1) \times (0, 1) \subset \mathbb{R}^2$. We assume that the time interval $(0, 1)$ is divided into equal subintervals of duration Δt. The finite element spaces are chosen to be piecewise biquadratic for the velocity and bilinear for the pressure, i.e., the Taylor–Hood finite element pair, based on a rectangular mesh. The mesh size is h and calculations with varying mesh sizes have been performed to make sure the results presented here are accurate. In this first test we are interested in the convergence history for the parameters involved and so a simple stationary target velocity $\mathbf{U} = (U, V)$ is chosen, where

$$U(x, y) = 10\frac{d\phi}{dy}(0.4, x, y) \qquad \text{and} \qquad V(x, y) = -10\frac{d\phi}{dx}(0.4, x, y)$$

with $\phi(t, x, y) = (1 - \cos(2\pi tx))(1 - x)^2(1 - \cos(2\pi ty))(1 - y)^2$.

Evolution of the velocity matching. We first consider the initial velocity $\mathbf{u}_0 = (u_0, v_0)$, where

$$u_0(x, y) = -10U(x, y) \qquad \text{and} \qquad v_0(x, y) = -10V(x, y)\,.$$

This initial velocity rotates in the opposite direction to that of the initial target velocity $\mathbf{U}|_{t=0}$. For this calculation, we use $\Delta t = 0.00625$, $h = 1/16$, and $\gamma = 30$.

Figure 7.1 provides the norm $\|\mathbf{u} - \mathbf{U}\|$ of the difference between the controlled flow $\mathbf{u}$ and the target flow $\mathbf{U}$ and the norm $\|\mathbf{f}\|$ of the control $\mathbf{f}$ vs. time. The norm $\|\mathbf{u} - \mathbf{U}\|$ of the difference quickly tends to zero, and a good match is achieved at $t = 0.2$. The control $\mathbf{f}$ works hard at the beginning in order to steer the controlled flow to the desired one and then remains flat. In fact, the norm of the control quickly approaches the constant (in time) value $\|\mathbf{F}\|$.[60]

The flow evolution is depicted in Figure 7.2. For the sake of clarity, the vector plots are normalized by the largest magnitude. We see that by $t = 0.2$ (and thereafter), we have a good match to the target flow.

Effects of changing γ. Next, we want to study the effect of changes in the parameter γ. The initial velocity is now set to zero and we set $\Delta t = 0.025$ and $h = 1/16$. Figure 7.3 provides the norm $\|\mathbf{u} - \mathbf{U}\|$ of the difference between the controlled flow $\mathbf{u}$ and the target flow $\mathbf{U}$ and the norm $\|\mathbf{f}\|$ of the control $\mathbf{f}$ vs. time. We note that by $t = 1$, the controlled flow matches the target flow very well for all values of γ. However, the match gets better more quickly for larger values of γ. Even for $\gamma = 1$, the controlled flow matches the target flow. For other, more energetic target flows in the Navier–Stokes setting,[61] it is possible that there is a limiting value for γ; for γ smaller than the limiting value, the controlled solution can be driven away from the target velocity. The norm $\|\mathbf{f}\|$ of the control $\mathbf{f}$ tends monotonically to $\|\mathbf{F}\|$, the norm of the forcing function corresponding to the target flow. For high values of γ, the control resembles a delta function in time (at $t = 0$) plus the body force $\mathbf{F}$ generated by the target velocity $\mathbf{U}$.

[60] In this example, the target flow $\mathbf{U}$ is a steady flow so that $\mathbf{F}$ defined by (7.2) is constant in time.

[61] For linear problems such as Stokes flows, γ need only be positive to obtain matching.

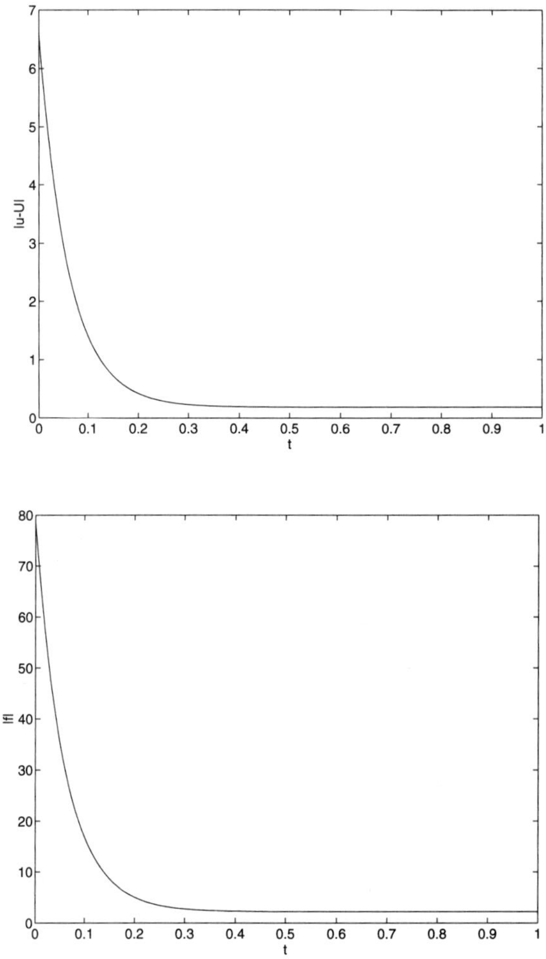

Figure 7.1. *The norm* $\|\mathbf{u} - \mathbf{U}\|$ *of the difference between the controlled flow and the target flow* (*top*) *and the norm* $\|\mathbf{f}\|$ *of the feedback control* (*bottom*) *vs. time.*

Velocity matching by feedback for an unsteady target flow

The computational set-up is similar to that for the previous example, except that we fix $\gamma = 10$ and now choose the time-dependent target velocity $\mathbf{U} = (U, V)$, where

$$\begin{aligned} U &= a(1, 0.4, x, y) + a(2, t, x, y)/(1 + 4\pi t)\,, \\ V &= b(1, 0.4, x, y) + b(2, t, x, y)/(1 + 4\pi t) \end{aligned}$$

with

$$a(k, t, x, y) = \frac{d}{dy}\big(\phi(k, t, x)\phi(k, t, y)\big)\,, \quad b(k, t, x, y) = -\frac{d}{dx}\big(\phi(k, t, x)\phi(k, t, y)\big)\,,$$

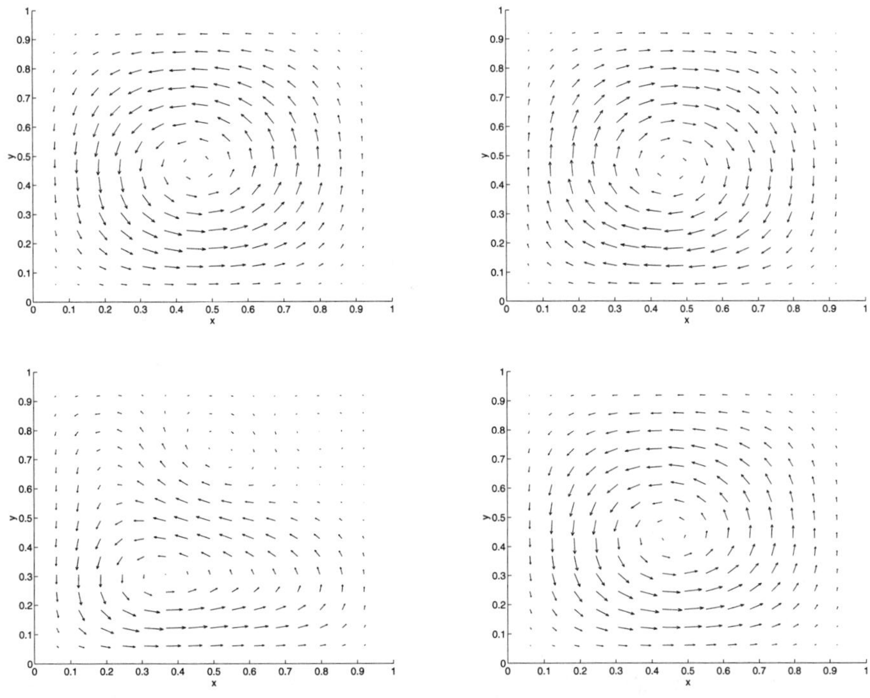

Figure 7.2. *The target flow* **U** *(upper left), the initial flow (upper right), and the controlled flow at* $t = 0.169$ *(lower left), and at* $t = 0.2$ *(lower right).*

and $\phi(k, t, z) = (1 - z)^2(1 - \cos(2k\pi tz))$. This target velocity field is the combination of two flows. One flow has a vortex at the center of the domain and another flow has four vortices. Only one, in the lower left corner, is visible due to the different magnitude of the flow in different portions of the domain. Each of these flows dominates at different times of the evolution. The initial velocity for the controlled flow is set to $\mathbf{u}_0 = (u_0, v_0)$, where

$$u_0(x, y) = -8U(1/4, x, y) \qquad \text{and} \qquad v_0(x, y) = -8V(1/4, x, y)\,.$$

Figure 7.4 provides the norm $\|\mathbf{u} - \mathbf{U}\|$ of the difference between the controlled flow $\mathbf{u}$ and the target flow $\mathbf{U}$ and the norm $\|\mathbf{f}\|$ of the control $\mathbf{f}$ vs. time. A very good match is achieved by $t = 0.3$. At the beginning, the control $\mathbf{f}$ works hard in order to steer the controlled flow to the desired one. Then, as the match improves, the control $\mathbf{f}$ tends to $\mathbf{F}$, the body force corresponding to the target flow. This last observation is made more obvious in Figure 7.5, where $\|\mathbf{f} - \mathbf{F}\|$ is plotted vs. time; we see clearly that $\mathbf{f}$ tends to $\mathbf{F}$ as the matching of the controlled flow $\mathbf{u}$ to the target flow $\mathbf{U}$ improves.

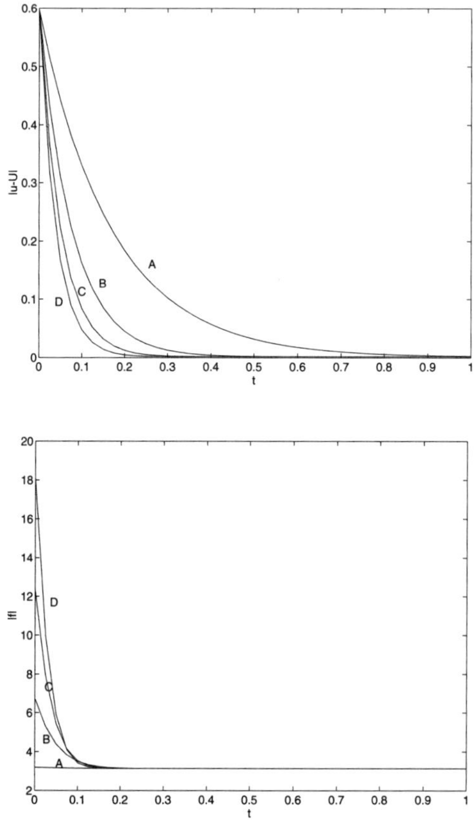

Figure 7.3. *The norm* $\|\mathbf{u}-\mathbf{U}\|$ *of the difference between the controlled flow and the target flow* (*top*) *and the norm* $\|\mathbf{f}\|$ *of the feedback control* (*bottom*) *vs. time for different values of* γ; $\gamma = 1$ (A), 10 (B), 20 (C), 30 (D).

7.2 Feedback control of lift oscillations in flow around a cylinder

We consider the classic problem of incompressible, viscous flow around a cylinder at a value of the Reynolds number that is high enough for a Karman vortex street to develop. Due to the asymmetric shedding of vortices, the lift on the cylinder is oscillatory in time; see Figure 7.6. Our goal is to reduce the amplitude of the oscillations in the lift coefficient. Details about this example and the approach used here may be found in [323].

The constraint equations, i.e., the flow model, are the time-dependent Navier–Stokes system for incompressible viscous flows. In what follows, we let (r, θ) denote a polar coordinate system with origin at the center of the cylinder and with the $\theta = 0$ line pointing upstream.

We want to sense the *asymmetry* in the flow at any instant of time. To do so, we use

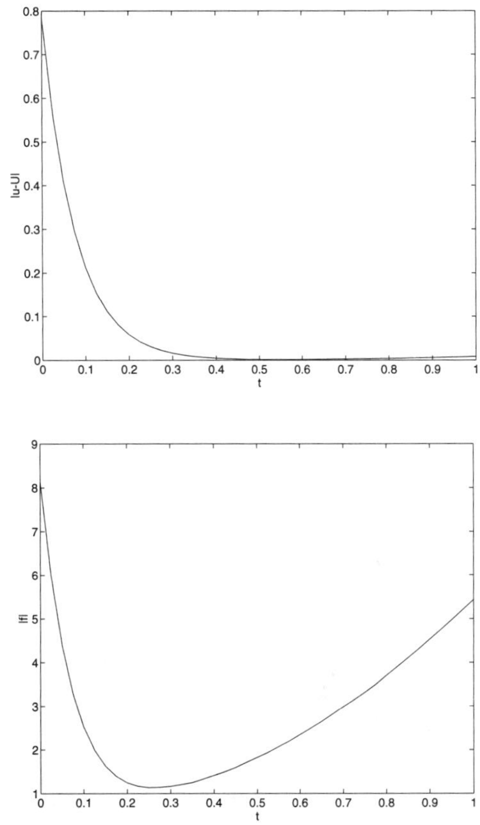

Figure 7.4. *The norm* $\|\mathbf{u} - \mathbf{U}\|$ *of the difference between the controlled flow and the target flow (top) and the norm* $\|\mathbf{f}\|$ *of the feedback control (bottom) vs. time.*

two pressure sensors placed on the upstream side of the cylinder and located symmetrically above and below the $\theta = 0$ point on the cylinder. Thus, we sense the pressure values $p(\theta_1)$ and $p(-\theta_1)$ with $0 < \theta_1 < \pi/2$. Actuation is effected by suction or injection of fluid through three fixed orifices located on the downstream side of the cylinder. We restrict our actuators so that each orifice is in either suction mode or injection mode, but cannot switch from one to the other; the total amount of mass injected equals the total amount of mass sucked so that the cylinder does not have to "carry" any injecting fluid. Note that all sensing and actuating is done on the surface of the cylinder, and not in the flow field itself. See Figure 7.7 for a schematic diagram of the placement of sensors and actuators.

The observations are the pressures at the two sensing locations on the cylinder.

We assume that at each orifice the velocity is normal to the cylinder surface, i.e., $u_\theta = 0$ at the orifices, where u_θ denotes the θ-component of the velocity vector. We choose the three orifices to have the same diameter and we choose the r-component of the velocity

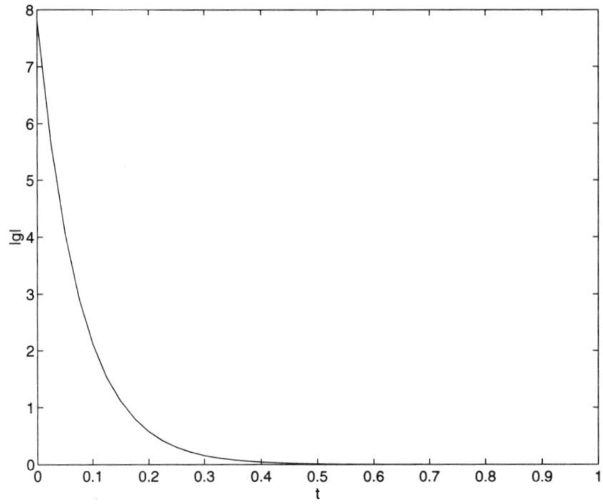

Figure 7.5. *The norm* $\|\mathbf{f}-\mathbf{F}\|$ *of the difference between the feedback control* $\mathbf{f}$ *and the body force* $\mathbf{F}$ *corresponding to the target flow vs. time.*

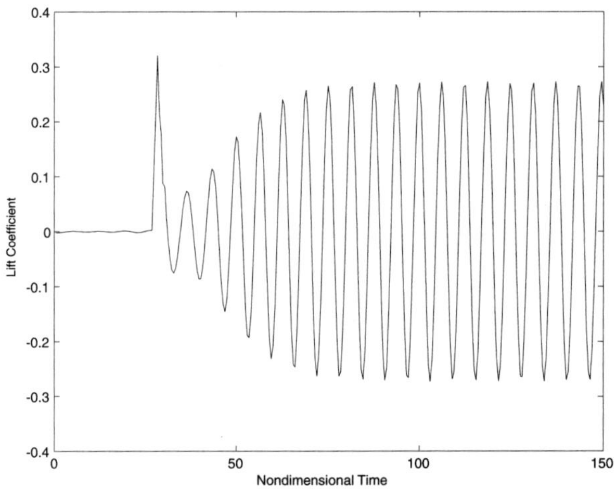

Figure 7.6. *Coefficient of lift vs. time for a flow around a cylinder at* $Re = 80$ *started impulsively from rest.*

at the orifices to be given by the *feedback law*

$$u_r = \begin{cases} -\min\Big(\alpha|p(\theta_1) - p(\theta_2)|, \beta\Big)\psi(\theta) & \text{at the suction orifices,} \\ +2\min\Big(\alpha|p(\theta_1) - p(\theta_2)|, \beta\Big)\psi(\theta) & \text{at the injection orifice,} \end{cases}$$

where α and β are positive constants and $\psi(\theta)$ is a given velocity profile (we take it to be a

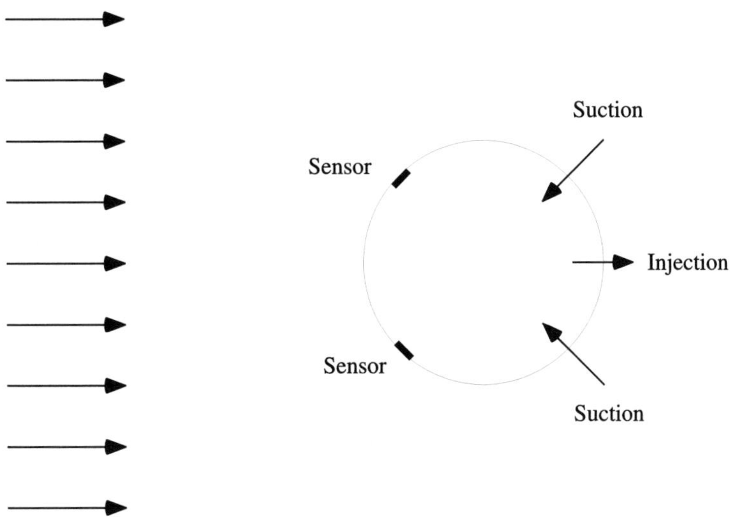

Figure 7.7. *Pressure sensor and injection/suction actuator locations.*

quadratic profile) at each orifice.

In this feedback law, β places a limit on the amount of fluid that can be injected or sucked; it plays the role of technological constraints on how much fluid can be pumped in or out through the orifices. Otherwise, the feedback law states that the amount of fluid injected or sucked through the orifices is proportional to the absolute value of the difference in the pressures at the two sensing locations. This simple, ad hoc feedback law is effective, at least for relatively low values of the Reynolds number, in reducing the oscillations in the lift; see Figure 7.8. Note that the feedback control reduces the amplitude of the oscillations but does not affect their frequency.

7.3 Suppression of 2D Tollmien–Schlichting instability through feedback

We again look at the problem of suppressing the two-dimensional Tollmien–Schlichting (TS) instability, this time through feedback.[62] Details about this example and the feedback approach used here may be found in [329].

Since everything, i.e., the phase, frequency, and amplitude, is known about the instability, one can use analytical information to control an actuator, in our case an injection/suction orifice, to suppress growth; this is the best case scenario. We will compare the effectiveness of the use of a simple feedback control law to the use of analytical information for suppressing the TS instability.

The instability is tripped by injection/suction through an orifice upstream of the sensors; both small and large amplitude tripping amplitudes are tested. We sense the pressure (or shear stress) at two nearby locations along the wall. The sensor data are put through a

[62]Recall that in Section 3.4 we treated this problem through optimal control.

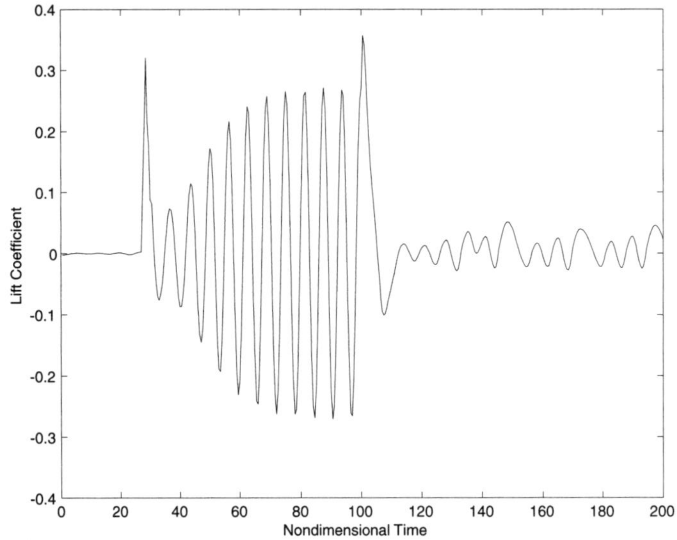

Figure 7.8. *Coefficient of lift vs. time for controlled flow at $Re = 80$ for $\alpha = 80$ and $\beta = 2$; control is turned on at $t = 100$.*

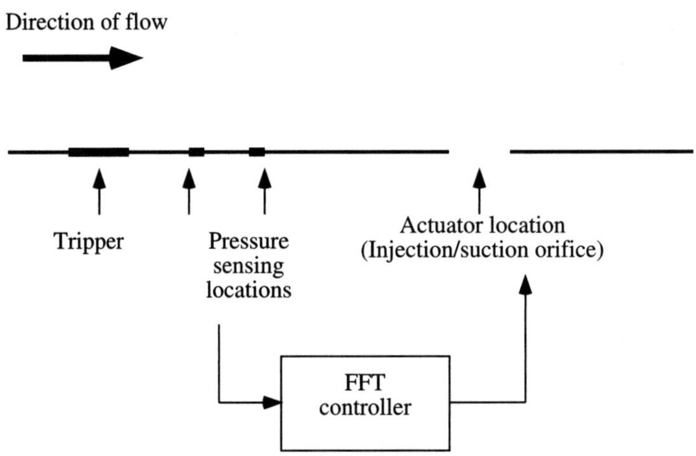

Figure 7.9. *Relative positions of the tripper injection orifice used to trigger the instability, the pressure sensors, and the actuation orifice.*

fast Fourier transform (FFT) to determine information about the instability, e.g., the frequency and growth rate. Actuation is effected by injection or suction of fluid through an orifice downstream of the sensors. The amplitude, frequency, and phase of the actuation, i.e., the feedback law, are determined from the output of the FFT. A sketch of the relative locations of the tripper, sensors, and actuator and of the relation between the actuator and sensor is given in Figure 7.9.

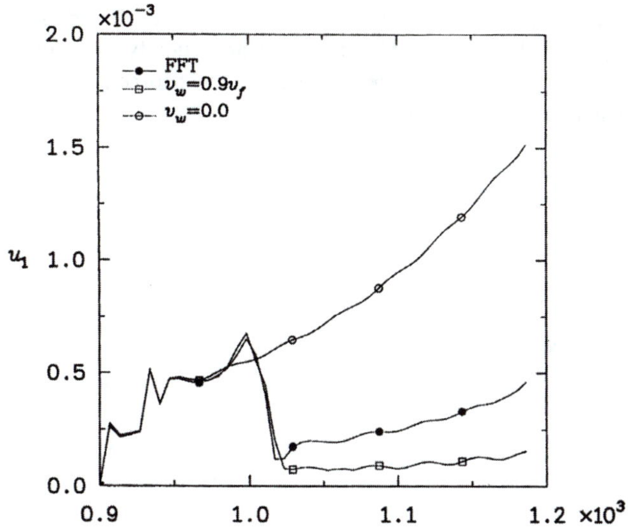

Figure 7.10. *Suppression of a small-amplitude TS wave (open circles) by using analytical information (squares) and by feedback control (filled circles).*

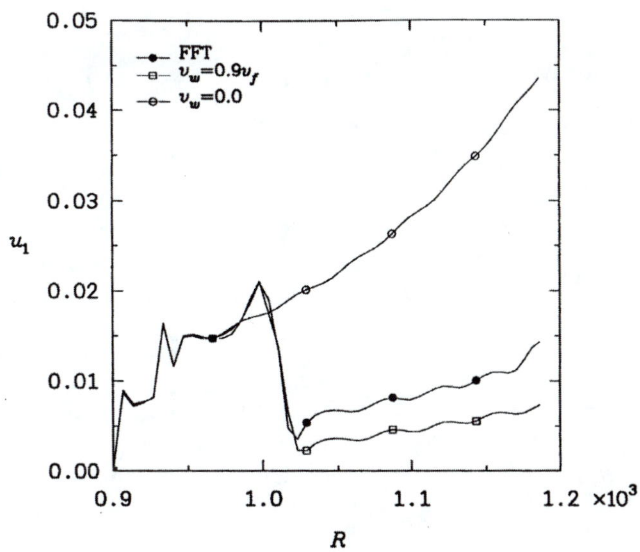

Figure 7.11. *Suppression of a large-amplitude TS wave (open circles) by using analytical information (squares) and feedback control (filled circles).*

In Figures 7.10 and 7.11, we compare the time history of the vertical velocity at a point near the wall for the uncontrolled flow, the controlled flow using analytical information to determine the actuation, and the feedback controlled flow using FFT-analyzed sensed data to determine the actuation. Figure 7.10 is for a small amplitude disturbance and Figure 7.11 is for a larger amplitude disturbance. Of course, the analytically determined controller does better at suppressing growth but we see that the FFT-determined feedback controller also does a good job at suppressing the instabilities.

Bibliography

[A] Books, edited books, proceedings volumes, review articles, and special journal issues devoted to flow control

[1] AGARD, *Computational Methods for Aerodynamic Design (Inverse) and Optimization,* AGARD-CP-463, Advisory Group for Aerospace Research and Development, Neuilly sur Seine, France, 1990.

[2] T. BEWLEY, *Flow control: New challenges for a new renaissance,* Progr. Aero. Sci. **37** 2001, pp. 21–58.

[3] J. BORGGAARD, J. BURKARDT, J. BURNS, E. CLIFF, M. GUNZBURGER, H.-C. KIM, H.-C. LEE, J. PETERSON, A. SHENOY, AND X. WU, *Algorithms for flow control and optimization,* in Optimal Design and Control, Birkhauser, Boston, 1995, pp. 97–116.

[4] J. BORGGAARD, J. BURNS, E. CLIFF, AND S. SCHRECK, Editors, *Computational Methods for Optimal Design and Control,* Birkhauser, Basel, 1998.

[5] J. BORGGAARD, J. BURKARDT, M. GUNZBURGER, AND J. PETERSON, Editors, *Optimal Control and Design*, Birkhauser, Boston, 1995.

[6] A. BOTTARO, J. MAUSS, AND D. HENNINGSON, Editors, *Flow, Turbulence and Combustion,* Special issue, **65**, Nos. 2 and 3, Kluwer, Dordrecht, 2000.

[7] A. FURSIKOV, *Optimal Control of Distributed Systems. Theory and Applications,* American Mathematical Society, Providence, RI, 2000.

[8] M. GAD-EL-HAK, *Flow control,* Appl. Mech. Rev. **42** 1989, pp. 262–292.

[9] M. GAD-EL-HAK, *Interactive control of turbulent boundary layers: A futuristic overview,* AIAA J. **32** 1994, pp. 1753–1765.

[10] M. GAD-EL-HAK, *Modern developments in flow control,* Appl. Mech. Rev. **49** 1996, pp. 365–379.

[11] M. GAD-EL-HAK, *Flow Control. Passive, Active, and Reactive Flow Management,* Cambridge University Press, Cambridge, 2000.

[12] M. GILES AND N. PIERCE, *An introduction to the adjoint approach to design,* Flow Turb. Comb. **65** 2000, pp. 393–415.

[13] M. GUNZBURGER, Editor, *Flow Control,* Springer, New York, 1995.

[14] M. GUNZBURGER, *A prehistory of flow control and optimization,* in Flow Control, Springer, New York, 1995, pp. 185–195.

[15] M. GUNZBURGER, *Flow control and optimization,* in Computational Fluid Dynamics Review, Wiley, West Sussex, 1995, pp. 548–566.

[16] M. GUNZBURGER, *Sensitivities in computational methods for optimal flow control,* in Computational Methods for Optimal Design and Control, Birkhauser, Boston, 1998, 197–236.

[17] M. GUNZBURGER, *Sensitivites, adjoints, and flow optimization,* Internat. J. Numer. Methods Fluids **31** 1999, pp. 53–78.

[18] M. GUNZBURGER, *Adjoint equation-based methods for control problems in viscous, incompressible flows,* Flow, Turb. Comb. **65** 2000, pp. 249–272.

[19] M. GUNZBURGER AND L. HOU, Editors, *International Journal of Computational Fluid Mechanics,* Special issue, **11**, Nos. 1 and 2, Gordon and Breach, New York, 1998.

[20] R. JOSLIN, *Aircraft laminar flow control,* Ann. Rev. Fluid Mech. **30** 1998, pp. 1–29.

[21] J. LUMLEY AND P. BLOSSEY, *Control of turbulence,* Ann. Rev. Fluid Mech. **30** 1998, pp. 311–327.

[22] V. MODI, *Moving surface boundary-layer control: A review,* J. Fluids Struc. **11** 1997, pp. 627–663.

[23] B. MOHAMMADI AND O. PIRONNEAU, *Applied Shape Optimization for Fluids,* Clarendon, Oxford, 2001.

[24] S. SRITHARAN, Editor, *Optimal Control of Viscous Flow,* SIAM, Philadelphia, 1998.

[25] R. VAN DEN BRAEMBUSSCHE AND M. MANNA, *Inverse Design and Optimization Methods,* Lecture Series 1997-05, von Karman Institute for Fluid Dynamics, Rhode Saint Genese, Belgium, 1997.

[B] Optimal flow control

[26] F. ABERGEL AND E. CASAS, *Some optimal control problems associated to the stationary Navier-Stokes equations,* in Mathematics, Climate and Environment, 1993, pp. 213–220.

[27] F. ABERGEL AND E. CASAS, *Some optimal control problems of multistate equations appearing in fluid mechanics,* Math. Model. Numer. Anal. **27** 1993, pp. 223–247.

[28] F. ABERGEL AND R. TEMAM, *On some control problems in fluid mechanics,* Theoret. Comput. Fluid Dynam. **1** 1990, pp. 303–325.

[29] N. AHMED, *Optimal control of turbulent flow as measure solutions,* Internat. J. Comput. Fluid Dynam. **11** 1998, pp. 169–180.

[30] P. ANDERSON, M. BERGGREN, AND D. HENNINGSON, *Optimal distrubances in boundary layers,* in Computational Methods for Optimal Design and Control, Birkhauser, Boston, 1998, pp. 1–26.

[31] P. ANDERSON, M. BERGGREN, AND D. HENNINGSON, *Optimal disturbances and bypass transition in boundary layers,* Phys. Fluids **11** 1999, pp. 134–150.

[32] T. BEWLEY AND O. AAMO, *A "win-win" mechanism for low-drag transients in controlled 2D channel flow and its implications for sustained drag reduction,* to appear in Phys. Fluids.

[33] T. BEWLEY AND S. LIU, *Optimal and robust control and estimation of linear paths to transition,* J. Fluid Mech. **365** 1998, pp. 305–349.

[34] T. BEWLEY AND P. MOIN, *Optimal control of turbulent channel flows,* in Active Control of Vibration and Noise, ASME DE **75** 1994, pp. 221–227.

[35] T. BEWLEY, P. MOIN, AND R. TEMAM, *Optimal and robust approaches for linear and nonlinear regulation problems in fluid mechanics,* AIAA paper no. 97-1872, AIAA, 1997.

[36] T. BEWLEY, R. TEMAM, AND M. ZIANE, *A general framework for robust control in fluid mechanics,* Physica D **138** 2000, pp. 360–392.

[37] J. BURKARDT, M. GUNZBURGER, AND J. PETERSON, *Insensitive functionals, inconsistent gradients, spurious minima, and regularized functionals in flow optimization problems,* Internat. J. Comput. Fluid Dynam. **16** 2002, pp. 171–185.

[38] J. BURKARDT AND J. PETERSON, *Control of steady incompressible 2D channel flows,* in Flow Control, Springer, New York, 1995, pp. 111–126.

[39] J. BURNS AND Y.-H. OU, *Optimal control of lift/drag ratios on a rotating cylinder,* Appl. Math. Lett. **5** 1992, pp. 57–62.

[40] E. CASAS, *The Navier-Stokes equations coupled with the heat equation: Analysis and control,* Control Cybernet. **23** 1994, pp. 605–620.

[41] E. CASAS, *Optimality conditions for some control problems of turbulent flows,* in Flow Control, Springer, New York, 1995, pp. 127–147.

[42] E. CASAS, *An optimal control problem governed by the evolution Navier–Stokes equations,* in Optimal Control of Viscous Flow, S. Sritharan, Editor, SIAM, Philadelphia, 1998, pp. 79–107.

[43] E. CLIFF, M. HEIKENSCHLOSS, AND A. SHENOY, *An optimal control problem for flows with discontinuities,* J. Optim. Theory. Appl. **94** 1997, pp. 273–309.

[44] S. Collis, K. Grayour, M. Heikenschloss, M. Ulbrich, and S. Ulbrich, *Numerical solution of optimal control problems governed by the compressible Navier-Stokes equations,* Internat. Series of Numerical Math. 139, Birkhauser, Basel, 2001, pp. 43–55.

[45] S. Collis, K. Grayour, M. Heikenschloss, M. Ulbrich, and S. Ulbrich, *Towards adjoint-based methods for aeroacoustic control,* AIAA paper no. 2001-0821, AIAA, 2001.

[46] C. Cuvelier, *Optimal control of a system governed by the Navier-Stokes equations coupled with the heat equation,* in New Developments in Differential Equations, North-Holland, Amsterdam, 1976, pp. 81–98.

[47] D. D'Alessandro, M. Dahleh, and I. Mezic, *Control of mixing in fluid flow: A maximum entropy approach,* IEEE Tran. Autom. Control. **44** 1999, pp. 1852–1863.

[48] M. Desai and K. Ito, *Optimal control of Navier-Stokes equations,* SIAM J. Control Optim. **32** 1994, pp. 1428–1446.

[49] Q. Du, M. Gunzburger, and L. Hou, *Analysis and finite element approximation of optimal control problems for a Ladyzhenskaya model for stationary, incompressible, viscous flows,* J. Comput. Appl. Math. **61** 1995, pp. 323–343.

[50] H. Fattorini and S. Sritharan, *Relaxation in semilinear infinite dimensional systems modelling fluid flow control problems,* in Control and Optimal Design of Distributed Parameter Systems, Springer, New York, 1995, pp. 93–111.

[51] H. Fattorini and S. Sritharan, *Existence of optimal controls for viscous flow problems,* Proc. Roy. Soc. London Ser. A **439** 1992, pp. 81–102.

[52] H. Fattorini and S. Sritharan, *Necessary and sufficient conditions for optimal controls in viscous flow problems,* Proc. Roy. Soc. Edinburgh **124A** 1994, pp. 211–251.

[53] H. Fattorini and S. Sritharan, *Optimal chattering control for viscous flows,* Nonlinear Anal. **25** 1995, pp. 763–797.

[54] H. Fattorini and S. Sritharan, *Optimal control problems with state constraints in fluid mechanics and combustion,* Appl. Math. Optim. **38** 1998, pp. 159–192.

[55] A. Fursikov, *On some control problems and results concerning the unique solvability of a mixed boundary value problem for the three-dimensional Navier-Stokes and Euler systems,* Soviet Math. Dokl. **21** 1980, pp. 889–893.

[56] A. Fursikov, *Control problems and theorems concerning the unique solvability of a mixed boundary value problem for the three-dimensional Navier-Stokes and Euler equations,* Math. USSR Sbornik **43** 1982, pp. 251–273.

[57] A. Fursikov, *Properties of solutions of control problems that are connected with the Navier-Stokes equations,* Soviet Math. Dokl. **25** 1982, pp. 40–45.

[58] A. FURSIKOV, *Properties of solutions of some extremal problems connected with the Navier-Stokes equations,* Math. USSR Sbornik **46** 1983, pp. 323–351.

[59] A. FURSIKOV, *Optimal control problems for Navier–Stokes system with distributed control function,* in Optimal Control of Viscous Flow, S. Sritharan, Editor, SIAM, Philadelphia, 1998, pp. 109–150.

[60] A. V. FURSIKOV, M. D. GUNZBURGER, AND L. S. HOU, *Boundary value problems and optimal boundary control for the Navier–Stokes system: The two-dimensional case,* SIAM J. Control Optim. **36** 1998, pp. 852–894.

[61] A. FURSIKOV, M. GUNZBURGER, AND L. HOU, *Optimal Dirichlet control and inhomogeneous boundary value problems for the unsteady Navier-Stokes equations,* in Control of Partial Differential Equations, ESAIM, Paris, 1998, pp. 97–116.

[62] A. FURSIKOV, M. GUNZBURGER, AND L. HOU, *Optimal boundary control of the Navier-Stokes equations with bounds on the control,* in Proc. Korean Advanced Institute for Science and Technology Mathematics Workshop on Finite Elements, 1999, pp. 41–60.

[63] A. FURSIKOV, M. GUNZBURGER, AND L. HOU, *Optimal boundary control for the evolutionary Navier–Stokes system: The three-dimensional case,* SIAM J. Appl. Dynam. Systems, submitted.

[64] O. GHATTAS AND J. BARK, *Optimal control of two- and three-dimensional incompressible Navier-Stokes flows,* J. Comput. Phys. **136** 1997, pp. 231–244.

[65] M. GUNZBURGER, L. HOU, S. MANSERVISI, AND Y. YAN, *Computations of optimal controls for incompressible flows,* Internat. J. Comput. Fluid Dynam. **11** 1998, pp. 181–191.

[66] M. GUNZBURGER, L. HOU, AND T. SVOBODNY, *Numerical approximation of an optimal control problem associated with the Navier-Stokes equations,* Appl. Math. Lett. **2** 1989, pp. 29–31.

[67] M. GUNZBURGER, L. HOU, AND T. SVOBODNY, *Optimal boundary control of nonsteady incompressible flow: An application to viscous drag reduction,* Proc. 29th IEEE Conf. Decision and Control, IEEE, 1990, pp. 377–378.

[68] M. GUNZBURGER, L. HOU, AND T. SVOBODNY, *A numerical method for drag minimization via the suction and injection of mass through the boundary,* in Stabilization of Flexible Structures, Springer, Berlin, 1990, pp. 312–321.

[69] M. GUNZBURGER, L. HOU, AND T. SVOBODNY, *Approximation of boundary control and optimization problems for fluid flows,* in 4th International Symposium on Computational Fluid Dynamics, U. California, Davis, 1991, pp. 455–460.

[70] M. GUNZBURGER, L. HOU, AND T. SVOBODNY, *Analysis and finite element approximations of optimal control problems for the stationary Navier-Stokes equations with distributed and Neumann controls,* Math. Comput. **57** 1991, pp. 123–151.

[71] M. GUNZBURGER, L. HOU, AND T. SVOBODNY, *Analysis and finite element approximations of optimal control problems for the stationary Navier-Stokes equations with Dirichlet controls,* Math. Model. Numer. Anal. **25** 1991, pp. 711–748.

[72] M. GUNZBURGER, L. HOU, AND T. SVOBODNY, *Control of temperature distributions along boundaries of engine components,* in Numerical Methods in Laminar and Turbulent Flow VII, Pineridge Press, Swansea, UK, 1991, pp. 765–773.

[73] M. GUNZBURGER, L. HOU, AND T. SVOBODNY, *Boundary velocity control of incompressible flow with an application to viscous drag reduction,* SIAM J. Control Optim. **30** 1992, pp. 167–181.

[74] M. GUNZBURGER, L. HOU, AND T. SVOBODNY, *Optimal control and optimization of viscous, incompressible flows,* in Incompressible Computational Fluid Dynamics: Trends and Advances, Cambridge University Press, Cambridge, 1992, pp. 109–150.

[75] M. GUNZBURGER, L. HOU, AND T. SVOBODNY, *The approximation of boundary control problems for fluid flows with an application to control by heating and cooling,* Comput. Fluids **22** 1993, pp. 239–251.

[76] M. GUNZBURGER, L. HOU, AND T. SVOBODNY, *Heating and cooling control of temperature distributions along boundaries of flow domains,* J. Math. Systems Estim. Control **3** 1993, pp. 147–172.

[77] M. GUNZBURGER, L. HOU, AND T. SVOBODNY, *Boundary control of incompressible flows,* in Computational Fluid Dynamics Techniques, Harwood, Harrow, UK, 1994, pp. 839–846.

[78] M. D. GUNZBURGER, L. S. HOU, AND T. P. SVOBODNY, *Optimal control problems for a class of nonlinear equations with an application to control of fluids,* in Optimal Control of Viscous Flow, S. Sritharan, Editor, SIAM, Philadelphia, 1998, pp. 43–62.

[79] M. GUNZBURGER, L. HOU, S. RAVINDRAN, AND J. TURNER, *Analysis and approximation of optimal control problems with nonlinear constraints,* in Proc. 33th IEEE Conf. Decision and Control, IEEE, 1994, pp. 299–304.

[80] M. GUNZBURGER AND O. IMANUVILOV, *Optimal control of stationary, low Mach number, highly nonisothermal viscous flows,* ESAIM: Cont. Optim. Calc. Var. **5** 2000, pp. 477–500.

[81] M. GUNZBURGER AND H.-C. LEE, *Analysis, approximation, and computation of a coupled solid/fluid temperature control problem,* Comput. Methods Appl. Mech. Engrg. **118** 1994, pp. 133–152.

[82] M. D. GUNZBURGER AND S. MANSERVISI, *The velocity tracking problem for Navier–Stokes flows with bounded distributed controls,* SIAM J. Control Optim. **37** 1999, pp. 1913–1945.

[83] M. D. GUNZBURGER AND S. MANSERVISI, *Analysis and approximation of the velocity tracking problem for Navier–Stokes flows with distributed control,* SIAM J. Numer. Anal. **37** 2000, pp. 1481–1512.

[84] M. D. Gunzburger and S. Manservisi, *The velocity tracking problem for Navier–Stokes flows with boundary control,* SIAM J. Control Optim. **39** 2000, pp. 594–634.

[85] M. Gunzburger, E. Ozugurlu, J. Turner, and H. Zhang, *Controlling transport phenomena in the Czochralski crystal growth process,* J. Crystal Growth **234** 2002, pp. 47–62.

[86] M. Heikenschloss, *Formulation and analysis of a sequential quadratic programming method for optimal boundary control of Navier-Stokes flows,* in Optimal Control: Theory, Algorithms, and Applications, Kluwer, Dordrecht, 1998, pp. 178–203.

[87] K. Heusermann and S. Manservisi, *On some optimal control problems for the heat radiative transfer equation,* ESAIM: COCV, 2000, pp. 425–444.

[88] K. Heusermann and S. Manservisi, *Numerical computations of some optimal control problems for the heat radiative transfer equation,* to appear in Comput. Methods Appl. Mech. Engrg.

[89] L. Hou and A. Meir, *Boundary optimal control for MHD flows,* Appl. Math. Optim. **32** 1995, pp. 143–162.

[90] L. Hou and J. Peterson, *Boundary optimal control for an electrically conducting fluid using boundary electrical potential controls,* J. Nonlinear Anal. - TMA **24** 1995, pp. 857–874.

[91] L. Hou and S. Ravindran, *Computations of boundary optimal control problems for ellectrically conducting flows,* J. Comput. Phys. **128** 1996, pp. 319–330.

[92] L. Hou and S. Ravindran, *Penalty methods for numerical approximations of optimal boundary flow control problems,* Internat. J. Comput. Fluid Dynam. **11** 1998, pp. 157–167.

[93] L. S. Hou and S. S. Ravindran, *A penalized Neumann control approach for solving an optimal Dirichlet control problem for the Navier–Stokes equations,* SIAM J. Control Optim. **36** 1998, pp. 1795–1814.

[94] L. S. Hou and S. S. Ravindran, *Numerical approximation of optimal flow control problems by a penalty method: Error estimates and numerical results,* SIAM J. Sci. Comput. **20** 1999, pp. 1753–1777.

[95] L. Hou, S. Ravindran, and Y. Yan, *Numerical solutions of optimal distributed control problems for incompressible flows,* Internat. J. Comput. Fluid Dynam. **8** 1997, pp. 99–114.

[96] L. Hou and W. Sun, *Numerical methods for optimal control of impressed cathodic protection systems,* Internat. J. Numer. Methods Engrg. **37** 1994, pp. 2779–2796.

[97] L. Hou and T. Svobodny, *Optimization problems for the Navier-Stokes equations with regular boundary controls,* J. Math. Anal. Appl. **177** 1993, pp. 342–367.

[98] L. S. Hou and Y. Yan, *Dynamics for controlled Navier–Stokes systems with distributed controls,* SIAM J. Control Optim. **35** 1997, pp. 654–677.

[99] K. Ito and S. S. Ravindran, *Optimal control of thermally convected flows,* SIAM J. Sci. Comput. **19** 1998, pp. 1847–1869.

[100] K. Ito, J. Scroggs, and H. Tran, *Optimal control of thermally coupled Navier-Stokes equations,* in Optimal Design and Control, Birkhauser, Boston, 1995, pp. 199–214.

[101] R. Joslin, M. Gunzburger, R. Nicolaides, G. Erlebacher, and M. Hussaini, *A methodology for the automated optimal control of flows including transitional flows,* in Proc. ASME Forum on Control of Transitional and Turbulent Flows, FED–237, ASME, 1996, pp. 287–294.

[102] R. Joslin, M. Gunzburger, R. Nicolaides, G. Erlebacher, and M. Hussaini, *Self-contained automated methodology for optimal flow control,* AIAA J. **35** 1997, pp. 816–824.

[103] H.-C. Lee, *Analysis and computation of optimal and feedback control problems for Navier-Stokes equations,* J. Korean Math. Soc. **34** 1997, pp. 841–857.

[104] H.-C. Lee and O. Imanuvilov, *Analysis of optimal control problems for the 2-D stationary Boussinesq equations,* J. Math. Anal. Appl. **242** 2000, pp. 191–211.

[105] H.-C. Lee and O. Y. Imanuvilov, *Analysis of Neumann boundary optimal control problems for the stationary Boussinesq equations including solid media,* SIAM J. Control Optim. **39** 2000, pp. 457–477.

[106] H.-C. Lee and Y.-H. Lee, *Finite element approximations and computations of boundary optimal contol problems for the Navier-Stokes flow through a channel with steps,* J. Korean Math. Soc. **36** 1999, pp. 195–214.

[107] K.-T. Li and A.-X. Huang, *Mathematical aspect of optimal control finite element method for Navier-Stokes equations,* J. Comput. Math. **2** 1984, pp. 139–151.

[108] Z. Li, M. Navon, M. Hussaini, and F. Dimet, *Optimal control of the unsteady Navier-Stokes equations,* Comput. Fluids **32** 2002, pp. 149–171.

[109] J. Loncaric, *Sensor/actuator placement via optimal distributed control exterior Stokes flow,* in Computational Methods for Optimal Design and Control, Birkhauser, Basel, 1998. pp. 303-322.

[110] Y.-C. Ma, *Optimal control finite element algorithm for penalty variational formulation of the Navier-Stokes equation,* J. Math. Res. Exposition **3** 1983, pp. 130–132.

[111] J. Malek and T. Roubicek, *Optimization of steady flows for incompressible viscous fluids,* in Applied Nonlinear Analysis, Kluwer, Dordrecht, 1999, pp. 355–372.

[112] M. MARIN AND M. KAWAHARA, *Optimal control of vorticity in Rayleigh-Bernard convection by finite element method,* Comm. Numer. Methods Engrg. **14** 1998, pp. 9–22.

[113] E. OZUGURLU, M. GUNZBURGER, J. TURNER, AND H. ZHANG, *Magnetic control of transport phenomena in CZ crystal growth processes,* Proc. IMECE'01, ASME, 2001.

[114] D. PELLETIER, J. BORGGAARD, AND J.-H. HETU, *A continuous sensitivity equation method for conduction and phase change problems,* AIAA report no. 2000-0881, AIAA, 2000.

[115] S. RAVINDRAN, *Numerical approximation of optimal flow control problems by SQP method,* in Optimal Control of Viscous Flow, S. Sritharan, Editor, SIAM, 1998, pp. 181–198.

[116] S. RAVINDRAN, *Numerical solution of optimal control for thermally convective fluid flow,* J. Numer. Methods Fluids **25** 1997, pp. 205–223.

[117] A. SAHRAPOUR AND N. AHMED, *Boundary control of Navier-Stokes equations with potential applications to artifical hearts,* in Proc. CFD Society of Canada, 1994, pp. 387–393.

[118] A. SHENOY, E. CLIFF, AND M. HEIKENSCHLOSS, *Thermal-fluid control via finite-dimensional approximations,* AIAA paper no. 96-1910, AIAA, 1996.

[119] S. SRITHARAN, *Dynamic programming of the Navier-Stokes equations,* System Control Lett. **16** 1991, pp. 299–307.

[120] S. SRITHARAN, *An optimal control problem in exterior hydrodynamics,* Proc. Roy. Soc. Edinburgh **121A** 1992, pp. 5–32.

[121] S. SRITHARAN, *An introduction to deterministic and stochastic control of viscous flows,* in Optimal Control of Viscous Flow, S. Sritharan, Editor, SIAM, Philadelphia, 1998, pp. 1–42.

[122] S. SRITHARAN, *Deterministic and stochastic control of Navier-Stokes equation with linear, monotone, and hyperviscosities,* Appl. Math. Optim. **41** 2000, pp. 255–308.

[123] H. WOOD AND M. GUNZBURGER, *Adjoint and sensitivity–based methods for optimization of gas centrifuges,* in Proc. 7th Work. Separation Phenomena in Liquids and Gases, Moscow Engineering Physics Institute, Moscow, 2000, pp. 89–99.

[124] N. ZABARAS, *Adjoint methods for inverse free convection problems with application to solidification processes,* in Computational Methods for Optimal Design and Control, Birkhauser, Basel, 1998. pp. 391–425

[125] N. ZABARAS AND G. YANG, *A functional optimization formulation and implementation of an inverse natural convection problem,* Comput. Methods Appl. Mech. Engrg. **144** 1997, pp. 245–274.

[C] Suboptimal flow control

[126] M. BERGGREN AND M. HEIKENSCHLOSS, *Parallel solution of optimal-control problems by time-domain decomposition,* in Computational Science for the 21st Century, Wiley, Chichester, 1997, pp. 102–112.

[127] M. HINZE AND K. KUNISCH, *On suboptimal control strategies for the Navier-Stokes equations*, in Control and Partial Differential Equations, ESAIM, Paris, 1998, pp. 181–198.

[128] M. HINZE AND K. KUNISCH, *Three control methods for time-dependent fluid flow,* Flow. Turb. Comb. **65** 2000, pp. 273–298.

[129] L. S. HOU AND Y. YAN, *Dynamics and approximations of a velocity tracking problem for the Navier–Stokes flows with piecewise distributed controls,* SIAM J. Control Optim. **35** 1997, pp. 1847–1885.

[130] C. LEE, J. KIM, AND H. CHOI, *Suboptimal control of turbulent channel flow for drag reduction,* J. Fluid. Mech. **358** 1998, pp. 245–258.

[131] H.-C. LEE AND B. SHIN, *Piecewise optimal distributed controls for 2D Boussinesq equations,* Math. Methods Appl. Sci. **23** 2000, pp. 227–254.

[D] Active flow control

[132] P. BALAKUMAR AND P. HALL, *Optimum suction distribution for transition control,* Theoret. Comput. Fluid Dynam. **13** 1999, pp. 1–19.

[133] H. BANKS AND R. SMITH, *Active control of acoustic pressure fields using smart material technologies,* in Flow Control, Springer, New York, 1995, pp. 1–34.

[134] L. BARAMOV, O. TUTTY, AND E. ROGERS, *Robust control of plane Poiseuille flow,* AIAA paper no. 2000-2684, AIAA, 2000.

[135] M. BERGGREN, *Numerical solution of a flow-control problem: Vorticity reduction by dynamic boundary action,* SIAM J. Sci. Comput. **19** 1998, pp. 829–860.

[136] W. BOWER, J. KEGELMAN, A. PAL, AND G. MEYER, *A numerical study of two-dimensional instability-wave control based on the Orr-Summerfeld equation,* Phys. Fluids **30** 1987, pp. 998–1004.

[137] J. BURNS AND Y.-H. OU, *Effect of rotation rate on the forces of a rotating cylinder: Simulation and control,* ICASE report no. 93-11, ICASE, 1993.

[138] D. BUSCHNELL AND C. MCGINLEY, *Turbulent control in wall flows,* Ann. Rev. Fluid Mech. **21** 1989, pp. 1–20.

[139] P. CATHALIFAUD AND P. LUCHINI, *Algebraic growth in boundary layers: Optimal control by blowing and suction at the wall,* European J. Mech. B - Fluids **19** 2000, pp. 469–490.

[140] Y. CHANG AND S. COLLIS, *Active control of turbulent channel flows based on large eddy simulation,* ASME Paper no. FEDSM-99-6929, ASME, 1999.

[141] S. CHERNYSHENKO, *Stabilization of trapped vortices by alternating blowing and suction,* Phys. Fluids **7** 1995, pp. 802–807.

[142] H. CHOI, P. MOIN, AND J. KIM, *Active turbulence control for drag reduction in wall-bounded flows,* J. Fluid. Mech. **262** 1994, pp. 75–110.

[143] S. COLLIS AND Y. CHANG, *Computer simulation of active control in complex turbulent flows,* in Modeling and Simulation Based Engineering I, Tech Science Press, Encino, CA, 1998, pp. 851–856.

[144] S. COLLIS AND Y. CHANG, *On the use of LES with a dynamic subgrid-scale model for optimal control of wall bounded turbulence,* in Recent Advances in DNS and LES, Kluwer, Dordrecht, 1999, pp. 99–110.

[145] S. COLLIS, Y. CHANG, S. KELLOGG, AND R. PRABHU, *Large eddy simulation and turbulence control,* AIAA paper no. 2000-2564, AIAA, 2000.

[146] S. COLLIS AND S. LELE, *A computational approach to swept leading-edge receptivity,* AIAA paper no. 96-0180, AIAA, 1996.

[147] S. COLLIS AND S. LELE, *Receptivity to surface roughness near a swept leading edge,* J. Fluid Mech. **380** 1999, pp. 141–168.

[148] B. FARREL AND P. IOANNOU, *Turbulence suppression by active control,* Phys. Fluids **8** 1996, pp. 1257–1268.

[149] J. FFOWCS-WILLIAMS AND B. ZHAO, *Active control of vortex shedding,* J. Fluids Struc. **S3** 1989, pp. 115–122.

[150] C. GMELIN AND U. RIST, *Active control of laminar-turbulent transition using instantaneous vorticity signals at the wall,* Phys. Fluids **13** 2001, pp. 513–519.

[151] E. HAMMOND, T. BEWLEY, AND P. MOIN, *Observed mechanisms for turbulence attenuation and enhancement in opposition-controlled wall bounded flows,* Phys. Fluids **10** 1998, pp. 2421–2423.

[152] J.-W. HE, R. GLOWINSKI, R. METCALFE, A. NORDLANDER, AND J. PERIAUX, *Active control and drag optimization of flow past a circular cylinder. I. Oscillatory cylinder rotation,* J. Comput. Phys. **163** 2000, pp. 83–117.

[153] J.-W. HE, R. GLOWINSKI, R. METCALFE, AND J. PERIAUX, *A numerical approach to the control and stabilization of advection-diffusion systems: Application to viscous drag reduction,* Internat. J. Comput. Fluid Mech. **11** 1998, pp. 131–156.

[154] M. HOGBERG AND M. BERGGREN, *Numerical approaches to optimal control of a model equation for shear flow instabilities,* Flow, Turb. Comb. **65** 2000, pp. 299–320.

[155] M. Hogberg and D. Henningson, *Linear optimal control applied to instabilities in spatial boundary layers,* to appear in J. Fluid Mech.

[156] C. Homescu, I. Navon, and Z. Li, *Suppression of vortex shedding for flow around a circular cylinder using optimal control,* Internat. J. Numer. Methods Fluids, **38** 2002, pp. 43–69.

[157] A. Iollo and L. Zannetti, *Optimal control of a vortex trapped by an airfoil with a cavity,* Flow Turb. Comb. **65** 2000, pp. 417–430.

[158] S. Joshi, J. Speyer, and J. Kim, *Finite-dimensional optimal control of Poiseuille flow,* J. Guidance Control Dynam. **22** 1999, pp. 340–348.

[159] R. Joslin, G. Erlebacher, and M. Hussaini, *Active control of instabilities in laminar boundary layers – Overview and concept validation,* J. Fluids Engrg. **118** 1996, pp. 494–497.

[160] S. Kang, V. Ryder, L. Cortelezzi, and J. Speyer, *State-space formulation and controller design for three-dimensional channel flows,* in Proc. 3rd American Control Conference **3** 1999, pp. 1627–1631.

[161] P. Koumoutsakos, *Active control of vortex-wall interactions*, Phys. Fluids **9** 1997, pp. 3808–3816.

[162] P. Koumoutsakos, *Vorticity control for a turbulent channel flow,* Phys. Fluids **11** 1999, pp. 248–250.

[163] K. Kwon and H. Choi, *Control of laminar vortex shedding behind a circular cylinder using splitter plates,* Phys. Fluids **8** 1996, pp. 479–485.

[164] E. Laurien and L. Kleiser, *Numerical solution of boundary-layer transition and transition control,* J. Fluid Mech. **199** 1989, pp. 403–440.

[165] L. Maestrello and L. Ting, *Analysis of active control by surface heating,* AIAA J. **23** 1985, pp. 1038–1045.

[166] K. McManus, T. Poinsot, and S. Candel, *Review of active control of combustion instabilities,* Prog. Energy Combust. Sci. **19** 1993, pp. 1–29.

[167] P. Moin and T. Bewley, *Application of control theory to turbulence,* in Proc. 12th Australian Fluid Mechanics Conference, University of Sidney, 1995, pp. 109–117.

[168] Y.-H. Ou, *Mathematical modeling and numerical simulation in external flow control,* in Flow Control, Springer, New York, 1995, pp. 219–255.

[169] B. Protas and A. Styczek, *Theoretical and computational study of the wake control problem,* Preprint, Department of Aerodynamics, Warsaw University of Technology, Warsaw, 1998.

[170] B. Protas and A. Styczek, *Optimal rotary control of the cylinder wake in the laminar regime,* to appear in Phys. Fluids.

[171] H. Rebbeck and K. Choi, *Opposition control of near-wall turbulence with a piston type actuator,* Phys. Fluids **13** 2001, pp. 2142–2145.

[172] M. Schumm, E. Berger, and P. Monkewitz, *Self-excited oscillations in the wake of two-dimensional bluff bodies and their control,* J. Fluid Mech. **271** 1994, pp. 17–53.

[173] R. Temam, *Remarks on the control of turbulence,* in Flow Control, Springer, New York, 1995, pp. 357–381.

[174] R. Temam, T. Bewley, and P. Moin, *Control of turbulent flows,* in Proc. of the 18th IFIP TC7 Conference on System Modelling and Optimization, Detroit, 1997.

[175] P. Tokumaru and P. Dimotakis, *Rotary oscillatory control of a cylinder wake,* J. Fluid Mech. **224** 1991, pp. 71–90.

[176] S. Walther, C. Airiau, and A. Bottaro, *Optimal control of Tollmien-Schlichting waves in a developing boundary layer,* Phys. Fluids **13** 2001, pp. 2087–2096.

[177] F. Zhu and J. Wu, *A numerical study of controlling stalled airfoil flow by boundary vorticity flux,* to appear in J. Fluid Mech.

[E] Shape optimization in fluids

[178] E. Arian and S. Ta'asan, *Shape optimization in one shot,* in Optimal Design and Control, Birkhauser, Boston, 1995, pp. 23–40.

[179] G. Armugan and O. Pironneau, *On the problem of riblets as a drag reduction device,* Optimal Control Appl. Methods **10** 1989, pp. 93–112.

[180] J. Borggaard, *On the presence of shocks in domain optimization of Euler flows,* in Flow Control, Springer, New York, 1995, pp. 35–48.

[181] J. Borggaard and J. Burns, *A sensitivity equation approach to shape optimization in fluid flows,* in Flow Control, Springer, New York, 1995, pp. 49–78.

[182] J. Brock and W. Ng, *Quasi-analytic shape modification for neighboring steady-state Euler solutions,* in Flow Control, Springer, New York, 1995, pp. 79–109.

[183] J. Burkardt, M. Gunzburger, and J. Peterson, *Discretization of cost functionals and sensitivities in shape optimization,* in Computation and Control IV, Birkhauser, Basel, 1995, pp. 43–56.

[184] H. Cabuk, C. Shung, and V. Modi, *Adjoint operator approach to shape design for internal incompressible flow,* in Proc. 3rd Intern Conf. on Inverse Design and Optimization in Engineering Sciences, 1991, pp. 391–404.

[185] R. Dziri and J.-P. Zolesio, *Dynamical shape control in non-cylindrical Navier-Stokes equations,* J. Convex Anal. **6** 1999, pp. 293–318.

[186] R. Glowinski and J. He, *On shape optimization and related issues,* in Computational Methods for Optimal Design and Control, Birkhauser, Basel, 1998. pp. 151–180.

[187] R. Glowinski, A. Kearsley, T. Pan, and J. Periaux, *Numerical simulation and optimal shape for viscous flow by a fictitious domain method,* Internat. J. Numer. Methods Fluids **20** 1995, pp. 695-711.

[188] R. Glowinski and O. Pironneau, *Toward the computation of minimum drag profile in viscous laminar flow,* Appl. Math. Modeling **1** 1976, pp. 58–66.

[189] M. D. Gunzburger and H. Kim, *Existence of an optimal solution of a shape control problem for the stationary Navier–Stokes equations,* SIAM J. Control Optim. **36** 1998, pp. 895–909.

[190] M. Gunzburger, H.-C. Kim, and S. Manservisi, *On a shape control problem for the stationary Navier-Stokes equations,* Math. Model. Numer. Anal. **34** 2000, pp. 1233–1258.

[191] M. Gunzburger and S. Manservisi, *A variational inequality formulation of an inverse elasticity problem,* Appl. Numer. Math. **34** 2000, pp. 99–126.

[192] M. Gunzburger and J. Peterson, *Issues in shape optimization for the Navier-Stokes equations,* in Proc. 31th IEEE Conf. Decision and Control, IEEE, 1992, pp. 3390–3392.

[193] T. Han, D. Hammond, and C. Sagi, *Optimization of bluff body for minimum drag in ground proximity,* AIAA J. **30** 1992, pp. 882–889.

[194] P. Hovland, B. Mohammadi, and C. Bischof, *Automatic differentiation and Navier-Stokes computations,* in Computational Methods for Optimal Design and Control, Birkhauser, Basel, 1998, pp. 265–284.

[195] W. Jou, W. Huffman, D. Young, R. Melvin, M. Bierterman, C. Hilems, and F. Johnson, *Practical considerations in aerodynamic design optimization,* AIAA paper no. 95-1730, AIAA, 1995.

[196] H.-C. Kim, *Penalized approach and analysis of an optimal shape control problem for the stationary Navier-Stokes equations,* J. Korean Math. Soc. **38** 2001, pp. 1–23.

[197] B. Mohammadi, *Optimal shape design, reverse mode of automatic differentiation and turbulence,* AIAA paper no. 97-0099, AIAA, 1997.

[198] B. Mohammadi, *Practical application to fluid flows of automatic differentiation for design problems,* in Inverse Design and Optimization Methods, Lecture Series 1997-05, von Karman Institute for Fluid Dynamics, Rhode Saint Genese, Belgium, 1997.

[199] B. Mohammadi, *Shape optimization for 3D turbulent flows using automatic differentiation,* Internat. J. Comput. Fluid Dynam. **11** 1998, pp. 27–50.

[200] B. Mohammadi, *Mesh adaptation and automatic differentiation for optimal shape design,* Internat. J. Comput. Fluid Dynam. **10** 1998, pp. 199–211.

[201] O. PIRONNEAU, *On optimal design in fluid mechanics,* J. Fluid Mech. **64** 1974, pp. 97–110.

[202] E. CLIFF, M. HEIKENSCHLOSS, AND A. SHENOY, *Airfoil design by and all-at-once method,* Internat. J. Comput. Fluid Mech. **11** 1998, pp. 3–25.

[203] J. SIMON, *Domain variation for drag Stokes flows,* in Lecture notes in Control and Information Sciences **114**, Springer, Berlin, 1987, pp. 277–283.

[204] J. SIMON, *Domain variation for Stokes flow,* in Lecture Notes in Control and Information Sciences **159**, Springer, Berlin, 1990, pp. 28–42.

[205] S. STOJANOVIC, *Nonsmooth analysis and free boundary poblems for potential and Stokes flow,* in Flow Control, Springer, New York, 1995, pp. 275–295.

[206] T. SVOBODNY, *Shape optimization and control of separating flow in hydrodynamics,* in Flow Control, Springer, New York, 1995, pp. 325–339.

[F] Optimal design in fluids

[207] N. ALEXANDROV, R. LEWIS, C. GUMBERT, L. GREEN, AND P. NEWMAN, *Optimization with variable fidelity models applied to wing design,* AIAA Paper 2000-0841, AIAA, 2000.

[208] W. ANDERSON AND D. BONHAUS, *Airfoil design on unstructured grids for turbulent flows,* AIAA J. **37** 1999, pp. 185–191.

[209] W. ANDERSON AND V. VENKATAKRISHNAN, *Aerodynamic design optimization on unstructured grids with a continuous adjoint formulation,* AIAA paper no. 97-0643, AIAA, 1997.

[210] J. APPEL, A. GODFREY, E. CLIFF, AND M. GUNZBURGER, *Optimization-based design in high-speed flows,* in CFD for Design and Optimization, ASME, New York, 1995, pp. 61–68.

[211] E. ARIAN, *On the coupling of aerodynamic and structural design,* J. Comput. Phys. **135** 1997, pp. 83–96.

[212] E. ARIAN, *MDO - A mathematical viewpoint,* in Computational Methods for Optimal Design and Control, Birkhauser, Boston, 1998, pp. 27–47.

[213] E. ARIAN AND S. TA'ASAN, *Analysis of the Hessian for aerodynamic optimization: Inviscid flow,* Comput. Fluids **28** 1999, pp. 853–877.

[214] O. BAYSAL AND M. ELESHAKY, *Aerodynamic design optimization using sensitivity analysis and computational fluid dynamics,* AIAA J. **30** 1992, pp. 718–725.

[215] J. BETTS, *The application of sparse least squares in aerospace design problems,* in Optimal Design and Control, Birkhauser, Boston, 1995, pp. 81–96.

[216] A. Booker, J. Dennis, P. Frank, D. Serafini, and V. Torczon, *Optimization using surrogate objectives on a helicopter test example,* in Computational Methods for Optimal Design and Control, Birkhauser, Boston, 1998, pp. 49–58.

[217] J. Borggaard and J. Burns, *A PDE sensitivity equation method for optimal aerodynamic design,* J. Comput. Phys. **136** 1997, pp. 366–384.

[218] J. Borggaard and D. Pelletier, *On optimal design using adaptive finite element methods,* in Proc. First International Conference on Nonlinear Problems in Aeronautics and Aerospace, Embry-Riddle Aeronautical University, 1997, pp. 33–40.

[219] J. Borggaard and D. Pelletier, *Observations in adaptive refinement strategies for optimal design,* in Computational Methods for Optimal Design and Control, Birkhauser, Basel, 1998, pp. 59–76.

[220] J. Borggaard and D. Pelletier, *Optimal shape design in forced convection using adaptive finite elements,* AIAA paper no. 98-0908, AIAA, 1998.

[221] E. Cliff, M. Heikenschloss, and A. Shenoy, *Adjoint-based method in aerodynamic design-optimization,* in Computational Methods for Optimal Design and Control, Birkhauser, Basel, 1998. pp. 91–112.

[222] A. Dadone and B. Grossman, *CFD design problems using progressive optimization,* AIAA paper no. 99-3295, AIAA, 1999.

[223] J. Elliot and J. Peraire, *Aerodynamic design using unstructured meshes,* AIAA paper no. 96-1941, AIAA, 1996.

[224] J. Elliot and J. Peraire, *Practical 3D aerodynamic design and optimization using unstructured meshes,* AIAA J. **35** 1997, pp. 1479–1485.

[225] P. Frank and G. Shubin, *A comparison of optimization-based approaches for a model computational aerodynamics design problem,* J. Comput. Phys. **98** 1992, pp. 74–89.

[226] M. Giles and N. Pierce, *Adjoint equations in CFD: Duality, boundary conditions and solution behaviour,* AIAA paper no. 97-1850, AIAA, 1997.

[227] M. Giles and N. Pierce, *On the properties of solutions of the adjoint Euler equations,* in Numerical Methods for Fluid Dynamics VI, ICFD, Oxford, UK, 1998.

[228] M. Giles and N. Pierce, *Improved lift and drag estimates using adjoint Euler equations,* AIAA paper no. 99-3293, AIAA, 1999.

[229] R. Hicks and P. Henne, *Wind design by numerical optimization,* J. Aircraft **15** 1978, pp. 407–412.

[230] D. Huddleston, *Development of a free-jet forebody design simulator design optimization method,* Report no. AEDC-TR-90-22, Arnold Engineering Development Center, Arnold AFB, 1990.

[231] D. HUDDLESTON AND C. MASTIN, *Optimization method applied to aerodynamic design problems in computational fluid dynamics,* in Proc. 7th International Conference on Finite Element Methods in Flow Problems, University of Alabama in Huntsville, 1989, pp. 899–907.

[232] W. HUFFMAN, R. MELVIN, D. YOUNG, F. JOHNSON, J. BUSSOLETTI, M. BIETERMAN, AND C. HIMES, *Practical design and optimization in computational fluid dynmaics,* AIAA paper no. 93-3111, AIAA, 1993.

[233] K. ITO, H. TRAN, AND J. SCROGGS, *Mathematical issues in optimal design of a vapor transport reactor,* in Flow Control, Springer, New York, 1995, pp. 197–218.

[234] A. JAMESON, *Aerodynamic design via control theory,* J. Sci. Comput. **3** 1988, pp. 233-260.

[235] A. JAMESON, *Computational aerodynamics for aircraft design,* Science **245** 1989, pp. 361-371.

[236] A. JAMESON, *Optimum aerodynamic design using CFD and control theory,* AIAA paper no. 95-1729, AIAA, 1995.

[237] A. JAMESON, *Re-engineering the design process through computation,* J. Aircraft **36** 1999, pp. 36–50.

[238] A. JAMESON, L. MARTINELLI, AND N. PIERCE, *Optimum aerodynamic design using the Navier-Stokes equations,* Theoret. Comput. Fluid Dynam. **10** 1998, pp. 213–237.

[239] V. KORIVI, A. TAYLOR, AND G. HOU, *Sensitivity analysis, approximate analysis and design optimization for internal and external viscous flows,* AIAA paper no. 91-3083, AIAA, 1991.

[240] G. KURUVILA, S. TA'ASAN, AND M. SALAS, *Airfoil design and optimization by the one-shot method,* AIAA paper no. 95-0478, AIAA, 1995.

[241] R. LEWIS AND R. AGARWAL, *Airfoil design via control theory using the full-potential and Euler equations,* in Forum on CFD Design and Optimization, IMECE 95, ASME, 1995.

[242] J. MALONE, J. NARRAMORE, AND L. SANKAR, *Airfoil design method using the Navier-Stokes equations,* J. Aircraft **28** 1990, pp. 216–224.

[243] S. NADARAJAH AND A. JAMESON, *A comparison of the continuous and discrete adjoint approach to automatic aerodynamic optimization,* AIAA paper no. 2000-0667, AIAA, 2000.

[244] R. NARDUCCI, B. GROSSMAN, AND R. HAFTKA, *Sensitivity algorithms for an inverse design problem involving a shock wave,* Inverse Problems Engrg. **2** 1995, pp. 49–83.

[245] J. NEWMAN, A. TAYLOR, R. BARNWELL, P. NEWRMAN, AND G. HOU, *Overview of sensitivity analysis and shape optimization for complex aerodynamic configurations,* J. Aircraft **36** 1999, pp. 87–96.

[246] P. NEWMAN, *Preparation of advanced CFD codes for use in sensitivity analyses and multidisciplinary design optimization,* in Optimal Design and Control, Birkhauser, Boston, 1995, pp. 241–274.

[247] P. NEWMAN, G. HOU, H. JONES, A. TAYLOR, AND V. KORIVI, *Observations on computational methodologies for use in large-scale, gradient-based, multidisciplinary design,* AIAA paper no. 92-4753, AIAA, 1992.

[248] E. NIELSEN AND W. ANDERSON, *Aerodynamic design optimization on unstructured meshes using the Navier-Stokes equations,* AIAA paper no. 98-4809, AIAA, 1998.

[249] O. PIRONNEAU, *On optimal design in fluid mechanics,* J. Fluid Mech. **64** 1974, pp. 97–110.

[250] J. REUTHER AND A. JAMESON, *Control theory based airfoil design for potential flow and a finite volume discretization,* AIAA paper no. 94-0499, AIAA, 1994.

[251] J. REUTHER AND A. JAMESON, *Control based airfoil design using the Euler equations,* AIAA paper no. 94-4272, AIAA, 1994.

[252] J. REUTHER, A. JAMESON, J. ALONSO, M. REMLINGER, AND D. SAUNDERS, *Constrained multipoint aerodynamic shape optimization and adjoint formulation and parallel computers. Part 1,* J. Aircraft **36** 1999, pp. 51–60.

[253] J. REUTHER, A. JAMESON, J. ALONSO, M. REMLINGER, AND D. SAUNDERS, *Constrained multipoint aerodynamic shape optimization and adjoint formulation and parallel computers. Part 2,* J. Aircraft **36** 1999, pp. 61–74.

[254] J. REUTHER, A. JAMESON, J. FARMER, L. MARTINELLI, AND D. SAUNDERS, *Aerodynamic shape optimization of complex aircraft configurations via adjoint formulation,* AIAA paper no. 96-0094, AIAA, 1996.

[255] R. SAMPATH AND N. ZABARAS, *Inverse thermal design and control of solidification processes in the presence of a strong external magnetic field,* Internat. J. Numer. Methods Engrg. **50** 2001, pp. 2489–2520.

[256] G. SHUBIN AND P. FRANK, *A comparison of the implicit gradient approach and the variational approach to aerodynamic design optimization,* Report number AMR-TR-163, Boeing Computer Services, Seattle, 1991.

[257] M. STANCIU AND B. MOHAMMADI, *Platform for multi-model design,* Flow Turb. Comb. **65** 2000, pp. 431–452.

[258] L. STANLEY AND D. STEWART, *A comparison of local and global projections in design sensitivity computations,* in Computational Methods for Optimal Design and Control, Basel, 1998, pp. 361–373.

[259] S. TA'ASAN, *Trends in aerodynamics design and optimization: A mathematical viewpoint,* AIAA paper no. 95-1731, AIAA, 1995.

[260] S. Ta'asan, G. Kuruvila, and M. Salas, *Aerodynamic design and optimization in one-shot method,* AIAA paper no. 92-0025, AIAA, 1992.

[261] A. Taylor, G. Hou, and V. Korivi, *A methodology for determining aerodynamic sensitivity derivatives with respect to variation in geometric shape,* AIAA paper no. 91-1101, AIAA, 1991.

[262] A. Taylor, G. Hou, and V. Korivi, *Sensitivity analysis, approximate analysis, and design optimization for internal and external flows,* AIAA paper no. 91-3083, AIAA, 1991.

[263] A. Taylor, V. Korivi, and G. Hou, *Sensitivity analysis applied to the Euler equations: A feasibility study with emphasis on variation of geometric shape,* AIAA paper no. 91-0173, AIAA, 1991.

[264] A. Taylor, V. Korivi, and G. Hou, *Approximate analysis and sensitivity analysis methods for viscous flows involving variation of geometric shape,* AIAA paper no. 91-1569, AIAA, 1991.

[265] E. Turgeon, D. Pelletier, and J. Borggaard, *A continuous sensitivity equation approach to optimal design in mixed convection,* AIAA report no. 99-3625, AIAA, 1999.

[266] X. Wu, E. Cliff, and M. Gunzburger, *An optimal design problem for a two dimensional flow in a duct,* Optimal Control Appl. Methods **17** 1996, pp. 329–339.

[267] G. Yang and N. Zabaras, *The adjoint method for an inverse design problem in the directional solidification of binary alloys,* Internat. J. Numer. Methods Engrg. **42** 1998, pp. 1121–1144.

[G] Flow sensitivity analysis

[268] K. Anderson, J. Newman, D. Whitfield, and E. Nielsen, *Sensitivity analysis for the Navier-Stokes equations on unstructured grids using complex variables,* AIAA paper no. 99-3294, AIAA, 1999.

[269] J. Appel and M. Gunzburger, *Sensitivity calculation in flows with discontinuities,* AIAA Paper no. 96-2471, AIAA, 1996.

[270] J. Appel and M. Gunzburger, *Difficulties in sensitivity calculations for flows with discontinuities,* AIAA J. **35** 1997, pp. 842–848.

[271] J. Appel and M. Gunzburger, *Approximate solutions via sensitivities,* in Inverse Design and Optimisation Methods, von Karman Institute for Fluid Dynamics, Rhode–Saint–Genèse, Belgium, 1997.

[272] J. Borggaard, J. Burns, E. Cliff, and M. Gunzburger, *Sensitivity calculations for a 2D, inviscid, supersonic forebody problem,* in Identification and Control in Systems Governed by Partial Differential Equations, SIAM, Philadelphia, H. T. Banks, R. H. Fabiano, and K. Ito, Editors, 1993, pp. 14–24.

[273] J. BORGGAARD AND D. PELLETIER, *Computing design sensitivities using an adaptive finite element method,* AIAA paper no. 96-1938, AIAA, 1996.

[274] J. BURKARDT AND M. GUNZBURGER, *Sensitivity discrepancy for geometric parameters,* in CFD for Design and Optimization, ASME, New York, 1995, pp. 9–15.

[275] O. GHATTAS AND X. LI, *Domain decomposition methods for sensitivity analysis of a nonlinear aeroelasticity problem,* Internat. J. Comput. Fluid Dynam. **11** 1998, pp. 113–130.

[276] A. GODFREY, *Using sensitivities for flow analysis,* in Computational Methods for Optimal Design and Control, Birkhauser, Basel, 1998. pp. 181–196.

[277] C. HOMESCU AND I. NAVON, *Numerical and theoretical consideration for sensitivity calculation of discontinuous flow,* to appear in Systems Cont. Lett.

[278] G. HOU AND J. SHEEN, *Numerical methods for second-order shape sensitivity analysis with application to heat conduction problems,* Internat. J. Numer. Methods Engrg. **36** 1993, pp. 417–435.

[279] G. HOU, A. TAYLOR, AND V. KORIVI, *Discrete shape senstivity equations for aerodynamic flow,* AIAA paper no. 91-2259, AIAA, 1991.

[280] V. KORIVI, A. TAYLOR, P. NEWMAN, G. HOU, AND H. JONES, *An approximately factored incremental strategy for calculating consistent discrete aerodynamic sensitivity derivatives,* AIAA paper no. 92-4746, AIAA, 1994.

[281] R. LEWIS, A. PATERA, AND J. PERAIRE, *A posteriori finite element bounds for sensitivity derivatives of partial-differential-equations outputs,* Finite Elem. Anal. Des. **34** 2000, pp. 271–290.

[282] L. MACHIELS, J. PERAIRE, AND A. PATERA, *Output bound approximations for partial differential equations: Application to the incompressible Navier-Stokes equations,* Comput. Methods Appl. Mech. Engrg. **190** 2001, pp. 3413–3426.

[283] M. PARASCHIVOIU AND A. PATERA, *A posteriori bounds for linear-functional outputs on Crouzeix-Raviart finite element discretizations of the incompressible Stokes problem,* Internat. J. Numer. Methods Fluids **32** 2000, pp. 823–849.

[284] M. PARASCHIVOIU, J. PERAIRE, Y. MADAY, AND A. PATERA, *Fast bounds for outputs of partial differential equations,* in Computational Methods for Optimal Design and Control, Birkhauser, Basel, 1998. pp. 323–360.

[285] D. PELLETIER, E. TURGEON, D. LACASSE, AND J. BORGGAARD, *Adaptivity, sensitivity and uncertainty: Towards standards in CFD,* AIAA report no. 2001-0192, AIAA, 2001.

[286] J. PRALITS, C. AIRIAU, A HANIFI AND D. HENNINGSON, *Sensitivity analysis using adjoint parabolized stability equations for compressible flows,* Flow Turb. Comb. **65** 2000, pp. 321–346.

[287] A. TAYLOR, G. HOU, AND V. KORIVI, *An efficient method for estimating neighboring steady-state numerical solutions to the Euler equations,* AIAA paper no. 91-1680, AIAA, 1991.

[288] G. SHUBIN, *Obtaining cheap optimization gradients from computational aerodynamic codes,* Report number AMR-TR-164, Boeing Computer Services, Seattle, 1991.

[289] A. TAYLOR, P. NEWMAN, G. HOU, AND H. JONES, *Recent advances in steady compressible aerodynamic sensitivity analysis,* in Flow Control, Springer, New York, 1995, pp. 341–355.

[290] E. TURGEON, D. PELLETIER, AND J. BORGGAARD, *A general continuous sensitivity equation formulation for complex flows,* AIAA report no. 2000-4732, AIAA, 2000.

[291] E. TURGEON, D. PELLETIER, AND J. BORGGAARD, *A continuous sensitivity equation method for flows with temperature dependent viscosities,* AIAA report no. 2000-4821, AIAA, 2000.

[292] E. TURGEON, D. PELLETIER, AND J. BORGGAARD, *Sensitivity and uncertainty analysis for variable property flows,* AIAA report no. 2001-0140, AIAA, 2001.

[H] Reduced-order methods for flow control

[293] K. AFANASIEV AND M. HINZE, *Adaptive control of a wake flow using proper orthogonal decomposition,* Report no. 648, Technische Universitat Berlin, Berlin, 1999.

[294] J. BURNS AND B. KING, *A reduced basis approach to the design of low order feedback controllers for nonlinear continuous systems,* J. Vib. Control **4** 1998, pp. 297–323.

[295] L. CORTELEZZI AND J. SPEYER, *Robust reduced-order controller of transitional boundary layers,* in Computational Methods for Optimal Design and Control, Birkhauser, Basel, 1998. pp. 127–136.

[296] L. CORTELEZZI AND J. SPEYER, *Robust reduced-order controller of laminar boundary layer transitions,* Phys. Rev. E **58** 1998, pp. 1906–1910.

[297] L. CORTELEZZI, K. LEE, J. KIM, AND J. SPEYER, *Skin-friction drag reduction via robust reduced-order linear feedback control,* Internat. J. Comput. Fluid Dynam. **11** 1998, pp. 79–92.

[298] L. CORTELEZZI, J. SPEYER, K. LEE, AND J. KIM, *Robust reduced-order control of turbulent channel flows via distributed sensors and actuators,* in Proc. 37th IEEE Conference on Decision and Control, **2**, IEEE, 1998, pp. 1906–1911.

[299] E. GILLIES, *Low-dimensional control of the circular cylinder wake,* J. Fluid Mech. **371** 1998, pp. 1157–178.

[300] W. GRAHAM, J. PERAIRE, AND K. TANG, *Optimal control of vortex shedding using low-order models. Part I - Open-loop model development,* Internat. J. Numer. Methods Fluids **44** 1999, pp. 945–972.

[301] W. GRAHAM, J. PERAIRE, AND K. TANG, *Optimal control of vortex shedding using low-order models. Part II - Model-based control,* Internat. J. Numer. Methods Fluids **44** 1999, pp. 973–990.

[302] K. ITO AND S. RAVINDRAN, *A reduced-order method for simulation and control of fluid flows,* J. Comput. Phys. **143** 1998, pp. 403–425.

[303] K. ITO AND S. RAVINDRAN, *Reduced basis method for optimal control of unsteady viscous flows,* Internat. J. Comput. Fluid Dynam. **15** 2001, pp. 97–113.

[304] K. ITO AND J. SCHROETER, *Reduced order feedback synthesis for viscous incompressible flows,* Math. Comput. Modelling **33** 2001, pp. 173–192.

[305] K. KUNISCH AND S. VOLKWEIN, *Control of Burgers' equation by a reduced order approach using proper orthogonal decomposition,* J. Optim. Theory Appl. **102** 1999, pp. 345–371.

[306] K. LEE, L. CORTELEZZI, J. KIM, AND J. SPEYER, *Application of robust reduced-order controller to turbulent flows for drag reduction,* Phys. Fluids **13** 2001, pp. 1321–1330.

[307] R. PRABHU, S. COLLIS, AND Y. CHANG, *The influence of control on proper orthogonal decomposition of wall-bounded turbulent flows,* Phys. Fluids **13** 2001, pp. 520–537.

[308] S. RAVINDRAN, *A reduced order approach for optimal control of fluids using proper orthogonal decomposition,* Internat. J. Numer. Methods Fluids **34** 2000, pp. 425–448.

[309] S. RAVINDRAN, *Reduced-order adaptive controllers for fluid flows using POD,* J. Sci. Comput. **15** 2000, pp. 457–478.

[310] S. S. RAVINDRAN, *Adaptive reduced order controllers for a thermal flow system using proper orthogonal decomposition,* SIAM J. Sci. Comput. **23** 2002, pp. 1924–1942.

[311] S. RAVINDRAN, *Proper orthogonal decomposition in optimal control of fluids,* Internat. J. Numer. Methods Fluids **34** 2000, pp. 425–448.

[312] K. TANG, W. GRAHAM, AND J. PERAIRE, *Active flow control using a reduced order model and optimal control,* AIAA paper no. 96-1946, AIAA, 1996.

[313] S. TANG AND N. AUBRY, *Suppression of vortex shedding inspired by low dimensional models,* J. Fluids Struc. **14** 2000, pp. 443–468.

[I] Feedback flow control

[314] V. BARBU, *Feedback control of time dependent Stokes flows,* in Optimal Control of Viscous Flow, S. Sritharan, Editor, SIAM, Philadelphia, 1998, pp. 63–77.

[315] V. BARBU AND S. SRITHARAN, *Flow invariance preserving feedback controllers for the Navier-Stokes equations,* J. Math. Anal. Appl. **255** 2000, pp. 281–307.

[316] T. BEWLEY, P. MOIN, AND R. TEMAM, *A method for optimizing feedback control laws for wall-bounded flows based on control theory,* in Proc. Fluids Engineering Division Conference, Part 2, FED-237, ASME, New York, 1999.

[317] T. BEWLEY, P. MOIN, AND R. TEMAM, *DNS-based predictive control of turbulence: an optimal benchmark for feedback algorithms,* J. Fluid Mech. **477** 2001, pp. 179–225.

[318] J. BURNS, B. KING, AND D. RUBIO, *Feedback control of a thermal fluid using state estimation,* Internat. J. Comput. Fluid Dynam. **11** 1998, pp. 93–112.

[319] H. CHOI, R. TEMAM, P. MOIN, AND J. KIM, *Feedback control for unsteady flow and its application to the stochastic Burgers equation,* J. Fluid Mech. **253** 1993, pp. 509–543.

[320] L. CORTELEZZI, *Nonlinear feedback control of the wake past a plate with suction points on the downstream wall,* J. Fluid Mech. **327** 1996, pp. 303–324.

[321] L. CORTELEZZI, Y. CHEN, AND H. CHANG, *Nonlinear feedback control of the wake past a plate: From a low order model to a higher order model,* Phys. Fluids **9** 1997, pp. 2009–2022.

[322] M. GUNZBURGER AND H.-C. LEE, *Feedback control of fluid flows,* in Proc. 14th IMACS World Congress on Computational and Applied Mathematics, Georgia Tech., Atlanta, 1994, pp. 716–719.

[323] M. GUNZBURGER AND H.-C. LEE, *Feedback control of Karman vortex shedding,* J. Appl. Mech. **63** 1996, pp. 828–835.

[324] M. GUNZBURGER AND S. MANSERVISI, *Analysis and approximation for linear feedback control for tracking the velocity in Navier-Stokes flows,* Comput. Methods Appl. Mech. Engrg. **189** 2000, pp. 803–823.

[325] H. HU AND H. BAU, *Feedback control to delay or advance linear loss of stability in planar Poiseuille flow,* Proc. Roy. Soc. London Ser. A **447** 1994, pp. 229–312.

[326] K. ITO AND S. KANG, *A dissipative feedback control synthesis for systems arising in fluid dynamics,* SIAM J. Control Optim. **32** 1994, pp. 831–854.

[327] K. ITO AND Y. YAN, *Viscous scalar conservation law with nonlinear flux feedback and global attractors,* J. Math. Anal. Appl. **227** 1998, pp. 271–299.

[328] S. JOSHI, J. SPEYER, AND J. KIM, *A systems theory appraoch to the feedback stabilization of infinitesimal and finite-amplitude disturbances in plane Poiseuille flow,* J. Fluid Mech. **332** 1997, 157–184.

[329] R. JOSLIN, R. NICOLAIDES, G. ERLEBACHER, M. HUSSAINI, AND M. GUNZBURGER, *Active control of boundary–layer instabilities: Use of sensors and spectral controller,* AIAA J. **33** 1995, pp. 1521–1523.

[330] P. Koummoutsakos, T. Bewley, E. Hammond, and P. Moin, *Feedback algorithms for turbulent control – Some recent developments,* AIAA paper no. 97-2008, AIAA, 1998.

[331] H.-C. Lee and B. Shin, *Dynamics for linear feedback controller for the two dimensional Benard equations with distributed control,* Appl. Math. Optim. **44** 2001, pp. 163–175.

[332] C. Min and H. Choi, *Suboptimal feedback control of vortex shedding at low Reynolds numbers,* J. Fluid Mech. **401** 1999, pp. 123–156.

[333] P. Moin and T. Bewley, *Feedback control of turbulence,* Appl. Mech. Rev. **47** 1994, pp. S3–S13.

[334] Y.-H. Ou, *Design of feedback compensators for viscous flow,* in Optimal Control of Viscous Flow, S. Sritharan, Editor, SIAM, Philadelphia, 1998, pp. 151–180.

[335] D. Park, *Theoretical analysis of feedback control of Karman vortex shedding at slightly supercritical Reynolds numbers,* European J. Mech. B Fluids **13** 1994, pp. 387–399.

[336] D. Park, D. Ladd, and E. Hendricks, *Feedback control of Karman vortex shedding behind a circular cylinder at low Reyolds numbers,* Phys. Fluids. **6** 1994, pp. 2390–2405.

[337] K. Roussopoulos, *Feedback control of vortex shedding at low Reynolds numbers,* J. Fluid Mech. **248** 1993, pp. 267–296.

[338] S. Sritharan, *Optimal feedback control of hydrodynamics: A progress report,* in Flow Control, Springer, New York, 1995, pp. 257–274.

[339] B. Ton, *An inverse problem for the Navier-Stokes equations: Feedback laws,* preprint, Department of Mathematics, University of British Columbia, Vancouver, 2001.

[J] Controllability of flows

[340] M. Berggren, R. Glowinski, and J.-L. Lions, *A computational approach to controllability issues for flow-related models (I): Pointwise control of the viscous Burgers equation,* Internat. J. Comput. Fluid Mech. **6** 1996, pp. 237–252.

[341] M. Berggren, R. Glowinski, and J.-L. Lions, *A computational approach to controllability issues for flow-related models (II): Control of two-dimensional, linear advection-diffusion and Stokes models,* Internat. J. Comput. Fluid Mech. **6** 1996, pp. 253–276.

[342] J.-M. Coron, *On the controllability of the 2-D incompressible Navier-Stokes equations with the Navier slip boundary conditions,* ESAIM: Cont. Optim. Calc. Var. **1** 1996, pp. 35–75.

[343] J.-M. CORON, *Controlabilite exacte frontiere de l'equation d'Euler des fluides parfaits incompressibles bidimensionnels,* C. R. Acad. Sci. Paris Ser. I Math. **317** 1993, pp. 271–276.

[344] J.-M. CORON, *On the controlability of 2-D incompressible perfect fluids,* J. Math. Pures Appl. **75** 1996, pp. 155–188.

[345] J.-M. CORON AND A. FURSIKOV, *Global exact controllability of the 2D Navier-Stokes equations on a manifold without boundary,* Russian J. Math. Phys. **4** 1996, pp. 429–448.

[346] A. FURSIKOV, *Exact boundary zero controllability of three-dimensional Navier-Stokes equations,* J. Dynam. Systems Control **1** 1995, pp. 325–350.

[347] A. FURSIKOV AND O. IMANUVILOV, *On approximate controllability of the Stokes system,* Ann. Fac. Sci. Toulouse **II** 1993, pp. 1–28.

[348] A. FURSIKOV AND O. IMANUVILOV, *On exact boundary zero-controllability of two-dimensional Navier-Stokes equations,* Acta Appl. Math. **36** 1994, pp. 1–10.

[349] A. FURSIKOV AND O. IMANUVILOV, *On controllability of certain systems simulating a fluid flow,* in Flow Control, Springer, New York, 1995, pp. 149–184.

[350] A. FURSIKOV AND O. IMANUVILOV, *Local exact controllability for 2-D Navier-Stokes equations,* Sbornik Math. **187** 1996, pp. 1355–1390.

[351] A. FURSIKOV AND O. IMANUVILOV, *Local exact controllability of Boussinesq equations,* Vestnik Ros. Univ. Dr. Nar. Ser. Math. **3** 1996, pp. 177–194.

[352] A. FURSIKOV AND O. IMANUVILOV, *Local exact controllability of the Navier-Stokes equations,* C. R. Acad. Sci. Paris **323** 1996, pp. 275–280.

[353] A. FURSIKOV AND O. IMANUVILOV, *Local exact controllability of the Navier-Stokes equations,* Control Math. **209** 1997, pp. 115–129.

[354] A. V. FURSIKOV AND O. Y. IMANUVILOV, *Local exact boundary controllability of the Boussinesq equations,* SIAM J. Control Optim. **36** 1998, pp. 391–421.

[355] O. IMANUVILOV, *On exact controllability for the Navier-Stokes equations,* ESAIM: Cont. Opt. Cal. Var. **3** 1997, pp. 97–131.

[K] Genetic and evolutionary algorithms for flow control

[356] Y. CRISPIN, *Aircraft conceptual optimization using simulated evolution,* AIAA paper no. 94-0092, AIAA, 1994.

[357] I. DE FALCO, *An introduction to evolutionary algorithms and their application to the airfoil design problem,* in Inverse Design and Optimization Methods, Lecture Series 1997-05, von Karman Institute for Fluid Dynamics, Rhode Saint Genese, Belgium, 1997.

[358] I. De Falco, A. Deua Cioppa, A. Iazzetta, and E. Tarantino, *Evolutionary algorithms for aerofoil design,* Internat. J. Comput. Fluid Dynam. **11** 1998, pp. 51–78.

[359] P. Gage and I. Kroo, *A role for genetic algorithms in a preliminary design environments,* AIAA paper no. 93-3933, AIAA, 1993.

[360] R. Gregg and K. Misegades, *Transonic wing optimization using evolution theory,* AIAA paper no. 87-0520, AIAA, 1987.

[361] S. Obayashi, *Aerodynamic optimization with evolutionary algorithms,* in Inverse Design and Optimization Methods, Lecture Series 1997-05, von Karman Institute for Fluid Dynamics, Rhode Saint Genese, Belgium, 1997.

[362] S. Obayashi, *Target pressure optimization using MOGA,* in Inverse Design and Optimization Methods, Lecture Series 1997-05, von Karman Institute for Fluid Dynamics, Rhode Saint Genese, Belgium, 1997.

[363] S. Obayashi and S. Takanashi, *Genetic optimization of target pressure distributions for inverse design methods,* AIAA J. **34** 1996, pp. 881–886.

[364] D. Quagliarella and A. Cioppa, *Genetic algorithms applied to the aerodynamic design of transonic airfoils,* AIAA paper no. 94-1896, AIAA, 1994.

[365] K. Yamamoto and O. Inoue, *Applications of genetic algorithms to aerodynamic shape optimization,* AIAA paper no. 85-1650, AIAA, 1985.

[L] General control, optimization, automatic differentiation, and other references

[366] N. Ahmed and K. Teo, *Optimal Control of Distributed Parameter Systems,* North Holland, New York, 1981.

[367] N. Alexandrov and J. Dennis, *Multilevel algorithms for nonlinear optimization,* in Optimal Design and Control, Birkhauser, Boston, 1995, pp. 1–22.

[368] N. Alexandrov, J. Dennis, R. Lewis, and V. Torczon, *A trust region framework for managing the use of approximation models in optimization,* Struc. Optim. **15** 1998, pp. 16–23.

[369] V. Alekseev, V. Tikhomirov, and S. Fomin, *Optimal Control,* Consultants Bureau, New York, 1987.

[370] R. Becker, H. Kapp, and R. Rannacher, *Adaptive finite element methods for optimal control of partial differential equations: Basic concept,* SIAM J. Control Optim. **39** 2000, pp. 113–132.

[371] J. Betts, *Practical Methods for Optimal Control Using Nonlinear Programming,* SIAM, Philadelphia, 2001.

[372] G. Biros, *VELTISTO: A parallel library for PDE-constrained optimization,* Technical report, Laboratory for Mechanics, Algorithms, and Computing, Carnegie Mellon University, Pittsburgh, 2000.

[373] G. Biros and O. Ghattas, *Parallel Lagrange-Newton-Krylov-Shur methods for PDE-constrained optimization. Part I: The Krylov-Schur solver,* Technical report, Laboratory for Mechanics, Algorithms, and Computing, Carnegie Mellon University, Pittsburgh, 2000.

[374] G. Biros and O. Ghattas, *Parallel Lagrange-Newton-Krylov-Shur methods for PDE-constrained optimization. Part II: The Lagrange-Newton solver and its application to optimal control of steady viscous flows,* Technical report, Laboratory for Mechanics, Algorithms, and Computing, Carnegie Mellon University, Pittsburgh, 2000.

[375] C. Bischof, A. Carle, G. Corliss, A. Griewnak, and P. Hoveland, *ADIFOR: Generating derivative codes from Fortran programs,* Sci. Prog. **1** 1992, pp. 11–29.

[376] C. Bischof, P. Khademi, A. Mauer, and A. Carle, *Adifor 2.0: Automatic differentiation of Fortran 77 programs,* Comput. Sci. Engrg. **3** 1996, pp. DG18–32.

[377] J. Borggaard and A. Verma, *On efficient solutions to the continuous sensitivity equation using automatic differentiation,* SIAM J. Sci. Comput. **22** 2000, pp. 39–62.

[378] M. C. Delfour and J.-P. Zolesio, *Shapes and Geometries: Analysis, Differential Calculus, and Optimization,* SIAM, Philadelphia, 2001.

[379] J. E. Dennis, M. Heikenschloss, and L. N. Vicente, *Trust-region interior-point SQP algorithms for a class of nonlinear programming problems,* SIAM J. Control Optim. **36** 1998, pp. 1750–1794.

[380] A. Faulds and B. King, *Sensor location in feedback control of partial differential equation systems,* in Proc. 2000 IEEE CCA/CACSD, IEEE, 2000, pp. 536–541.

[381] C. Faure, *Generating adjoints of industrial codes with limited memory,* Flow Turb. Comb. **65** 2000, pp. 453–467.

[382] D. Gay, *Algorithm 611. Subroutines for unconstrained minimization using a model/trust-region approach,* ACM Trans. Math. Software **9** 1983, pp. 503–525.

[383] P. Gill, W. Murray, and M. Wright, *Practical Optimization,* Academic Press, Boston, 1981.

[384] A. Griewank, *Achieving logarithmic growth of temporal and spatial complexity in reverse automatic differentiation,* Optim. Methods Software **1** 1992, pp. 35–54.

[385] A. Griewank, *On automatic differentiation,* in Mathematical Programming (Tokyo 88), Kluwer, Dordrecht, 1989, pp. 83–107.

[386] A. Griewank, *Evaluating Derivatives: Principles and Techniques of Algorithmic Differentiation,* SIAM, Philadelphia, 2000.

[387] M. D. GUNZBURGER AND L. S. HOU, *Finite-dimensional approximation of a class of constrained nonlinear optimal control problems,* SIAM J. Control Optim. **34** 1996, pp. 1001–1043.

[388] M. GUNZBURGER AND H.-C. LEE, *Analysis and approximation of optimal control problems for first-order elliptic systems in three dimensions,* Appl. Math. Comput. **100** 1999, pp. 49–70.

[389] M. GUNZBURGER AND H.-C. LEE, *A penalty/least–squares method for optimal control problems for first-order elliptic systems,* Appl. Math. Comput. **107** 2000, pp. 57–75.

[390] W. HAGER, *Runge-Kutta methods in optimal control and the transformed adjoint system,* Numer. Math. **87** 2000, pp. 83–117.

[391] J. HASLINGER, *Comparison of different fictitious domain approaches used in shape optimization,* Report no. 15/1996, Department of Mathematics, University of Jyvaskyla, Jyvaskyla, Finland, 1996.

[392] J. HASLINGER AND P. NEITTAANMAKI, *Finite Element Approximation for Optimal Shape Design,* Wiley, Chichester,1988.

[393] M. HEIKENSCHLOSS, *Projected sequential quadratic programming methods,* SIAM J. Optim. **6** 1996, pp. 373–417.

[394] M. HEIKENSCHLOSS, *A trust region method for norm constrained problems,* SIAM J. Numer. Anal. **35** 1998, pp. 1594–1620.

[395] M. HEIKENSCHLOSS AND L. VICENTE, *TRICE: A package for the solution of optimal control and engineering design problems. User's guide,* Technical report, Department of Computational and Applied Mathematics, Rice University, Houston, 1996.

[396] C. T. KELLEY, *Iterative Methods for Optimization,* SIAM, Philadelphia, 1999.

[397] C. T. KELLEY AND E. W. SACHS, *Quasi-Newton methods and unconstrained optimal control problems,* SIAM J. Control Optim. **25** 1987, pp. 1503–1516.

[398] R. LEWIS, *Numerical computation of sensitivities and the adjoint approach,* in Computational Methods for Optimal Design and Control, Birkhauser, Basel, 1998, pp. 285–302.

[399] X. LI AND J. YONG, *Optimal Control Theory for Infinite Dimensional Systems,* Birkhauser, Boston, 1995.

[400] J.-L. LIONS, *Optimal Control of Systems Goverened by Partial Differential Equations,* Springer, Berlin, 1971.

[401] J.-L. LIONS, *Control of Distributed Singular Systems,* Dunod, Paris, 1985.

[402] D. LIU AND J. NOCEDAL, *On the limited memory BFGS method for large optimization,* Math. Prog. **45** 1989, pp. 503–528.

[403] D. LUENBERGER, *Introduction to Linear and Nonlinear Programming,* Addison-Wesley, Reading, MA, 1973.

[404] O. PIRONNEAU, *Optimal Shape Design for Elliptic Systems,* Springer, Berlin, 1984.

[405] R. RAUSCH, J. BATINA, AND H. YANG, *Three dimensional time-marching aeroelastic analyses using an unstructured-grid Euler method,* AIAA J. **31** 1993, pp. 1626–1633.

[406] J. M. RESTREPO, G. K. LEAF, AND A. GRIEWANK, *Circumventing storage limitation in variational data assimilation studies,* SIAM J. Sci. Comput. **19** 1998, pp. 1586–1605.

[407] N. ROSTAING, S. DALMAS, AND A. GALLIGO, *Automatic differentiation in Odysee,* Tellus **45a** 1993, pp. 558–568.

[408] J. SOKOLOWSKI AND J. ZOLESIO, *Introduction to Shape Optimization: Shape Sensitivity Analysis,* Springer, Berlin, 1992.

[409] W. SQUIRE AND G. TRAPP, *Using complex variables to estimate derivatives of real functions,* SIAM Rev. **40** 1998, pp. 110–112.

[410] V. TICHOMOROV, *Fundamental Principles of the Theory of Extremal Problems,* Wiley, Chichester, 1982.

[411] M. ULBRICH, S. ULBRICH, AND M. HEIKENSCHLOSS, *Global convergence of trust-region interior-point algorithms for infinite-dimensional nonconvex minimization subject to pointwise bounds,* SIAM J. Control Optim. **37** 1999, pp. 731–764.

[412] X. ZOU, I. M. NAVON, M. BERGER, K. H. PHUA, T. SCHLICK, AND F. X. LE DIMET, *Numerical experience with limited-memory quasi-Newton and truncated Newton methods,* SIAM J. Optim. **3** 1993, pp. 582–608.

[M] Other references cited

[413] R. ADAMS, *Sobolev Spaces,* Academic Press, New York, 1975.

[414] I. BABUSKA, *The finite element method with Lagrangian multipliers,* Numer. Math. **16** 1973, pp. 179–192.

[415] I. BABUSKA AND A. AZIZ, *Survey lectures on the mathematical foundations of the finite element method,* in The Mathematical Foundations of the Finite Element Method with Applications to Partial Differential Equations, Academic Press, New York, 1972, pp. 3–359.

[416] F. BAUER, P. GARABEDIAN, AND D. KORN, *Supercritical Wing Sections,* Springer, Berlin, 1972.

[417] F. BREZZI, *On the existence, uniqueness, and approximation of saddle-point problems arising from Lagrange multipliers,* RAIRO Anal. Numer. **8** 1974, pp. 129–151.

[418] F. Brezzi and M. Fortin, *Mixed and Hybrid Finite Element Methods,* Springer, New York, 1991.

[419] F. Brezzi, J. Rappaz, and P. Raviart, *Finite-dimensional approximation of non-linear problems. Part I: Branches of nonsingular solutions,* Numer. Math. **36** 1980, pp. 1–25.

[420] M. Bristeau, R. Glowinski, J. Perrier, O. Pironneau, and G. Poirer, *Application of optimal control and finite element methods to the calculation of transonic flows and incompressible viscous flows,* in Numerical Methods in Applied Fluid Mechanics, Academic Press, London, 1980, pp. 203–312.

[421] P. Ciarlet, *The Finite Element Method for Elliptic Problems,* North-Holland, Amsterdam, 1978.

[422] R. Courant and D. Hilbert, *Supersonic Flow and Shock Waves,* Wiley, New York, 1948.

[423] M. Crouzeix and J. Rappaz, *On Numerical Approximation in Bifurcation Theory,* Masson, Paris, 1990.

[424] R. Dautray and J.-L. Lions, *Mathematical Analysis and Numerical Methods for Science and Technology,* Vol. 1–6, Springer-Verlag, Berlin, 1990–1993.

[425] Q. Du and M. Gunzburger, *Centroidal Voronoi tessellation based proper orthogonal decomposition analysis,* to appear.

[426] Q. Du, M. Gunzburger, and J. Lee, *Optimization-based methods for multidisciplinary simulation and optimization,* in Proc. 8th Annual Conference of the CFD Society of Canada, CERCA, Montreal, 2000, pp. 689–694.

[427] G. Farin, *Curves and Surfaces for Computer Aided Geometric Design: A Practical Guide,* Academic Press, San Diego, 1988.

[428] V. Girault and P.-A. Raviart, *Finite Element Methods for Navier-Stokes Equations*, Springer, New York, 1986.

[429] G. Grubb, *Solution of parabolic pseudo-differential initial-boundary value problems,* J. Differential Equations **87** 1990, pp. 256-304.

[430] M. Gunzburger, *Finite Element Methods for Incompressible Viscous Flows: A Guide to Theory, Practice, and Algorithms*, Academic, Boston, 1989.

[431] M. D. Gunzburger and H. K. Lee, *An optimization-based domain decomposition method for the Navier–Stokes equations,* SIAM J. Numer. Anal. **37** 2000, pp. 1455–1480.

[432] M. D. Gunzburger and L. S. Hou, *Treating inhomogeneous essential boundary conditions in finite element methods and the calculation of boundary stresses,* SIAM J. Numer. Anal. **29** 1992, pp. 390–424.

[433] C. Hirsch, *Numerical Computation of Internal and External Flows,* Vols. 1 and 2, Wiley, Chichester, 1988 and 1990.

[434] P. Holmes, J. Lumley, and G. Berkooz, *Turbulence, Coherent Structures, Dynamical Systems and Symmetry*, Cambridge University Press, Cambridge, 1996.

[435] M. Kline, *Mathematical Thought From Ancient to Modern Times,* Oxford University Press, New York, 1972.

[436] H. Liepmann and A. Roshko, *Elements of Gas Dynamics,* John Wiley, New York, 1957.

[437] H. Manouzi and L. Hou, *Resolution of the Navier-Stokes equations in vorticity-velocity formulation via control techniques,* Physica D **60** 1992, pp. 185–193.

[438] A. Noor, C. Andersen, and and J. Peters, *Reduced basis technique for collapse analysis of shells,* AIAA J. **19** 1981, pp. 393–397.

[439] A. Noor and J. Peters, *Reduced basis technique for nonlinear analysis of structures,* AIAA J. **18** 1980, pp. 455–462.

[440] A. Noor and J. Peters, *Tracing post-limit-point paths with reduced basis technique,* Comput. Methods Appl. Mech. Engrg. **28** 1981, pp. 2217–2240.

[441] J. Ortega and W. Rheinboldt, *Iterative Solution of Nonlinear Equations in Several Variables,* Academic, New York, 1970.

[442] J. S. Peterson, *The reduced basis method for incompressible flow calculations,* SIAM J. Sci. Stat. Comput. **10** 1989, pp. 777–786.

[443] H. Schlichting and K. Gersten, *Boundary Layer Theory,* Springer, Berlin, 2000.

[444] R. Temam, *Navier-Stokes Equations,* North-Holland, Amsterdam, 1979.

Index